Robert Winston is one of the country's best-known scientists. As Professor of Fertility Studies at Imperial College, University of London, and Director of NHS Research and Development and Consultant Obstetrician and Gynaecologist at Hammersmith Hospital, he has made advances in fertility medicine and been a leading voice in the debate on genetic engineering. His television series include *Making Babies*, *The Human Body*, *Superman*, *Human Instinct* and *Walking with Cavemen*, and have made him a household name across Britain. He became a life peer in 1995. His previous book, *Human Instinct*, is also published by Bantam Books.

the human**mind**
and how to make the most of it

ROBERT WINSTON

BANTAM BOOKS

LONDON • TORONTO • SYDNEY • AUCKLAND • JOHANNESBURG

THE HUMAN MIND
A BANTAM BOOK: 0 553 81619 5

Originally published in Great Britain by Bantam Press,
a division of Transworld Publishers

PRINTING HISTORY
Bantam Press edition published 2003
Bantam edition published 2004

1 3 5 7 9 10 8 6 4 2

Copyright © Professor Robert Winston 2003
By arrangement with the BBC.
The BBC logo is a trademark of the British Broadcasting Corporation
and is used under licence.
BBC logo © BBC 1996
The Human Mind © BBC 2003
Illustrations by Peter Gardiner

The right of Robert Winston to be identified as the author of
this work has been asserted in accordance with sections 77
and 78 of the Copyright Designs and Patents Act 1988.

The quotation from 'Dockery and Son' by Philip Larkin is taken from his
Collected Poems and is reproduced by permission of Faber & Faber.
The quotation from Luigi Pirandello's *Six Characters in Search of an
Author* (translated by John Linstrum) is reproduced by permission of Methuen
Publishing Limited and the Pirandello Estate, Toby Cole, Agent.

The publishers have made every effort to contact the copyright owners of
the extracts reproduced in this book. In any case where they have been
unsuccessful they invite copyright holders to contact them direct.

Set in 11/13pt Sabon by
Falcon Oast Graphic Art Ltd.

Bantam Books are published by Transworld Publishers,
61–63 Uxbridge Road, London W5 5SA,
a division of The Random House Group Ltd,
in Australia by Random House Australia (Pty) Ltd,
20 Alfred Street, Milsons Point, Sydney, NSW 2061, Australia,
in New Zealand by Random House New Zealand Ltd,
18 Poland Road, Glenfield, Auckland 10, New Zealand
and in South Africa by Random House (Pty) Ltd,
Endulini, 5a Jubilee Road, Parktown 2193, South Africa.

Printed and bound in Great Britain by
Cox & Wyman Ltd, Reading, Berkshire.

Papers used by Transworld Publishers are natural, recyclable products
made from wood grown in sustainable forests. The manufacturing
processes conform to the environmental regulations of the
country of origin.

To
Professor Colin Blakemore
and
Professor Richard Dawkins

This book is dedicated with respect and admiration to two distinguished scientists who have done such outstanding work in promoting science in society. They have led where others of us hope to follow.

Contents

	Acknowledgements	9
	Foreword	13
One	Body, brain and mind	17
Two	How does your brain work?	55
Three	Coming to your senses	86
Four	Paying attention	131
Five	The emotional mind	178
Six	The learning mind	239
Seven	A question of character	308
Eight	The loving mind	355
Nine	The amazing mind: intelligence, creativity and intuition	414
	References	473
	Glossary	485
	Index	493

Acknowledgements

This book has many imperfections of which I am only too aware. If there are errors, let me clearly state – and this is not an excuse – I am the only person to be blamed. None of the kind, gifted and intelligent people who examined bits of the manuscript have seen the whole book in its entirety – there simply was not time.

I owe a huge debt to a number of people. Particular thanks go to John Lynch, Head of Science at the BBC. He has been very kind and supportive throughout and I owe the opportunity to work on this project to him. He had sufficient faith to encourage me to present material related to cognitive neuroscience on mainstream BBC television and his enthusiasm was essential to start this project and, indeed, to write this somewhat personal account of the mind.

The book could not have been completed in the time available to me without the solid work, dedication and advice of Matt Baylis. We first met nearly nine years ago when he helped me with the book *Making Babies* and with background to various articles I was writing at the time. His towering intelligence and ability to research, ferret out, and sift information was vital for this book and his constant advice on structure and history has been critical. He has always been available when I was in a panic; it has been a pleasure to work with him.

Secondly, thanks go to Joel Winston, who read most of the manuscript, made numerous corrections and amendments, and prevented me from falling flat on my face at several stages during the writing. I fear though at times he has not been as critical as he should for fear of embarrassing his father. My thanks also go to Sarah-Jayne Blakemore, who has an extraordinarily incisive, analytical brain. Her lucid criticism, kindness and advice have been enormously helpful. Any neuroscience errors in this book remain only because they are in a section I never discussed with her. As an advisor on the television series, too, the quality of her guidance and research has been pivotal in making the programmes. My old friend Alison Dillon and I have worked together previously, and her help with editing part of the manuscript is appreciated.

Mark Hedgecoe was series producer for *The Human Mind*. His acumen, considerate advice, military planning and, especially, his friendship are so appreciated. Many times during the preparation of the manuscript his enthusiastic response to my imperfect writing was a key reason for continuing with this project. I am also grateful to Jessica Cecil, the magnificent executive producer, whose warm support and careful discernment have meant much.

Three people, above all, stand out in the ways ideas have been expressed in both the programmes and this book. The BBC has always ensured that I have been extraordinarily fortunate in working with genuinely gifted producers, whose mixture of scientific sense, calm appraisal, artistic skill and film-making ability has been wonderfully stimulating. The producers on this series, Johanna Gibbon, Diana Hill and Nick Murphy, have all been exceptional (even by the standards of the BBC) and it has been a great privilege, a pleasure and fun to work with each of them. I am also grateful to the powerhouse of assistant producers and researchers who were so diligent in digging out and collating background research – they are Miranda Eadie, James Marshall, Nicola Cook, Stian Reimers and, of course, Sarah Blakemore.

The actual process of filming has been important in writing this book – it gave pause for thought, a change of environment, and the stimulus of working with a crew. It is invidious to pick out individual members of a film crew, but it would be remiss not to mention three extraordinarily talented cameramen who did the bulk of the filming, and with their artistry and commitment to the project set the tone for the whole endeavour. They are Chris Hartley, Rob Goldie and Paul Jenkins.

Maggie Pearlstine, my remarkable agent, and Jamie Crawford, her brilliant assistant, have been the usual tower of strength. Their regular comments on the manuscript and enthusiasm for its content have spurred me throughout. I am so grateful for their confidence in me as I am for that shown by my publishers, in particular the keen interest of Sally Gaminara and her outstanding team at Transworld, and the concentrated work of the copy editor, Mari Roberts, and the Managing Editor at Bantam Press, Katrina Whone.

One huge advantage I have is in belonging to a great university, Imperial College. It has a wealth of expertise, excellent libraries, and strong backing for the communication of the ideas behind science to a wider audience. I am grateful for the support of my senior colleagues there, the Rector, and the Principal of Medicine, and my Divisional Chairman, David Edwards, and for the facilities of the university. Also, so many students from the university – the physicists, the drama club, the orchestra and the rowing club, amongst many others – put up with the imposition of a film crew with patience and good humour. It was a real pleasure to meet and work with them.

Finally, as any writer will probably admit, the act of setting down a manuscript is usually miserable for much of the time. My family, as always, have suffered (mostly stoically) and the love and support of my wife, Lira, and her very many helpful suggestions about the text, are deeply appreciated.

Foreword

The first few chapters in this book attempt to describe the basic functioning of the brain and how scientists have used various tools to measure this. Such an introduction is inevitably somewhat technical, but it is important to have a loose idea of the complexity of the anatomical regions of the brain and how it functions. Without this we cannot begin to understand how the mind works. I hope the reader will accept the necessity of this and be encouraged to persevere with the book by my use of anecdote and personal experience. In later chapters there is much more discussion of some of the current problems in our exploration of the science of the mind. One thing I hope will be obvious: there is a considerable amount that is controversial and a great deal that we do not know. Science is about uncertainty, and the topics of the human mind and human consciousness are some of the greatest unsolved scientific issues. Perhaps because, paradoxically, we are using an imperfect tool – the human mind itself – to understand the mind, there will always be unresolved questions. Finally, in the last part of this book I deal with issues like personality and intelligence; and I hope it will become obvious that a determinist view of the mind is not, in my view, at all adequate to explain what makes each of us who we are. We are, of course, invested with genes that direct many broad aspects of character and ability, but the environment in which we are placed can have the most profound influence. I hope, therefore, that it will become clear to those who persist beyond the first two or three chapters that there are indeed numerous ways in which we can improve our minds.

It is highly presumptuous for me, a mere reproductive biologist, to have attempted to write any book of this sort. Cognitive neuroscience is not my field of biology, and it is one of the most complex, rapidly changing and contentious areas of modern science. Moreover, a huge number of people, many of them outstanding scientists and a great number excellent writers, have written very good books on this subject. So I have little excuse. But over recent years I have become so interested in this field that if I had my time again as a medical scientist I think this is the area that would now hold the greatest attractions.

In many ways for me this subject is a natural progression from my general scientific and broadcasting interests. These have increasingly touched on various aspects of brain function. Nearly all the TV series and programmes with which I have been involved recently, from *The Human Body*, *Superhuman* and *Instincts* to *Threads of Life*, *Walking with Cavemen* and *Child of Our Time*, have had the brain – with its development, adaptations and workings – as their unifying interest. Though I genuinely never intended to write such a book, I found myself increasingly persuaded that it might be worth doing so during the filming of my latest BBC project. This book was largely stimulated by my presenting *The Human Mind* series and simple intellectual curiosity. When I was encouraged by numerous colleagues at the BBC to write a book around that subject, I took up the challenge with diffidence but growing enthusiasm. I hope its readers will find my particular approach fresh and interesting.

This is not simply the book of the BBC television series. Of course, it refers to it in many places, but it uses only a limited amount of the material seen on the screen. The book is intended to go further and to act as an independent companion, one that will stand alone. It tries to give the background – scientific, historical, social and philosophical – to what we in the production team found interesting or extraordinary about many aspects of human consciousness. So it is focused on the things that I find interesting and is not intended to be a comprehensive account. Nevertheless, I hope that its readers will enjoy reading this attempt to explain some aspects of what is most topical, controversial and intriguing in this extraordinary field.

chapter one

Body, brain and mind

If I were to die later today before finishing this chapter and my brain was removed from my skull, it would weigh about 1,400 grams – roughly the same as a bag and a half of granulated sugar. Before being preserved for posterity by being marinated in a jar of formalin (better still, perhaps, strong alcohol with some flavouring), about 75 to 80 per cent of my brain would be made up of water, with just over 10 per cent fat and about 8 per cent protein. If, once it was fixed, people came to examine it and poke it about a bit, it would appear rather crinkled and whitish, and have the slightly rubbery consistency of a large mushroom. And what is more, if you, dear reader, were to die at the same time and have your brain treated in the same peremptory fashion, it would be so similar to mine that any difference would almost certainly be undetectable.

The chances are that no matter how closely our respective brains were viewed, there would be hardly anything obvious to show that what amused passers-by were gazing at were two totally different specimens of the most complicated structure on this planet. There would be nothing to reveal that these two rubbery objects, which to some bystanders would seem faintly disgusting, respectively comprised the sum total of our very being and personality. Nothing to show that at some time we had both loved in different ways, had known different pains, ambitions and disappointments, and had been angry and taken pleasure at different things. Nor that we had learnt different physical and intellectual skills, had mind-bending experiences in different parts of the world, had totally different memories, liked different food or music, and that each of us had quite different human strengths and failings.

Perhaps it is not so surprising then that it has taken humans such a long time to understand the complex nature of the brain, and that it is the very centre of what makes us who we are. Although surgical drilling of holes in the skull (for whatever now mysterious purpose) goes back to Cro-Magnon man some 40,000 years ago, and knowledge of the mind-altering nature of alcohol and the sap from the poppy plant is longstanding, most old civilizations regarded the heart, not the brain, as the centre of the soul. Ancient Egyptians, when embalming human bodies, religiously preserved the heart but destroyed the brain – because otherwise it would rot – by scraping it out of a hole they drilled in the bones at the back of the nose and palate. But it was an Egyptian surgeon who left the first written descriptions that give evidence of some basic insight into neuroscience.

The Edwin Smith Papyrus is one of the oldest known written documents. It is around 3,700 years old and is a surgical treatise describing injuries, mostly to the head, in forty-eight different patients. The Egyptologist, Edwin Smith, who first handled this extraordinary manuscript brought it back from Luxor in 1862, but he did not understand the remarkable nature of the text. Its real significance was recognized by James Breasted, director of the Chicago Oriental Institute, in 1930, who realized it was a scribe's copy of a treatise from an even earlier time – possibly some 5,000 years ago. The horrifying injuries of Case Number Six give a description of the pulsating brain under the surgeon's hands:

> If thou examinest a man having a gaping wound in his head, penetrating to the bone and smashing his skull, and rending open the brain, thou shouldst palpate the wound. Shouldst thou find that smash which is in his skull those corrugations* which form in molten copper, and something therein throbbing and fluttering under thy fingers – like the weak place in an infant's fontanelle before it becomes whole . . . then if he suffers blood from both his nostrils and stiffness in his neck . . . thou shouldst say concerning him 'An ailment not to be treated'. Thou shouldst anoint that wound with grease but not bind it; thou shalt not apply two strips upon it until thou knowest he has reached a decisive point.

So even Egyptian physicians knew when it might be more prudent not to treat a patient actively. These hieroglyphics go on to describe the delicate membranes lining the injured brain, the meninges, and the discharge of cerebro-spinal fluid from inside the head. Elsewhere the papyrus records the symptoms of a patient unable to move one limb after severe head injuries on one side, and loss of speech resulting from injuries to the temple – presumably damage to the frontal lobe and Broca's area – several thousand years before Dr Paul Pierre Broca described the speech centre in the 1860s.

---

* The 'corrugations' almost certainly refers to the convolutions seen in the cerebral cortex.

The mind/body debate

Thousands of years elapsed before the brain, rather than the heart, was universally recognized as the most important organ in the body. Alcmaeon, who around 500BC was one of the earliest to see the brain's importance, regarded it as the centre of sensation – he removed an animal's eye and noted the tracts leading to the brain, recording that 'all senses are connected to the brain'. Plato believed in the soul – the essence of ourselves, and what we might in modern times call the 'mind' – and he thought that it had a separate existence from the body, to the extent that it could survive after the body had expired. He believed that the centre of the intellect was in the head.

But Aristotle, who lived from 384 to 322BC, appears to have disagreed with his teacher, Plato. He seems to have regarded the heart as more important. All the lower animals he examined – worms, insects and shellfish – had a pulsating organ resembling a heart but they did not have an obvious brain. All blood vessels led towards the heart, and he describes how the heart twitched when touched while the brain of higher animals remained inert. The fact that a chicken ran about after its head was cut off helped Aristotle to the view that 'the seat of the soul and the control of voluntary movement – in fact of nervous functions in general – are to be sought in the heart. The brain is an organ of minor importance, perhaps necessary to cool the blood.' Aristotle hugely influenced the medieval scholars who came later; after all, his view of the importance of the heart fitted with biblical accounts. The notion of the heart as the centre of human behaviour survived until the sixteenth century. 'Faith sits under the left nipple,' said Martin Luther.

A little earlier than Aristotle, though, the philosopher Democritus argued against the heart being the centre of human functions. He writes: 'The brain watches over the upper limbs like a guard, as citadel of the body, consecrated to its protection,' and adds that 'the brain, guardian of thoughts or intelligence', contains the principal 'bonds of the soul'.

Hippocrates, the father of medical practice, recognized the unique nature of the brain: 'Men ought to know that from the human brain and from the brain only arise our pleasures, joys, laughter, and jests as well as our sorrows, pains, grief and tears . . . It is the same thing which makes us mad or delirious, inspires us with dread and fear, whether by night or by day, brings us sleeplessness, in-opportune mistakes, aimless anxieties, absent-mindedness and acts that are contrary to habit . . .'

In the third century BC, Herophilus and Erasistratus, both human anatomists, dissected thousands of bodies and demonstrated that nerves were different from blood vessels and that they originated not in the heart, as Aristotle thought, but in the brain or the spinal cord. Then, almost five hundred years after Herophilus' day, the Greek physician Galen (AD130–200) dissected pigs, cattle and monkeys and wrote meticulous accounts of what he had seen. By cutting various nerves, such as those coming from the spinal cord, he established the lack of function caused by their damage. He also demonstrated that severing the laryngeal nerve resulted in the loss of the ability to make noise. During his career he was a physician to gladiators in Rome. Seeing many head injuries presumably gave him an insight into the working of the nervous system and the understanding that the brain played a central role in controlling bodily and mental activity.

When I was a medical student, one of the organs I found most mysterious in the brain was the pineal gland. It is a tiny rounded protuberance, pretty much deep in the middle of the brain, and is unique because, being in the centre, it is the only structure in the brain that is not paired. It has always been associated with mystery – the ancient Hindu mystics thought of it as the seat of the human soul. In the early seventeenth century, the French philosopher René Descartes took the debate further. He concluded that the pineal gland was vital because it functioned by connecting our physical being with our spiritual side. He decided that these two separate and apparently contradictory attributes – the corporeal body and the entirely non-physical mind – could communicate and act as one using the pineal as a conduit. This idea of a conduit seems odd and I am unsure how Descartes came to believe it. To me, the pineal gland seems rather puzzling, if not totally useless. I know that it is supposed to respond to the light of day and the dark of night via the optic nerves and thus regulate the body clock. Everybody repeatedly tells me how it does this by producing the hormone melatonin. Nowadays, travellers can buy melatonin in airports to combat jet lag. But numerous transatlantic flights to the second laboratory in which I work, in California, have never been made even slightly easier by my taking melatonin. I still wake up in Los Angeles feeling like death at three in the morning.

Though he never sat for ten hours in an aircraft, Descartes felt that the pineal was highly significant. So-called Cartesian dualism, the idea of a physical body and abstract mind (communicating via the pineal), endured for a long time because it helped Science and Faith to co-exist. But historical events lent strength to an opposite point of view. The French Revolution, with the infamous apparatus 'popularized' by Dr Joseph Ignace Guillotin, provided the curious with an ample supply (it is calculated at about 40,000) of freshly decapitated heads – and hence brains – to view. And just over a century later, the horrors of the first world war provided doctors and scientists with a steady stream of conscious but brain-damaged young men to observe. Between these two great historical events, advances in physics and better methods of measurement gave scientists the means to observe fluctuations in electricity and pressure. Thanks to these developments, the early neurologists were able to begin to amass a body of evidence to challenge the Cartesian point of view. Rather than the mind and its faculties floating formlessly in the ether and communicating with the brain and body through the pineal gland, it began to seem that the mind was a product of the brain. Moreover, it became increasingly clear that different and distinct regions of the brain were responsible for different aspects of 'mind'.

One lump or two?

In medieval times, various attributes of the self were thought to reside in different bodily fluids. It was generally believed that the cerebrospinal fluid, the watery fluid in which the brain and spinal cord are bathed, provided the faculty of thought. Then in the early nineteenth century, before science could really prove its claims, phrenology became popular. It was argued by its many adherents that differences in brain architecture – and hence in the character of individuals – could be felt and seen as differences in the shape of the skull.

Dr Franz Joseph Gall, a Viennese doctor, seems to have been the father of phrenology in the late 1790s. Gall, together with his pupil Johann Spurzheim, established many of the basic notions behind this 'science'. Gall recognized that the brain was where the mind was situated. The essential features of his thinking were that the mind is made up of distinct faculties and attributes. He postulated that because these faculties were distinct, they each should occupy a different location within the brain. The bigger the brain, the bigger a person's mental capacity. It therefore followed, in his thinking, that those parts of the brain that were identifiably larger were larger because the attribute associated with them was particularly well developed. But the key to the science of phrenology was that the skull takes its shape from the brain – so the surface of the skull, and its lumps and protuberances, gave a clue as to what lay underneath. Although this is ridiculed now, these ideas are not so silly. For example, we now know that people with trained memory, such as London cab drivers, sometimes have an area – the posterior hippocampus, a structure deep inside the brain associated with memory – which is larger than average. Another example is that of the estimations made by palaeo-anthropologists – they frequently use skull shape to deduce information about the characteristics of our early, ape-like, hominid ancestors.

By the end of the Napoleonic Wars, phrenology was getting established as a way of telling a person's character and personality. Phrenologists palpated the head and even made complicated measurements with specially designed calipers to find out whether someone was likely to be conscientious or a convict. Soon, phrenology was out of the medical and scientific journals and discussed in the serious literary papers; not all comment was glowing. In June 1815, *The Edinburgh Review*, the serious journal of its day, reported:

The writings of Drs. Gall and Spurzheim have not added one fact to the stock of our knowledge respecting either the structure or the functions of man; but consist of such a mixture of gross errors, extravagant absurdities, downright mis-statements, and unmeaning quotations from Scripture as can leave no doubt, we apprehend, in the minds of honest and intelligent men as to the real ignorance, the real hypocrisy, and the real empiricism of the authors . . .

Such is the trash, the despicable trumpery, which two men, calling themselves scientific inquirers, have the impudence gravely to present to the physiologists of the nineteenth century, as specimens of reasoning and induction.

In spite of this sometimes hostile press, phrenology became increasingly popular, particularly in Europe, Britain and then in the United States. Cesare Lombroso, the Italian physician and criminologist, gained widespread attention when he published his book *L'uomo delinquente* (*The Criminal Man*) in 1876. His view of criminal attributes was partly stimulated by Darwinian ideas and the notion that primitive man was descended from apes. He believed that measuring the heads of people gave a clue about their criminal tendencies. He based his observations on measurements of both living and executed criminals. He compared these with the skulls of non-human primates and came to the view that criminals were 'throwbacks' – the consequence of atavism.

In Britain, phrenology was used to reinforce conventional arguments and prejudices about class, race and criminality. Soon employers were seeking information from their local phrenologist about the predilections of candidates for a post they might be offering. For some gullible members of the public, a visit to a phrenologist was rather similar to a modern-day visit to a clairvoyant or astrologer. And in addition to the thirty-five areas of the head responsible for such attributes as Fidelity, Musical Talent, and Verbal Memory, several devout Christian theologians saw cranial organs for Veneration and Wonder – clear evidence of divine design.

Phrenology was responsible for even more thoroughly laughable claims, such as the idea that Eastern peoples were less warlike because they had smaller heads. And though phrenology has been long discredited, it still had some uncomfortable echoes in the twentieth century. There seems to be a strong element of this stereotypic thinking behind the detailed skull measurements used for some of the Nazi

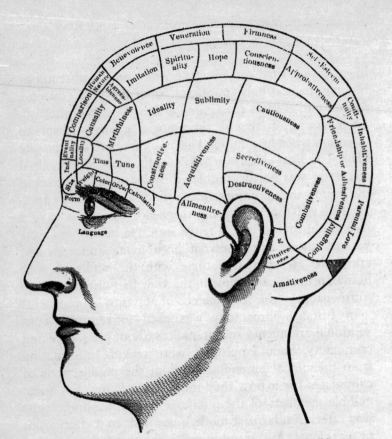

NUMBERING AND DEFINITION OF THE ORGANS.

1. AMATIVENESS, Love between the sexes.
A. CONJUGALITY, Matrimony—love of one. [etc.
2. PARENTAL LOVE, Regard for offspring, pets,
3. FRIENDSHIP, Adhesiveness—sociability.
4. INHABITIVENESS, Love of home.
5. CONTINUITY, One thing at a time.
E. VITATIVENESS, Love of life.
6. COMBATIVENESS, Resistance—defense.
7. DESTRUCTIVENESS, Executiveness—force.
8. ALIMENTIVENESS, Appetite—hunger.
9. ACQUISITIVENESS, Accumulation.
10. SECRETIVENESS, Policy—management.
11. CAUTIOUSNESS, Prudence—provision.
12. APPROBATIVENESS, Ambition—display.
13. SELF-ESTEEM, Self-respect - dignity.
14. FIRMNESS, Decision—perseverance.
15. CONSCIENTIOUSNESS, Justice. equity.
16. HOPE, Expectation—enterprise.
17. SPIRITUALITY, Intuition—faith—credulity.
18. VENERATION, Devotion—respect.
19 BENEVOLENCE, Kindness—goodness.
20. CONSTRUCTIVENESS, Mechanical ingenuity
21. IDEALITY, Refinement—taste—purity.
B. SUBLIMITY, Love of grandeur—infinitude.
22. IMITATION, Copying—patterning.
23. MIRTHFULNESS, Jocoseness—wit—fun.
24. INDIVIDUALITY, Observation.
25. FORM, Recollection of shape.
26. SIZE, Measuring by the eye.
27. WEIGHT, Balancing—climbing.
28. COLOR, Judgment of colors.
29. ORDER, Method - system—arrangement
30. CALCULATION, Mental arithmetic.
31. LOCALITY, Recollection of places.
32. EVENTUALITY, Memory of facts.
33. TIME, Cognizance of duration.
34. TUNE, Sense of harmony and melody.
35. LANGUAGE, Expression of ideas.
36. CAUSALITY, Applying causes to effect. [tion
37. COMPARISON, Inductive reasoning—illustra
C. HUMAN NATURE, Perception of motives.
D. AGREEABLENESS, Pleasantness—suavity.

Phrenological map (Wellcome Library, London)

pronouncements on race. In Hitler's Germany, the shape of the skull was frequently evaluated to help determine true Aryan origins, or to detect underlying Jewish racial degeneracy.

In spite of all this, phrenology was vaguely groping in the right direction in a few instances. The so-called 'Organ of Mirthfulness' is a region slightly north-west of the crown of the head. A University of California Medical School surgeon recently applied an electrical current to this region of a patient's brain as part of a procedure to reduce epilepsy, and she burst out laughing.[1] The Organ of Mirthfulness was a happy accident for phrenology, and it shows that there is an extent to which the brain is a box of separate functions. However, these distinct brain regions do not cause protuberances in the skull.

So Gall was not entirely on the wrong track after all. Sadly for him, he lived well before any of the modern-day methods of proving brain function existed. There are many ironies in the history of neurology. In 1828, Gall suffered a fatal stroke and his own head ended up as the last, rather morbid addition to his personal skull collection.

Nowadays, we have clearly arrived at a point where we understand something vital about humankind: that the brain is the key organ responsible for who we are – that the 'mind' is created by the brain. But, as we have seen, this was by no means something we had always understood, or a view that was universally held. In reaching this understanding, the key process, which developed with improved scientific measurement from the middle of the nineteenth century onwards, was brain mapping.

The process of mapping has provided medical science with some of its most crucial information. Thanks to pioneers in this field, we now know that there are regions of the brain that are responsible, at least in part, for movement, sensation and vision, and others even for such things as some religious experiences, and speech. We know that there are sites which store our vocabulary for topics as specific as the naming of vegetables and gemstones. And still others which, when damaged, cause responsible people to become rash, impulsive, promiscuous risk-takers.

The neuron – the basic unit in the brain

A key development was an understanding of the intricate internal anatomy of the brain. Among the first anatomists to look at the brain through a microscope was the Czech professor Johannes Purkinje (1787–1869) who, among other achievements, first classified human fingerprints. Purkinje is credited with describing the first cell identified in the central nervous system in the cerebellum. The cerebellum is the folded structure at the back of the brain that deals largely with balance and coordinated movements. The nerve cell or neuron named after him, the Purkinje cell, is gigantic and it has a most complex web of delicate fronds at its end which connect with other nerve cells. It is almost as thick as a human hair (very large indeed for a single cell) and about ten times bigger than most other cells in the cerebellum, the part of the brain where it mostly resides. It may be of interest to know that somebody calculated that there are around 26 million of these cells in the brain.

Presumably the very size of these large neurons enabled Purkinje to identify them under the microscope and to make drawings. Remarkably, one Purkinje cell with its huge network has been calculated to receive inputs from up to 200,000 other cells. Jim Bower, a neuroscientist from California Institute of Technology, waxes lyrical about the Purkinje cell: 'They are continuously active at night . . . awake, asleep, or anesthetized,' he writes. 'They

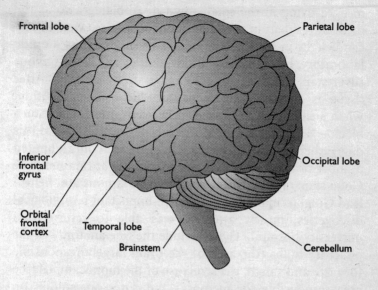

Frontal lobe

Parietal lobe

Inferior frontal gyrus

Occipital lobe

Orbital frontal cortex

Temporal lobe

Brainstem

Cerebellum

The brain seen from the left side

are physically stunning, absolutely beautiful. And the ones in your brain are the same shape as the ones in the brain of a fish, bird, reptile, or earliest vertebrate.'

Towards the end of the nineteenth century, the use of tissue stains made it increasingly clear that some cell types were more prevalent in one region of the brain and other types in another. Many great anatomists, among them Nobel Prize winners such as Santiago Cajal from Spain and Camillo Golgi from Italy, were responsible for this key progress, and the basic cell unit in the brain, the neuron, was described in increasing detail.

The respective stories of Golgi and Cajal are interesting and poignant, and eventually clash with a bitter scientific rivalry – not unlike, sadly, scientific jealousies of today. Human nature does not change. Golgi came from Milan, studied medicine at the University of Pavia, but, because he was financially desperate, left academic work to earn a meagre living in the small town of Abbiategrasso as the medical resident in the Home for Incurables. Working by candlelight in his kitchen, which he modified as a crude laboratory, he eventually discovered, in 1873, the famous silver stain that is specifically taken up by neurons. After tissue is hardened in potassium bichromate solution, when a silver salt is applied, nerve cells immediately turn black – 'la reazione negra'. This black reaction is, to this day, rather mysterious because it happens in only a limited number of neurons, and those apparently at random. The Golgi stain, as it is still called, became famous and helped lead Golgi to making a number of important pathological discoveries, mostly after he was able to return to a university hospital. These include the fine anatomy of the olfactory bulbs (the area of the brain largely responsible for taste and smell), the structure of the hippocampus (the area responsible for memory) and, later on, studies on brain tumours, the sensory apparatus in tendons and the internal structure in cells called the Golgi apparatus, which has become increasingly important in modern biology because it is this structure that helps the cell to process proteins. He even used his techniques of staining to help in the understanding of the parasite Plasmodium – the organism responsible for malaria. This last achievement alone is of signal importance and it is not surprising that a kind of shrine devoted to his memory was established at the University of Pavia.

Santiago Ramón y Cajal was born in a small village in northern Spain and studied medicine in the University of Saragossa. He had an ambition to become a painter, but his father was determined that his son should be a doctor. As it happened, his artistic skill was to be of huge benefit. He eventually became Professor of Anatomy in Valencia. Soon after Golgi published his silver stain method, Cajal started to make detailed, quite beautiful drawings of different parts of the brain, viewed down a microscope. In those days, photography down a microscope was of course virtually impossible, so accurate drawing was crucial. His meticulous drawings are still a source of wonderment. He eventually built up a complete picture of the whole of the brain, area by area. As his work developed, Cajal became increasingly convinced that the neuron, the single nerve cell with its minute, fine and delicate processes connecting with other nerve cells, was the key unit in the brain. This is where he had a fundamental disagreement with Golgi. Golgi and his followers were convinced that the brain's tissue was a diffuse net of nerves that were interconnected – a kind of miniature mesh. Of course, the lack of photography meant that the evidence Cajal could produce, a mere drawing, was open to scepticism and ridicule. Many scientists, including Golgi, regarded Cajal's drawings as mere artistic interpretations.

It turns out that Cajal was right, and Golgi wrong. Cajal's genius lies partly in the fact that even though he was using the same kind of microscope and similar staining methods, he was rigorous in drawing what he could

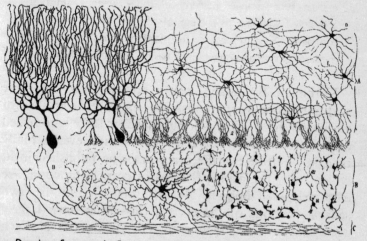

Drawing of neurons by Santiago Ramón y Cajal (Wellcome Library, London)

31

see and did not overinterpret what he could not see. But, in 1906, they were jointly awarded the Nobel Prize for their contribution to neuroanatomy, a matter that caused fury and bitterness for Golgi, who was a supreme egotist. To be fair, presumably the appalling way he behaved to Cajal was at least in part because of the difficulties he had repeatedly experienced in getting his earlier methods accepted. It seems they met only once – at the Nobel Prize ceremony in Stockholm. Golgi gave his Nobel lecture first, in which he re-emphasized his belief in a neural mesh – a theory that was then completely contradicted by Cajal's lecture only minutes later. It is interesting to me that, in spite of their rivalry, Santiago Cajal must have been a generous man. Even though his own work had been criticized and ridiculed by some of his opponents, he felt able to write of Golgi's method in 1917:

I expressed in former paragraphs the surprise and wonder which I experienced upon seeing with my own eyes the wonderful revelatory powers of the chrome-silver reaction and the absence of any excitement in the scientific world by its discovery. How can one explain such strange indifference? Today, when I am better acquainted with the psychology of scientific men, I find it very natural … Out of respect for the master, no pupil is wont to use methods of investigation which he has not learnt from him.

Korbinian Brodmann was a contemporary of these two famous neuroanatomists. In one way, he is a bit of an unsung hero. His book published in 1909 is regarded as a classic work, but few people have ever actually read it. Nonetheless, the brain maps he published in it, which show the principal areas of cortical function, are extremely well known. He identified around fifty-two separate regions of the cortex, classifying them largely according to the cell types seen in the different areas. Each of the areas was numbered by him, and some of these numbers are still in use today – however, as modern neuroscientists are starting to understand more about brain function, many of Brodmann's areas are controversial.

Korbinian Brodmann was born on 17 November 1868, and studied medicine in Munich, Würzburg and Berlin. He worked in Switzerland and then Berlin for a while, and he became interested in neurology after he contracted diphtheria. His boss, Oscar Vogt, to whom he owed much of his later career, described him as having 'broad scientific interests, a good gift of observation and great diligence in widening his knowledge'.

Brodmann wrote a thesis on chronic changes in the blood vessels of the brain, in Leipzig in 1898. His meeting with the famous Alzheimer there inspired an interest in the anatomy of brain tissue that he pursued for much of the rest of his scientific career. Even with Vogt's support, he was passed over for promotion in Berlin – the establishment there did not recognize the real value of his meticulous observations. So, again in a relatively short career, he moved institutions. After working in Tübingen, in 1916 Brodmann took over as Director of the Nietleben Mental Asylum in Halle. He left getting married – to Margarete Francke – until the age of forty-eight and by then he was financially secure for the first time in his life, happy and settled.

On 17 August 1918, shortly after the birth of his daughter, he had flu, and septicaemia followed. Brodmann was generally very fit, and was quite pleased to be ill because it gave him a chance to catch up with paperwork. He did not understand how ill he was becoming. One day, his wife saw him making writing motions on his bed with his finger. He then sat back in the bed and died, at the age of forty-nine. Part of the reason I find his death so poignant is the memory of the death of my father, who contracted influenza, did not take it seriously, was quite happy to catch up with work and died of septicaemia – which led of all things to an abscess on the brain – at the age of just forty-two. I was nine when he died, so, unlike Brodmann's baby, at least I was a child who knew something of his father's attributes – but it was a death that now, with better antibiotic treatment, would certainly not have happened.

Brodmann's work is fundamental because it not only involved human brain mapping, it involved work comparing these maps with those in rodents and primates. It led to a much better understanding of how different areas in the cortex might work. Of course, it is only with the development of techniques such as MRI, PET scanning and identification of chemicals in different regions of the brain – mostly eighty or more years after his death – that the principles behind Brodmann's ideas, which were not always appreciated at the time, have been vindicated. But new maps are being produced now that are much closer to how we believe the brain works.

Localizing the workings of the brain

Just before scientists like Brodmann were mapping the detailed microscopic structure of brain tissue, others had started to look at the brain in a more general way. Paul Pierre Broca was one of the most brilliant young men at his medical school in Paris, graduating at the age of twenty. He also held a degree in mathematics, and others in physics and French literature. He was soon appointed Professor of Pathology in Paris at the Hôtel Dieu and Bicêtre hospital. His career was extremely distinguished; he described the spread of cancers through the veins of the body, the effect of dietary deficiency in causing rickets and congenital muscle wasting caused by muscular dystrophy. He also used volunteer medical students to test the temperature of the scalp during various mental activities. Being something of a polymath, he became fascinated by anthropology and this interest landed him in considerable trouble. He was on record in the police files as a 'suspicious person'; the Parisian authorities, particularly the Prefect of Police and the Catholic Church, had him followed and all his movements monitored. At this time discussion of knowledge about human beings was considered to be against the public interest and hence against the interest of the state. His attempts to popularize this science led him to be dogged constantly by plain-clothes police and denounced as a subversive, and the public anthropological meetings he organized were infiltrated by agents and informers.

On 4 April 1861, Broca attended a lecture by Ernest Aubertin at the Anthropology Society that was to change his life. Aubertin was a pupil of the phrenologist Gall, and was convinced that the power of speech was localized in the front part of the brain. Paul Broca determined to find a suitable brain on which to work and looked for a

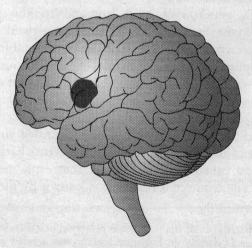

The left hemisphere of the brain, showing Broca's area

patient who had a severe speech disorder. A week later, Monsieur Leborgne, a patient who had been incapacitated in Broca's hospital ward for many years, died of septicaemia and gangrene of the right leg and buttock. Leborgne had been paralysed down his right side for some thirty years and was only able to say the words 'tan – tan' in response to any question. His plight was so well known to those who nursed him that they called him 'Tan'. At a post-mortem, Broca found a lesion on the surface of the left side of the frontal lobe of his brain.* Leborgne's pickled brain can now be viewed in Paris at the Dupuytren Museum, a short distance from Odéon Métro station, at no. 15, rue de l'Ecole de Médecine. By 1863, Broca had managed to identify similar lesions in the same part of the brain in no fewer than twenty-five patients who had lost their powers of speech. This area of the brain, the left inferior frontal gyrus, became known as Broca's area, and his reputation was made. The recognition that specific areas of the brain were important for particular functions was an important advance in the history of neurology.

Broca generously paid homage to the maligned phrenologist Franz Gall – at least in general terms. In 1861 he wrote, 'Gall ... had the undisputable merit of proclaiming the great principle of cerebral localization, which – it may be said – was the starting point of the discoveries of our century concerning the physiology [of the brain].' Despite this compliment, Broca seemed less inclined to credit Gall with having identified virtually the same location in the brain some fifty years earlier, which Gall had called 'Organ 14', the centre of 'memory for words'; its connected region was known by the father of

---

* Recent scans of this brain, which is preserved in a glass jar, show that the area of damage in Leborgne's brain was actually considerably bigger than that described by Broca.

phrenology as 'Organ 15' – 'the sense of language, of speech'.

Broca was not always quite so blessed in his researches. Later in his anatomical studies, he compared the sizes of brains taken from male and female corpses. The average weight of a man's brain turned out to be 1,325 grams, while women, on average, had a brain that weighed 180 grams less. He was led to the obvious (but wrong) conclusion that men were more clever than women, a view shared by many other scientists – particularly Darwinians – of his time. Once the political climate in France changed, Broca ended his career in 1880 as an elected member of the upper chamber of the French parliament, a senator. He was widely regarded as a fine speaker and raconteur, and a generous, compassionate Christian. By some cruel irony, he died shortly after his elevation, following a ruptured aneurysm in the brain – bleeding from the brain vessels being something he had most carefully documented in his nine-hundred-page monograph on the subject some twenty years earlier.

Measuring brain function

One of my finest moments was when I acquired my first stethoscope. Go to any teaching hospital and you can still spot the proud beginners and the newer students because they will be displaying a form of designer wear – prominently wearing their stethoscope around the neck or more casually thrown over the shoulder. As students, our chiefs were always calling on us to listen to remote, faint and totally obscure sounds down the tubing, and the usual temptation was to pretend to hear a noise when you almost invariably couldn't. As we progressed in my teaching hospital, we weren't required merely to listen to the sounds of breathing or the heart. Advanced students even listened to the neck, and I remember on one occasion mystifying both my elderly patient and myself by listening to his knee. I couldn't hear a thing but it impressed my class greatly.

In 1928, John Fulton broke new ground when he listened to the back of a man's head. While a doctor in training in Boston, a mere resident at Peter Bent Brigham Hospital, he heard of a patient with gradually decreasing vision on whom surgery to remove a collection of abnormal vessels at the back of the brain had failed. The surgery had left the patient with a gap in his skull where bone had been removed. The man complained that whenever he opened his eyes, he heard a roaring sound. The noise became a bit more intense, the patient said, whenever he was doing something difficult. Fulton wasn't interested in trying to impress his colleagues but he picked up his stethoscope and identified what clinicians technically refer to as a faint 'bruit' or a 'murmur' – a rushing, pulsating sound caused by turbulent blood flow. Fulton decided that the tangle of blood vessels in the visual area of his brain was the cause, and that the roaring noise they could both hear was the noise of chaotic blood flow as it rushed to the area of the brain doing the work when he opened his eyelids and used his remaining vision. The noise was particularly intense when he attempted to read. So the more intense the mental activity, the more the blood flow, because the brain was using more oxygen to complete its task. The remarkably resourceful Fulton even developed a very primitive 'imaging' machine. He rigged up a stethoscope so that the amplified sound vibrations could record changes in the sound of blood flow by dragging a needle across a smoked drum. Thus equipped, he would appear to have provided the first ever 'picture' of brain activity.

But although in recent years measurements of blood flow in the brain have become important in monitoring its activity, it was a long time before suitable instruments for measurement were available. Around the 1880s there was a growing realization that the brain might be producing electrical activity, and the most promising experiments of the time were to try to capture these faint signals. In Liverpool, a physician at the medical school, Richard Caton, was examining what happened to the brains of monkeys and rabbits when he shone lights into their eyes. Electric light wasn't yet available, so a naked flame sufficed, and he needed to take off the top of their skulls in order to apply his crude probes to the surface of the brain. He recorded electrical impulses by means of a mirror attached to a fine coil of wire – the so-called mirror galvanometer invented by the physicist Lord Kelvin. Removing the cranium was hardly a procedure, in those days of relatively crude anaesthetics, that could be applied to humans. It wasn't until the 1930s that we developed electrodes sensitive enough to routinely register changes in charge through the bone and the scalp.

The first person to record electrical activity in the brain of human subjects was probably the Austrian Dr Hans Berger. He started animal experiments before the first world war, but these were mostly inconclusive and unhelpful. However, as early as 1902 he had found a number of human subjects who had defects in the skull bones, and he used primitive electrodes and galvanometers to see whether he could detect faint electrical activity. After his return from the war, he set about using the rather more sophisticated equipment that had become available. Once again, he found a few people with defects in the skull – presumably not so difficult in Germany after the ravages of battle. He tried various approaches to measuring activity in these people, one of which involved stimulating one side of the brain with an electrode and seeing if there were any demonstrable pulses coming through on the opposite side. But he soon realized that it was possible to measure the slight fluctuations in electrical signals that were occurring spontaneously when no artificial stimulus was being given. He first published this work, coining the word 'electroencephalogram', in 1929. This is still regarded as an important milestone in the history of neuroscience because what Berger described were the first rhythms in human brain physiology, the alpha waves, which change in pattern and frequency during activity.

Berger was an odd man. He was demanding and reserved. His experiments were conducted in the strictest secrecy. He refused access to his laboratory and never mentioned to any colleagues what he was researching. He was obsessed with the idea that others might take credit for his work, and was haunted by the many false turns he had made in the thirty years of interest he had had in this work. In 1937, when international recognition was finally accorded to him at the Parisian Psychology Congress, he cried when he said, 'But in Germany I am not so famous.' He was an opponent of the Nazi regime, and watched the rise of Hitler with growing despair and depression, being forced to retire from the University of Jena in 1938. In 1941, he took his own life by hanging himself in the hospital where he had worked.

Electroencephalography – EEG – has limitations when trying to evaluate the function of the mind. It is a rather unselective form of monitoring the brain's electrical signals – like a TV that manages to receive all channels at once. Until relatively recently it has been thought nearly useless for researchers trying to isolate the brain activity associated with a particular thought or feeling or action. However, in the last twenty-five years electroencephalography has become more refined and has been used extensively in neuropsychology. It is undergoing something of a renaissance for this kind of investigation.

The use of radioisotopes

In the 1970s, a team of Danish and Swedish researchers hit upon a new technique. Using Geiger counters to measure radiation, they injected a tiny amount of a radioactive gas into the bloodstream of the subject: the more blood flow, the more radiation capable of detection. The gas they used was xenon, which is completely inert and harmless, but several of its isotopes are radioactive. Radioactive isotopes decay over time, eventually becoming devoid of radioactivity. The radioactivity of a substance is normally measured by its half-life, which is the time it takes for half the radioactive atoms to decay. Compounds with a short half-life are valuable for medical measurement because the patient receives a rapidly diminishing amount of radiation, and the dose of radiation is greatly limited. Where the radioactive isotopes of xenon that were once used produced radiation that lasted for a few hours or several days, nowadays only isotopes with a very short half-life are used – a few minutes at most.

Once the xenon had been injected by the Scandinavian scientists, its progress could be measured by the Geiger counters close to the scalp as it travelled in the bloodstream of the brain. The injection was not exactly pleasant – to get enough into the head entailed direct injection into the carotid artery in the neck using a hand-held syringe, and the only subjects hardy enough to withstand it were, rather surprisingly, big, beefy men drawn from the local fire station in Lund. (In my experience big beefy males would most probably be the worst candidates for this kind of procedure, likely to faint at the mere sight of the needle.) Crude though this procedure might have been, the researchers were able to produce pictures of the blood flow inside the brain as the subjects performed tasks such as listening to speech, tapping fingers, even

41

imagining making sudden movements. From these early eye-watering experiments, the technique of Positron Emission Tomography, or PET scanning, was born.

In PET, a fast-decaying radioactive tracer is injected into the blood, and the resultant gamma rays are read while the subject undertakes the activity being studied – counting to ten, for instance, or thinking happy thoughts. By analysing a few million of these readings at a time using sophisticated computing, it is possible to build a three-dimensional image of the brain as it goes to work. While PET was a step forward, it was undoubtedly a large-scale procedure. The machines were huge and costly – the PET unit at the Hammersmith Hospital in London, for instance, required an army of physicists, engineers, chemists and computer operators. It also required a small nuclear reactor, a cyclotron, to generate the rapidly decaying isotopes with a short half-life that would not give too much radiation to the subject. The Hammersmith cyclotron, which largely started as a possible way to give deep X-ray therapy to cancer patients, was also used to develop radioactive nucleotides for injection into patients having scans other than those done by PET.

When I first arrived at the Hammersmith in 1969, as a very junior physician and research worker, it seemed not so much a proper hospital as a collection of disparate buildings that did not match each other, dotted around in an almost random fashion next to Wormwood Scrubs prison. From time to time the more adventurous among those held as involuntary guests of the Crown would slip over the wall next to the high-security wing and into the grounds of the hospital. One such prisoner was George Blake, the atom spy. The collection of brick, concrete and wooden buildings, old workhouse premises, isolated out-patient facilities, Nissen huts and Portakabins (or whatever was the equivalent in those days) that made up the hospital was ideal territory for a hunted man looking for temporary cover, but not so good for someone trying to get from place to place in a hurry. It was especially difficult when attempting to use short-half-life isotopes for injection into patients who were not in the PET scanner, which was carefully sited in the cyclotron building. One of my colleagues, a junior house physician, was instructed to scan the neck of a patient on the second floor of the main hospital to identify how her parathyroid glands were working. The isotope that would be taken up in this gland and was therefore deemed suitable had a half-life of seven minutes. Several practice runs were made by the overworked houseman so that this highly valuable isotope, freshly made up and incorporated into the carrier compound, could be taken safely from the cyclotron unit, about a quarter of a mile distant. Two aluminium ladders, a huge keep net, a bicycle and a fishing rod with a large reel were prepared. The practice runs went without difficulty and it was found that an ampoule, resembling in size and shape the real container of the isotope, could be delivered from the cyclotron in four and a half minutes.

On the morning of the intended injection, the Geiger counters and the patient were prepared. Her bed was placed close to a window on the second floor. The fishing rod, complete with reel, was placed next to the window. Once the isotope had been delivered from the machine, prepared for injection and placed in a suitable vial and then into a canvas bag, the nearest window, on an upper floor of the cyclotron building, was opened and the vial was dropped vertically into the large-aperture keep net held by the waiting houseman in the car park. He mounted his bicycle, pedalled furiously to the first perimeter wall, shinned up the aluminium ladder, mounted the wall and gingerly felt his way over to the other ladder held firm by a waiting assistant, then sprinted 200 yards to the back of the ground floor of the hospital, two floors below the Endocrine Unit. Right on time, the line from the fishing rod was unreeled, the houseman gave the signal after tying on the bag containing the vial, and the isotope inside it was wound in to the window above, in the ward sister's office. But the patient had felt a call of nature and had locked herself into the lavatory. After various negotiations through the solid wooden door, the patient emerged and a sweating member of the medical team injected the isotope into the patient with just over two minutes to spare – still a new Hammersmith record. I have no idea whether the patient's parathyroid gland showed positive on the Geiger counter.

Picture the brain at work

The presence of the cyclotron at the Hammersmith was extremely important in establishing one of the first PET scanning units in Europe looking at the brain. The use of compounds such as those containing radioactive oxygen or radioactive glucose meant that both blood flow and brain metabolism could be measured. One powerful technique introduced fairly soon was substraction scanning. The brain might be scanned while the volunteer in the scanner is gazing at a blank screen, and then scanned again, for example when the screen contains a word. Thus physiologists and psychologists could learn what part of the brain is actively associated with reading. But one problem was that all the attention given to generating isotopes and working out the physics of the scanning process meant that the people best able to operate the machines were not the same individuals conducting research into the function of the brain. At that time, psychology was somewhat in thrall to the newly emergent discipline of computer science. The accepted view was that the human brain was no more than a very complex logic machine, one that could be replicated as our skills in programming grew sharper. PET was never going to be the technology that knocked this view from its throne, partly because it was so expensive. But what is more important is that any process that involves injecting radioactive substances into a person is likely to be limited – particularly if its purpose is research, rather than cure.

The resolution in PET scanning is limited, so structures in the brain do not show up very sharply. PET also has the severe disadvantage that volunteers cannot be repeatedly scanned. Repeated exposure to radiation would be harmful, and this is a major limitation, particularly as so much neurological investigation requires repeated evaluation in the same subject under differing conditions. And it is especially unwise to use women and children for this kind of research – women because they might be pregnant, and children because their threshold of what is acceptable and safe is much reduced. Ethics committees are properly cautious about giving approval to techniques involving PET that might expose the subject to any extra radiation risk. So while PET provided some very exciting three-dimensional images of the brain at work, it was never likely to enable us to build a detailed map of consciousness. Consequently, PET scanning is now largely used for specific indications where actual measure of brain blood flow is required, for looking for tumours (which tend to have a high metabolic activity), and in scanning people who might have pieces of metal in their skull when magnetic resonance imaging (MRI) cannot be used. PET is also used extensively for research into drugs that may be active in the brain. PET can evaluate very specific markers for brain activity – not, for example, just blood flow. So, if a researcher is interested in, say, dopamine function, PET can be of considerable value. PET is useful, too, for psychological evaluation of auditory function. This is because the PET machine is very quiet, unlike the MRI machine that has tended to supersede it.

Magnetic resonance imaging has proved to be a hugely important tool for brain and cognitive research. In another book, *Human Instinct*, I have explained in detail how MRI images are obtained. Essentially, if a person is placed in an intense magnetic field, the molecules in the body line up like minute compass needles. Thereafter, the molecules can be knocked off their axis by exposing them to a pulse of radiowaves. The molecules in different tissues spin at slightly different rates, emitting a signal that is detectable by arrays of sensors connected to a computer. Remarkable three-dimensional images can be generated and, as far as we know, without risk. It seems that repeated exposure to even very powerful magnetic fields is harmless and this has led to MRI being used widely to investigate the brains of willing subjects repeatedly. More recently, fMRI, or functional magnetic resonance imaging scanning, has been an important step forward. This utilizes a property of haemoglobin, the oxygen-carrying pigment in red blood cells. The amount of oxygen carried by haemoglobin changes the degree to which the haemoglobin, flowing through the brain, disturbs a magnetic field. By taking many readings from slightly different angles and building up a computed picture, it is possible to construct a three-dimensional image of great clarity.

While writing this book, I had the privilege of meeting Professor Richard Ernst from Switzerland, who won the Nobel Prize in 1991 for his remarkable work on improvements in magnetic resonance technology. This great scientist rather wistfully told the story of how he had recently sat in an MRI scanner to have his own brain imaged. With all the photographs taken, the technician came up to him and said, 'Professor, your brain is completely normal.' Had a technician said this to me I would have been reassured. But then I haven't won a Nobel Prize; it does rather illustrate that a brain that produces great intellectual achievements does not generally stand out from the run of the mill.

MRI not only delivers an image of great clarity in a short period of time, it is cheaper than PET technology. Nearly every major hospital now has an MRI scanner, and in recent years MRI has become increasingly used to replace other kinds of imaging, such as conventional X-rays. Moreover it does not require a large cohort of staff to operate it. As the *New Scientist* writer and author of *Going Inside*, John McCrone, describes it, MRI is 'brain scanning for the masses' – a reasonably accessible technology that nearly any scientist can use to test his or her thesis.

MRI is not the last word in brain mapping technology. In recent years, the technique of magnetoencephalography (MEG) has offered a few further refinements. MEG is a sensitive technology, able to read the minute traces of magnetic activity given off; however, the spacial resolution is limited and not substantially different from EEG. Because the signals are so small and detection of them difficult, it can only be used in specially constructed shielded rooms, and even then experiments can be hampered by the faintest movement. But one of MEG's great advantages is its speed. PET scanners could measure what happened in the brain over a period of minutes, fMRI over a few seconds. But MEG can deal in thousandths of a second. Given that our thoughts take fractions of a second to appear and disappear, any technology that can keep up with them is likely to be a useful tool. Another significant advantage over EEG (when background electrical activity can confuse the picture) is that the scalp and skull bones do not interfere with the underlying changes in the magnetic field occurring in the brain.

Watching single neurons at work

One of the most important advances in brain science has been the development of the ability to assess what an individual neuron is doing at a particular moment. Most of this work has been done in laboratory animals. Studying single cells by assessing their electrical activity is an important alternative to looking at regional activity in the brain, and makes it possible to look at what individual elements are doing within a region of interest.

Interest in single neuron measurement originates from the 1940s and a wonderfully productive partnership by two remarkable scientists. Andrew Huxley and Alan Hodgkin won the Nobel Prize for their work in 1963. Hodgkin is regarded as one of the great physiologists of the last century. Together with Huxley, he illustrated how nerves work electrically and how they 'fire' impulses. Hodgkin and Huxley did something revolutionary by inserting a tiny electrode directly into a nerve cell. Part of their genius was to use an animal with unusually large nerve cells – the squid – which they caught offshore in Plymouth. When I say these nerve cells are big, I am still referring to a cell around the thickness of a human hair – about 0.5mm across. They made a surprising discovery. Far from what was expected, they found that the resting potential inside the cell was –60 millivolts, and during a nerve impulse this changed to reach a peak of around +40 millivolts. They were unlucky in one sense – the discovery was made in July 1939 and a month later Britain was at war with Germany.

Alan Hodgkin was brilliant (and also something of an amateur mathematician), and the War Office commissioned him to work with radar. He was highly instrumental in developing radar in fighter planes so that they might detect Luftwaffe bombers. He worked on the spot and flew many hours strapped into the tiny space in the tail of fighter planes. His experiences must have been hair-raising – literally – because his equipment needed 15,000 volts and in the rare atmosphere 10,000 to 16,000 feet above sea level was prone to spark massively without warning. From time to time he found himself strapped right next to an electrical fire in the equipment surrounding him.

Andrew Huxley, who comes from an extraordinarily gifted scientific family, was an equally talented Cambridge undergraduate. But like many good scientists he kept a sense of proportion about his achievements. When speaking in Stockholm at the time of his being the joint recipient of the Nobel Prize, he modestly paid tribute to Hodgkin as his teacher. His war was also spent involved in radar, but more in the development of radar for gunnery and for use in naval ships. Like Hodgkin, his wartime work must have saved countless lives and was one of the most important defensive advances in recent military history. Also like Hodgkin, he eventually became President of the Royal Society, the most influential position in British science. In recent years his work at University College, London, concentrated more on the action of muscles than nerves.

The award of the Nobel Prize to these two great men somewhat overshadows important earlier work on micro-electrodes. Ida Hyde in 1902 at Harvard constructed tiny glass tubes of about three microns' diameter that were filled with salt water and therefore good electrical conductors. She was interested in cells that contracted, like muscle cells. More relevant to neuroscience was the pioneering work of Angelique Arvanitaki, who dissected giant cells from molluscs, initially the sea hare and the land snail. She showed, a little earlier than Huxley and Hodgkin, that nerve cells rhythmically produced electrical impulses and that one nerve cell in close proximity to another could influence the way it discharged.

There have been many refinements in electrodes, making the study of smaller neurons possible. Glass has a valuable property – fine glass tubing can be drawn easily into very thin tubes after being briefly heated, under tension, in a laser beam – a technique I use in my own work studying eggs and embryos. But modern single-cell research does not necessarily require the glass electrode to be inserted into the neuron itself. The electrode can be inserted painlessly into the brain of an animal, and, providing the tip of a very fine glass electrode is in close proximity to a neuron, its activity can be accurately measured. Important observations can be made by watching the change in electrical potential of single cells in this way, for example when an animal moves a particular limb, sees something or performs a given task – or possibly when it learns something new. This approach has been expanded in an exciting way recently. A number of physiologists have made simultaneous recordings from more than one single cell in animals. It is reported that one recent study of the hippocampus involved recordings from no fewer than 150 cells, taken simultaneously. Single-cell science, though, by no means solves all the puzzles of brain function. Studying very many neurons working together in harmony must hold many of the answers to understanding the nature of the human mind.

Animal research

No potted history of neuroscience is possible without a discussion on the use of experimental animals. As we shall see again and again throughout this book, animal experimentation has been central to the understanding of brain function ever since Galen in the second century. A surprising amount of brain research, valuable in understanding human nervous function, can be done in relatively 'primitive' creatures, such as fruit flies and, as we've seen, molluscs. But the mammalian brain has a unique cortex and much of the exploration of mind – cognitive neuroscience – requires work in mammals. Make no mistake about it. The study of a whole range of serious or fatal human diseases, for example cranial tumours, brain damage in babies, depression, schizophrenia and CJD, has required work with mammals, and this work has to continue, in my opinion, if human lives are to be saved. There is a strong moral imperative here. The central tenet of human ethics is the notion of the sanctity of life – human life above all. But we have important moral obligations to the lives and welfare of animals, too. These responsibilities must be taken with absolute seriousness. Not to do so would be to brutalize our society. So the codes of conduct in the human treatment of animals – which are now more rigorously enforced than ever before – must be taken seriously. We have no right to make animals suffer more than is absolutely necessary; they must be treated with humanity and respect at all times.

There are certain basic principles that experimenters must take into account. First, on conducting any research with animals there must be recognition that the studies are worthwhile and likely to lead to measures that have a reasonable chance of benefiting health. In this respect, repeating experiments that have already been done by others purely for the sake of minor advantage is not ethical. Secondly, all experiments must be conducted throughout with the minimum of pain or distress, with anaesthesia or analgesics used if there is any pain, and animals kept clean, comfortable, well housed and respected. Very few experiments cause any pain. Where this does happen – such as when finding out how to avoid or minimize human pain, or how to improve anaesthesia – it is kept to a minimum and pain relief is given. If there is evidence of prolonged suffering, there is a need to destroy the animal humanely and swiftly. Thirdly, the number of experiments must be kept to a minimum and alternatives that might be valuable should always be preferred even if they are more expensive or difficult. Cell culture techniques, human tissue experiments, studies on simple single-celled or invertebrate species are to be preferred if they are a realistic alternative.

Animal research has been attacked with massive hostility in recent years. Most scientists find this puzzling when the majority of the population eat meat and wear leather shoes and belts or carry leather handbags. Animal farming is far less rigorously regulated than laboratory animal usage. Leaving aside the question of food – and an animal diet is certainly not essential for human health – there are the issues raised by therapeutic drugs. All of us in modern society take drugs, use vaccines and employ pharmaceuticals that protect and maintain health. These must be tested for safety on animals, and usually more than one species will be required – sometimes in large numbers. Any alternative, with the attendant risks to human life, would be unthinkable to all but a tiny group of people. It is no accident that a recent House of Lords Select Committee, a body known for its independence of mind, reported of drug testing that it is '. . . essential for medical practice and the protection of consumers and the environment, as it provides information that is not currently available from any other source'. Perhaps, before the animal protesters persuade people that animal research is wrong, every drug bottle and preparation in this and other European countries should be clearly stamped: 'This preparation has been produced, and your safety improved, by the use of animal testing.'

What should be deeply worrying to people living in a democracy is the lengths to which some animal rights groups and protesters will go in their attempts to ban animal research. Their attacks on scientists and on scientific institutions risk human life and have barbaric consequences. Much of their propaganda, put out on the internet for example, is misleading and inflammatory. Their exhortations to criminal behaviour are clearly illegal and often violent. Their boycott of, or attacks on, companies that support animal work, however loosely – such as banks, insurance companies and pharmaceutical companies – are wicked because they frighten and intimidate. It should be a source of deep concern to every democrat in a modern society that governments do not do far more to protect individuals and institutions who are targeted by these protest groups, and by the violent action they frequently use. Until proper and effective measures are taken about criminal action and criminal threats, neuroscience and the health of people who suffer from brain diseases, such as stroke and Alzheimer's, will continue to be in serious jeopardy.

Let me conclude here with a rather surprising statistic. If you can bear it, visit some of the more virulent websites that contain material which violently condemns animal research and those who undertake it. Many canards are repeated: that animal research is done to promote the careers of scientists, that there are strong commercial motives behind it, that research has seldom achieved any scientific knowledge which could not have been achieved by other means. Throughout this book there are references to great scientists who have made significant contributions to our understanding of the brain. The highest accolade of all is reserved generally for the Nobel Prize winners – these are men and women whose science is considered of the highest importance. When a Nobel Prize is awarded for medicine or physiology, it is because the contribution is regarded as of extreme importance for health. Since 1901, no fewer than 132 individuals have won Nobel Prizes whose work would not have been possible had all research on animals been banned. Seven of the last ten Nobel prizewinners have relied at least partly on animal work. Of these 132 Nobel laureates, at least thirty-two individuals have made a direct contribution to neuroscience – indeed, many of them are mentioned in this book. These people did not use animals to further their career. They did so because they believed, and have demonstrated, that their work was important in the battle to save lives and to promote health.

chapter two

How does your brain work?

It is remarkable how controversial the study of the brains of dead people has become in Britain. Since the recent publication of the Isaacs Report by the government, there has been an extraordinary outcry. This is because pathologists have taken it upon themselves to store brains after post-mortem examinations to further knowledge of their possible abnormalities. The British press have made claims that such brains 'were illegally taken' or even that they were 'stolen'. These statements, designed to shock, have not been denied by the NHS, Department of Health officials or ministers of government. This irresponsible failure on the part of officials who should know better is having a devastating effect on brain research in Britain. It is widely agreed among neurologists that detailed study of brain tissue after death is of major medical importance. However, very soon after life ceases, the brain decomposes so fast that its structure can no longer be usefully studied unless it is first rapidly preserved by fixation in formalin. Inevitably, detailed study at the time of a post-mortem is simply impossible, so pathologists – completely responsibly in my opinion – have fixed brains to store them for later assessment when there is time to do detailed analysis. Because government officials have been so unsupportive of brain researchers undertaking this kind of study, there is a serious restriction on our knowledge of one of the most important and expensive issues facing the NHS and our society – namely, mental ill-health.

I hold strong religious views and certainly most religious people rightly desire that respect should be accorded to dead bodies. These religious views are essentially based on a conviction of the sacredness of life. But to preserve life, it is sometimes necessary to study death. Of course, I hold that the religious views of others should be respected deeply. But I do not consider that retarding the progress of society at large by overreacting to the religious sensitivities of a small number of people is acceptable. Of course, the relatives of somebody who may have had misgivings about a post-mortem have a right to refuse that examination and for the body to be buried intact. That is perfectly reasonable. But if relatives give permission for a post-mortem, then it should be understood that certain tissues may be required to be stored for future study – as part of the original post-mortem.

Neural networks

What is it we can see when we look at a brain after its preservation in formalin? Externally, not much. That unprepossessing object weighing roughly three pounds (or 1,400 grams) looks mostly like a gigantic, convoluted fungus. But under the microscope we can see it contains an extraordinary number of cells. Some are neurons, the cells that do the work of thinking: roughly 100 billion of them. Only a small proportion of the brain cells – probably around 10 per cent – are neurons, however. They need a supporting structure of other cells, and these are

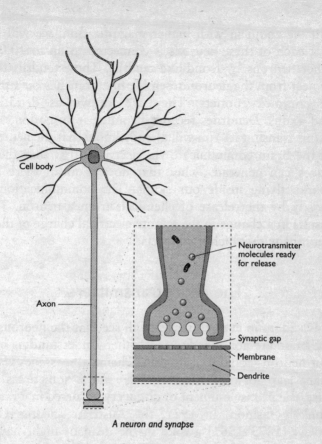

Cell body

Axon

Neurotransmitter
molecules ready
for release

Synaptic gap

Membrane

Dendrite

A neuron and synapse

called the glia. The early anatomists thought of the glia as a 'glue', hence the name.

Glial cells are of various types. Some perform the function of protecting the neurons from undesirable chemical compounds, such as many drugs. Other cells in the glia are used in the healing process – if neurons die, these cells invade the damaged areas and eat up the dead neuronal tissue by engulfing it. Lastly, other cells of the glia form the substance myelin, a fatty sheath wrapped around the neurons that acts as a kind of electrical insulator.

If we zoom in with higher magnification, we will see that each of these neurons is connected to up to 10,000 other neurons by frond-like tendrils. These tendrils that project from the neuron resemble fine branches on a tree, or complex rootlets. They are known as dendrites and axons. Dendrites receive incoming information, and axons transmit it. You will already know that the neurons of the brain communicate via electricity that they themselves can generate – but that doesn't mean there are sparks flying inside our skulls. The communication is mostly by the release of chemicals from a neuron. This results in a change in the minute electrical charge of those neurons with which it connects.

The neurotransmitters

If we zoom in even closer, we will see that the neurons of the brain are not joined to each other. Understanding how our neurons communicate using chemical messages was a giant leap forward for neurology. As is sometimes the case, the precise moment of discovery occurred in a rather unlikely fashion. In 1921, the Austrian scientist Otto Loewi (1873–1961) awoke in the dead of night having had an inspirational dream. He scrawled it down on a piece of paper next to his bed and promptly went back to sleep. He woke with a start at six o'clock in the morning, realizing that he had dreamed of something most important, but turning to his notebook he was completely unable to decipher his dead-of-night scrawl. He described the following day as 'the most desperate day of my scientific life'. But the next night, the dream came to him again and, taking no chances, he immediately got up, rushed to his lab at three in the morning, purportedly still in his nightshirt, and began working on the experiment that had come to him in his sleep.

He used two frog hearts which both had their vagus nerve still attached; the hearts were still beating. He placed one heart into a chamber filled with salt water (at a concentration similar to normal body fluid), and connected this chamber by a fluid-filled tube to a second chamber containing the other heart. He then began to stimulate the vagus nerve of the first heart with an electrical current that immediately caused its beat to slow down. Then came the amazing moment of truth. After a short delay of no more than a minute or two, Loewi noticed that the second heart also began to slow down. Loewi drew the obvious and correct conclusion. He realized that his electrical stimulation of the vagus nerve had caused the production of a chemical message. The chemical had passed from the first chamber into the second chamber in the saline, so causing the second heart to slow down without being directly electrically stimulated. By five o'clock that very morning, just two hours after his dream, Loewi had the first proof that nerve impulses were carried by chemical messengers – later known as neurotransmitters – around the brain and body. For this work, Loewi was awarded the Nobel Prize in 1936. But just two years later, being a Jew, he was arrested by stormtroopers. In order to escape their clutches, he had to instruct his bank to transfer all his Nobel Prize money to a Nazi account; and, as a refugee, all his other possessions were subsequently confiscated. He was lucky to arrive in England, penniless.

While Loewi's work showed the chemical message passed between a nerve cell and a muscle cell, what he had observed was precisely the same mechanism by which chemical messages are passed between the neurons in the brain. Neurons aren't joined to each other, so how do they communicate? There is a tiny gap between each axon transmitter site and the dendrite receiver sites of its neighbours. This gap is called the synapse. When communication between neurons occurs, this gap is filled up with the chemicals called neurotransmitters – these carry the electrochemical 'message'. These messages travel at a variety of speeds – between 2 and 20 miles an hour – depending on the size of the axon and the density of that fatty myelin sheath enveloping the neurons. But the human brain, as I say, is the most complicated object in the known universe. There are estimated to be an amazing 1,000 million synapses in a piece of your brain the size of a single grain of sand.

To date, around fifty different neurotransmitters have been identified – though we shall mainly be concerned here with only a few of them, mostly active particularly in the cortex. Most neurons are capable of producing more than one neurotransmitter, depending on the degree of stimulation they receive.

Glutamate is one of the most important and widespread 'general-purpose' neurotransmitters found everywhere in the brain. It is a chemical closely related to monosodium glutamate (MSG), widely used in cooking – particularly in Chinese restaurants. Because monosodium glutamate is so closely related to this ubiquitous neurotransmitter there has been some concern in the past that too much of it added to food might affect the way the nervous system functions. But the latest studies in America tend to confirm that MSG is harmless.

Gamma aminobutyric acid or GABA is the most important inhibitory neurotransmitter in the brain. To prevent the brain being overloaded by too many neurons and all their synapses firing at once, the neuronal circuits have to 'decide' which messages to pass on. GABA is the neurotransmitter that most frequently acts as a controller. If, for some reason, there is insufficient GABA, the system reacts chaotically, and many thousands of neurons send messages simultaneously, which can result in a fit. I have heard neuroscientists describe an epileptic seizure as losing the battle between stimulation by glutamate and inhibition by GABA. GABA helps neurons discriminate about which signals they respond to. Evidence from Rhesus monkeys suggests there may be inadequate GABA in old age. This could be one of the reasons why there is mental deterioration as we get older, and the findings imply that treatment with this compound or others that increase its action may improve well-being in old people.

Dopamine is associated particularly with voluntary movements – and people with low concentrations of it, such as sufferers from Parkinson's disease, generally have problems with initiating action. Low dopamine levels are also found in people with Attention Deficit Hyperactive Disorder, and in depression. Schizophrenia is associated with high dopamine levels. The dopamine pathways of the brain are well defined, and we can see how variations of dopamine secretion are responsible for certain disorders and traits. For instance, one key dopamine pathway travels from an area of the brainstem called the substantia nigra pars compacta up to the basal ganglia, enabling smooth performance of automatic movements – for instance, walking.

Among other things, as we shall see, dopamine is also linked to certain behaviour – such as alcoholism, criminality, thrill-seeking and drug taking. This is not surprising, given that dopamine is particularly important in the pleasure centres of the brain. The American geneticists David Blum and Kenneth Comings believe they have identified a specific change in one gene that is found in 50–80 per cent of alcoholics, drug addicts, compulsive eaters and compulsive gamblers. This gene prevents dopamine from binding to neurons in the 'reward-motivation' pathways of the brain. As a result, people with this gene do not feel sufficiently satisfied when they have engaged in behaviour calculated to give them a sense of reward. So they do it again and again.

Serotonin, among other functions, plays a role in appetite regulation – people with high levels of serotonin may, for example, have a poor appetite. It is also important in the regulation of mood. The anti-depressant Prozac is one of a class of drugs called Specific Serotonin Re-uptake Inhibitors. These work by preventing the action whereby the neuron reabsorbs serotonin after it has done its 'message-relay' job. The result is high levels of serotonin sloshing around the brain – which results in improved alertness and mood.

Adrenaline (or *epinephrine*, its alternative name) is chiefly found elsewhere in the body, though small amounts also act within the brain. Not surprisingly, it is associated with high levels of alertness and excitement. *Noradrenaline* (or *norepinephrine*) is more concentrated within the brain, and its levels increase dramatically once we wake up. This is primarily a vigilance neurotransmitter – not only does it help us focus attention, it also performs the same function at the molecular level within the brain – ensuring that weak neural connections wither away while strong ones are nourished.

We will come across several other neurotransmitters along our journey – among them the *endorphins*, which control pain and stimulate reward centres within the brain, and *adrenocorticotrophin*, which appears to inhibit the analytical left brain and enable creative thought processes.

The regions of the brain

If we now step back once more and look at the brain with the naked eye, we see that this remarkable object is not just composed of two halves; each half is split into four further components (see pages 29 and 64). Right at the back is the occipital lobe, largely responsible for visual processing. Further forward, just around the left and right ears, are the temporal lobes, which deal with, among many other things, language and sound processing. At the top of the brain are the parietal lobes responsible for perception of space and where you are in it, and partially at least for movement and touch. Right at the front are the frontal lobes – responsible for what we call the 'higher functions' of mankind, such as thought and planning. There is also another 'lobe' or region, the insula, deep between the temporal and parietal areas, which is important in processing visceral sensation – for example taste, and temperature.

The frontal lobes play a key role in regulating our behaviour. We shall explore the development of the human brain from conception to old age in a later chapter, but it is worth noting here that the frontal lobes take the longest to mature, in some cases not completing their growth until a person is in their mid-twenties. It is clear that children's capacity for learning to control themselves is something that can only improve with time – a ray of hope for exasperated parents and teachers!

Going in underneath the surface of the brain – that characteristic formation of convoluted ridges (known as gyri) and grooves (known as sulci) – we see a further set of structures, each with a twin in the opposite hemisphere. (The only item not replicated in both hemispheres, as we saw, is the pineal gland.) Four of these units – the *amygdala*, *hypothalamus*, *thalamus* and *hippocampus* – are connected to form what is sometimes called the *limbic system*. The limbic system is thought of as the seat of our emotions, along with many of the instincts, appetites and drives that help us to survive. The amygdala, a tiny, almond-shaped structure, is the area active when we feel emotions. If your amygdala was damaged or absent, you would be likely to lack the capacity to feel fear.

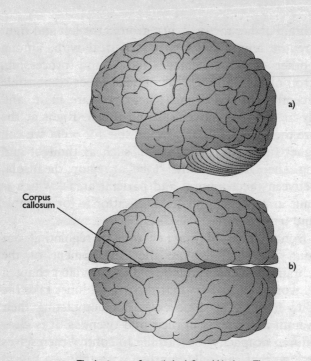

*The brain seen from a) the left and b) above. The two
hemispheres of the brain are joined by the corpus callosum*

The hypothalamus acts as a processing centre for many of the other regions of the brain and the body, receiving information and sending out appropriate responses. It's the hypothalamus, and in turn its mastery of the pituitary gland, that causes our hearts to race and our blood pressure to rise when we are afraid, for example. People with impaired hypothalamic function may have a tendency towards anorexia or overeating, as if they are unable to communicate with their own appetites. They may also be infertile because if the wrong signals are sent by the hypothalamus to the pituitary gland, the hormonal instructions to produce eggs or sperm may be inadequate. The thalamus, meanwhile, acts as the connection site between the sensory components and the higher functions of the brain. All incoming sensory information – with the exception of smell, the most primitive sense – passes through the thalamus before being sent upstairs to the cortex. The hippocampus is responsible for the storage and retrieval of long-term memory. People with hippocampal damage experience severe memory problems, and it has been shown that people who have been through serious trauma, and subsequently have only fragmentary recall, may have reduced hippocampal tissue.

Above and slightly encircling the amygdala and hippocampus are the basal ganglia. These act as an interface between the rest of the brain and the motor centres of the brainstem below them. Hence they have a distinct role in the planning and production of movements. But, as with many regions of the brain, their function cannot be so neatly labelled. People with damage to the basal ganglia can also have difficulties in understanding the non-verbal communications of other people. The same is true of the structure that sits at the very back of the brain, just above the stem. This area, the cerebellum (which means 'little brain' in Latin), has long been seen as a purely primitive motor, movement and balance centre. But recent research has demonstrated the cerebellum's role in the planning and timing of action, judging the shape of objects and creating properly proportioned drawings. And there is increasing evidence of its involvement in language and reading and many other aspects of cognition.

Together with the cerebellum, the bottom-most region, the brainstem, forms the oldest section of the human brain – the part we have in common with our reptilian ancestors. The brainstem is the site of the most basic survival programmes – those that control eating, sleeping, defecation, orgasm and breathing. It does enough to keep an organism alive in the absence of any other functions. Cats whose brainstem has been detached from the rest of their brains can still move around, pounce if they

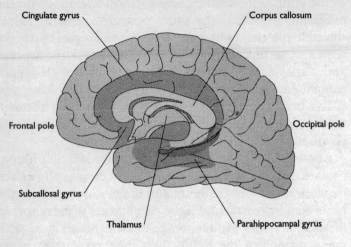

The limbic system

hear a noise and eat food if it is placed in front of them.

These various modules of the brain are connected to one another by densely packed, criss-crossing ropes of axons. There is also a thick band of connecting material that joins the left and right hemispheres of the brain – known as the *corpus callosum*. As its position would suggest, the corpus callosum allows the two halves of the brain to communicate with one another. As we shall see, the left and right hemispheres, and their interaction, play a vital role in the creation of who we are.

A brain of two halves

While the left and right hemispheres of our brain each have identical structures and in normal circumstances are in constant communication with one another, each half also houses somewhat different functions. Having said that, the two sides of the brain do not work in isolation from each other. In most people, the left side tends to do

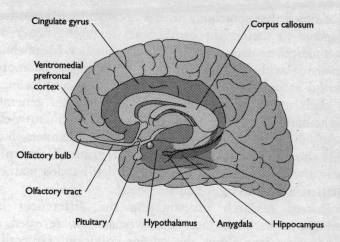

Right hemisphere seen after bisection of brain (inner surface)

Labels in figure:
Cingulate gyrus
Corpus callosum
Ventromedial prefrontal cortex
Olfactory bulb
Olfactory tract
Pituitary
Hypothalamus
Amygdala
Hippocampus

more analytical processing. As well as being the site of language faculties in most people, it is also often responsible for aspects of reason and deduction. The right side, in contrast, tends to be a much more holistic machine.

Much has been written, often without good evidence, about the difference between the function of the right side of the brain and the left side. Certainly, under a microscope, a difference between the left and right brain can be seen. Our right hemisphere is made up of more white matter than grey matter. The grey matter is essentially where the processing gets done; the white matter provides the channel of communications. The right hemisphere with its greater amount of white matter is sometimes stated to have more density of connections between the neurons. The connecting axons in the right brain are longer than those in the left brain and they can connect with neurons that are further away. Some writers have claimed, without very good evidence, that the left brain is better at associating logic, ideas and concepts and that the right brain is the seat of creative and intuitive thought. It is claimed that somebody with a dominant left brain (or damage to their right hemisphere) might be better at methodical, analytical work – like accountancy. Meanwhile someone with a dominant right brain might be more skilful at linking seemingly disparate ideas and feelings – skills useful in art, music and literature. In general, right-handed people tend to have a dominant left hemisphere, and the reverse is true for left-handers. But this shouldn't be taken as a universal truth. If this rule was always consistent there'd be no left-handed mathematicians and no right-handed poets!

Recent research investigating the differences in behaviour between men and women has revealed a fascinating fact about the corpus callosum. Generally speaking, it tends to be proportionately larger in women. Ruben and Raquel Gur at the Pennsylvania Medical Center in Philadelphia have confirmed that women tend to be more able to 'multi-task'.[2] Women regularly carry out a number of tasks, say in the office or around the home, all at the same time. Men, on the other hand, are more inclined to focus on and complete each individual task they're given. Ruben and Raquel Gur suggest that the corpus callosum is larger on average in women because it has more neural connections, enabling the brain to communicate more quickly and easily between its different regions. In men, there are fewer connections across the two hemispheres, so neuronal communication is somewhat slower and therefore more effective when a function is taking place in one particular brain region rather than across the two halves. They suggest that this difference in the corpus callosum might even explain some well-known emotional differences between men and women. There is much speculation about this, and some sources claim emotional processing is largely located in the right hemisphere, and that the power to express our feelings through language lives in the left. Could it be that for men, with fewer connections between the two halves, it is harder to talk about emotions? For women, on the other hand, with all those extra fibres in the corpus callosum, expressing emotions is second nature. So, at least, we men perhaps have some sort of excuse for our behaviour.

A splitting headache

In the 1960s, surgeons tried a last-ditch treatment for severe epilepsy. They set about severing the corpus callosum. Often the reason for a major fit with loss of consciousness (and the risk of death) is that the electrical discharges from the damaged area of the brain spread through the cortex, involving both sides. Cutting the connections between the two halves is intended to stop the seizures from spreading to involve the whole brain. More often than not it helped dramatically. However, it wasn't long before the patients began demonstrating some rather odd side effects.

The neurobiologist Roger Sperry, who won the Nobel Prize in 1981, was one of many illustrious scientists to spend part of his career at the California Institute of Technology. He was something of a polymath – one of his colleagues there, for example, records what a fine sculptor he was. But in the annals of neuroscience, he is perhaps best known for his work on these split-brain patients. He once famously declared, 'What it comes down to is that modern society discriminates against the right hemisphere.' In a noted case, he asked a woman to fix her eyes on a screen that had a small dot in the centre. There is a crossover between the eyes and the parts of the visual cortex that receive the information – so that under normal circumstances, information from the right side of the visual fields goes to the left side of the brain, and vice versa. The same is true of hands – the left hand is wired directly to the right brain, and vice versa. But clearly in the split-brain patients, most of the crossovers have been severed. Sperry tested the effects of this by flashing up an image of a cup for one twentieth of a second, just to the right of the dot. This is the amount of time it takes for the eye to register an object but not to focus on it and thereby send information across the two hemispheres. So in this woman's case, thanks to her severed corpus callosum, the image went only to the left side of the brain. Because that is also the language region of the brain, she could identify and speak the name of what she had just seen. A cup.

But then an image of a spoon was flashed up to the left of the dot, so that it would enter her right hemisphere. Asked to report what she had seen, she answered 'nothing'. Sperry then asked her to reach underneath the table without looking and select, with her left hand using touch alone, from a number of objects, the one that she had just seen. She immediately picked the spoon. But when asked to identify the object in her hand, she defined it as 'a pencil'.

In this experiment, we can see a clear example of the left and right hemispheres at work. When the cup image was on the right, her left brain could not only identify it, but also allow her to say what it had seen. But when the spoon image was on the left, and went to her right brain, she couldn't tell Sperry about it – because the left brain controls language, and the left brain had not seen it. The only thing her left brain could say was 'I see nothing.' But the image of the spoon was absorbed by her right brain – so it could use the left hand, without language, to select the correct object, even though it couldn't name it properly.

Not everyone has his or her language faculty situated solely in the left brain. A very few rare split-brain patients have been able to communicate from their right hemisphere – giving rise to some eerie results. Michael

Gazzaniga, whose fine books on neuroscience I strongly recommend, worked at the University of California and now works in Dartmouth, New Hampshire. He came across a student, a patient whose right hemisphere had the ability to communicate in short phrases. It wasn't an easy task, because spoken questions – unlike images – can't be sent to one hemisphere in isolation. When the left brain hears language, it takes over. So Dr Gazzaniga and his co-researcher Robert LeDoux devised a new method: they would speak the bulk of the question, but flash the key-word up on a screen to be read, thereby letting the right brain absorb what the left brain could not. For instance, they might ask – 'Tell us what are your . . .' and then flash the word 'hobbies' on the screen. The patient would then answer questions directed to his right hemisphere by arranging Scrabble letters.

The patient's right brain began to emerge as a distinct personality with its own likes and dislikes. When asked to rate a long string of items such as foods, colours and names, the right hemisphere consistently judged them to be worse than the left. Even more bizarrely, when asked what he wanted to do after graduation, the left brain vocally answered 'draughtsman' while the right brain spelt out the words 'automobile racer'. So are we all walking around with two separate personalities inside our heads, one of them rendered mute by its lack of language? That would certainly be the most extreme (and somewhat worrying) conclusion to draw here. It might just be the case that your consciousness moves continually from one hemisphere to the other, unless interrupted. Or that there are many different 'versions' of yourself inside your brain, usually integrated and harmonious thanks to your very own corpus callosum.

Professor Gazzaniga also writes of an intriguing episode in which a short piece of distressing film was shown to the right hemisphere of a split-brain patient (in other words, through her left visual field – the left side of both eyes). Having no language capacity in her right hemisphere, she couldn't report what she'd seen – and in fact said that she'd seen nothing at all. But nevertheless she did mention that she felt scared and anxious. Later on, she let slip to another member of the research team, 'I know I like Dr Gazzaniga, but right now, I'm scared of him for some reason.' It seemed that the emotional content of what she'd seen had been picked up, so she had tried to find some explanation for the feeling. The same kind of phenomenon can be seen when more neutral images are shown to split-brain patients – such as ocean scenes and waterfalls. The patients report feelings of tranquillity; they remain unaware that those feelings have been caused by the footage they've been shown.

How do split-brain patients go about living a normal everyday life? Some of them report being unable to perform even the simplest tasks because, while one hand acts according to their will, the other hand carries out all manner of Faustian mischief. Alan J. Parkin, Professor of Experimental Psychology at the University of East Sussex, has reported on patients with so-called 'Alien Hand Syndrome' (AHS), whose one hand takes off their clothes as fast as the other hand can dress them.

Alien Hand is the stuff of nightmares and horror movies. About two years ago, Dr Feinberg, from the Albert Einstein Medical College, on the television Learning Channel described a woman who, awakening from a troubled sleep, found in the pitch black of her bedroom a hand tightening around her neck, vicious and implacable. Using all the strength in her right hand, she tried desperately to pull it off. Then, struggling, she realized that the hidden assailant was her own left hand. So it is hardly surprising that, for the rest of that night, and most nights to follow, she was too scared to fall asleep unless totally exhausted.

While many people with AHS have had radical brain surgery to treat severe epilepsy, Feinberg says it can happen to anyone who has had brain damage. The case of the woman whose hand attempted to strangle her goes back to the earliest study of the disorder, written in Germany in 1908; but in matter-of-fact fashion, Larry, one of the patients on the show, also reported waking up one night to find his left hand around his neck. Sometimes when he is trying to eat, recalled Larry, his left hand will grab his right hand, 'and just won't let me get a fork – or spoonful – of food'.

I quite like this story, which I found on a motorbikers' website:

I met this Japanese rider yesterday. Nice guy, maybe in his mid-40s. Well, he tells us that he's not working right now. Says he had a medical condition: 'Baka Te'. I'm totally confused. That means crazy hand. What in the world is crazy hand? So he goes on to tell us that at times he loses control of his right hand. It just does what it wants. One night he was coming down the Daishein Keihin something. All of a sudden his right hand locks up on the throttle. Scary, 'cause this guy rides a 1300 Suzuki Hayabusa. This guy accelerated out of control. He told us that he was doing somewhere around 270kph, struggling and beating his right hand the whole way. Eventually, he was caught by the police running over 250kph with his mad hand. He was taken to the police station but later all charges were dropped against him because of his medical condition – Alien Hand Syndrome. He's STILL riding. Can you imagine that?

Clowns to the left of me, mopers to the right

We all differ to some extent in the degree of dominance of one hemisphere over the other – though, as we shall see later on, there need be nothing permanent about this arrangement. But differences across this axis can be responsible for key facets of our temperaments. A 1994 study by Richard Davidson at the University of Wisconsin involved showing film clips to various subjects, and then evaluating the blood flow to their left and right hemispheres.[3] The film clips were of two types – positive, happiness-inducing, and negative, sadness-inducing. Davidson noted that people with more left-side activity seemingly responded more positively to the positive film clips. And those with more right-brain activity responded more negatively to the negative film clips. It seems that, in short, people with more active left brains set up a loop of feedback between themselves and positive stimuli in the world around them, ensuring a more constant positive mood. And people with more right-brain activity do exactly the opposite – as if they have a tendency to be more moved by the sad things they see. This is certainly interesting. People with depression often say that they 'see the world in a different way' to everybody else. Perhaps there is a lot to be said for not always grasping the big picture.

We can see this difference if we look at people who have suffered a stroke that has effectively 'knocked out' one or the other hemisphere. People with damaged left brains are more prone to depression; they tend to treat their problem as a major disaster. In contrast, people with damaged right brains sometimes even fail to recognize that they have a problem.

The left/right split is seen in our reactions to art – quite often we say that we like paintings but cannot explain why. Indeed, art teachers encourage students raised in an academic environment where thoughts are rationalized to leave their preconceptions behind at the door. They ask them to start connecting with the feelings that art produces. This may seem a throwback to 'method acting' and other movements in the arts in the Sixties, but our left brain is where we use language to define and pin things down, and our right brain is where creativity is most at work. That is why advertising makes so much more use of imagery and music than it does of words. If our left brain got in the way and started analysing whether we really needed this revolutionary new nasal hair trimmer, we might not buy it . . .

A similar process can be seen if a written command is quickly sent to the right hemisphere of split-brain patients. In one case, after the command 'laugh' was flashed up, a patient promptly fell about chuckling. When asked why, she responded that the researchers were making her laugh with their absurd experiments. The left brain was seeking to make sense of actions produced by the disconnected right brain.

The Alien Hand Syndrome described earlier does not result solely from damage to the corpus callosum. An area at the top of the brain called the SMA (supplementary motor area) is also implicated. This area springs into action whenever we want to undertake some action of our own free will. It does not initiate the action so much as send signals to parts of the surrounding motor cortex. As we have seen with many functions of the brain, there is cross-wiring, so that the right part of the SMA controls the left side of the body and vice versa. But notably, whichever side of the body is used in the action, areas on both sides of the SMA are active. The action on the right side of the SMA for a right-handed action is fairly weak, but, Professor Alan Parkin believes, enough to initiate some action if not inhibited. Normally this inhibition comes directly from the side of the SMA opposite the side of the body being moved. It's as if a message is sent saying, 'Do not move – leave this to me.' To get from one side of the SMA to the other, an intact corpus callosum is needed. Hence, in some split-brain patients, there is a tendency for left and right hands to do contradictory tasks. Both sides of the SMA are sending 'Do it' messages to the motor areas and thence on to the limbs themselves.

In Parkin's view, there is nothing mischievous about the work of Alien Hands, or anything to suggest that we are all inhabited by impish, playful alternative personalities. In his view, Alien Hands are most often to be observed undoing the work that the other hand has done, because that's the only job that seems to need doing. For instance, say, in a case of split brain, the left hand opens the door. The right hand arrives on the scene and sees an open door, but carrying the crucial message 'Door – Handle – Do'. Not being able to communicate with the right brain and thus know that the relevant job has been done, the left brain assumes that the task is to shut the door and duly does so. Frustrating if you are in a hurry.

Whether or not we are seduced by the appeal of plural personalities residing in our right and left brain, it is becoming clear that the view of the brain as a series of separate modules is an oversimplification, for in the execution of the simplest tasks, we call upon a whole array of faculties. Imagine buying a pie in a baker's shop, for instance. We use our motor functions to walk in and not bang into the counter. We use our visual systems to see a pie and our memory to line up former pies with the pies we see in front of us. We call our language and social behaviour faculties into action in order to behave appropriately in our interactions with the counter assistant, and our reasoning capacities to work out how much money we need to hand over. In so much of our daily lives, our ancestors are working for us – the 'reptilian' and 'early mammalian' brain as much as the higher cortex. It might be more appropriate to think of the brain as operating like a large city, where the pie manufacturer needs his neighbours at the paper mill to provide the packaging and the oil refinery to provide petrol for transport. And as we shall see shortly, the brain is not a closed system. Like a city, in which businesses and public institutions and family homes grow and move in and out of different premises, the brain is undergoing constant change. In many cases, a faculty impaired is not a faculty lost for all time, but one that may be taken over elsewhere. But before we can explore this concept, we need to understand in a little greater depth precisely how the different units of the brain communicate with one another.

You might have noticed something unusual about one of the above statements. If not, read the information about noradrenaline again. A key aspect – indeed asset – of the human mind is contained within that paragraph. Namely that the brain which creates it, or at least many areas in it, is subject to a state of constant change. Synaptic connections between neurons grow and become more established. Connections that are seldom used grow weak and eventually wither away, the neurons responsible for them taking on new routes and new tasks. In short, the brain is not something formed during the first twenty years of our lives and in a state of decay ever after. It is a moving, changing, fluid organ that interacts with its environment.

The plastic brain

Do you remember being told that you were born with a finite number of brain cells and that, from the moment you were born, these were dying away, never to be replaced? It's a common misconception. The brain is not only the most complex organ in the body, it is also arguably one with great capacity to reinvent itself. That's why the mind can be the product of a flesh-and-blood organ, and yet be subject to change over time – why we can learn new skills, why we forget others we once knew well, why stroke victims can recover lost faculties with patient practice, even why teenage communists can become high Tories in their thirties. The discovery that this is the case has had far-reaching consequences for medicine and for the whole 'nature versus nurture' debate.

The tale of the plastic brain began in the 1940s with a Canadian psychologist called Donald O. Hebb. At that stage, brain researchers had already begun to discover that communication between neurons was no one-way street. On the contrary, it seemed that whenever a neuron was receiving input from another, it would not only pass its message on to other neurons, but send feedback to the neuron from which it had received the original message. This created a loop, which in turn could amplify the signalling of the original cell, or alternatively dampen it down.

Further to this, Hebb discovered that cells involved in a pattern of communication were changed by it. After the initial process of electrical firing had died down, a secondary activity took place, during which neurons strengthened their connections with one another, growing new axons and increasing their supplies of neuro-transmitters, so that it became easier for them to fire again in the same pattern in the future. In other words, the net-work of neurons in the brain regrouped itself according to what information came in, and it remained primed to regroup if that information should come in again.

This feedback and learning mechanism within the brain means that each synaptic connection is 'told' whether it has contributed to the final outcome of the communi-cation. If it has, it is encouraged to react a little more strongly next time, by growing new connection sites and by sending more neurotransmitters across the synaptic cleft. If it hasn't, it will inhibit its actions.

We will deal in greater depth with the brain chemistry of learning and memory at a later date, but for now, let us use an example from daily life. Imagine, for instance, I start working in a bar. Let's also imagine, for the sake of illustration, that it is an old-fashioned bar without a fancy modern electronic till. And that my mental arithmetic is a little rusty. The first time I am called upon to add up the cost of three pints and then deduct this from a twenty-pound note, the calculation may take up some considerable length of time. Inside the deductive regions of my brain, neurons are firing together, perhaps for the first time in years. But as they fire together, a pattern is established – in the same way as generations of footsteps will beat a path through thick undergrowth, making it smoother and clearer with each journey. When I serve the next customer, and the next, the pathway is that little bit more pronounced as the neurons involved become more tightly enmeshed, and the increasing quantities of neurotransmitter promote ever more vigorous firing. Over time, I become proficient at addition and subtraction – faster, more accurate and more confident. I am Barman of the Year. To some, I may seem to be overstating the obvious maxim that practice makes perfect. But what is significant about it is that this process is built into the chemistry of the human brain. As we learn, the circuitry of our brain learns, too.

The best neuron for the job

Our emergent understanding of this process gave birth to a theory known as Neural Darwinism – coined by Gerald Edelman, an American who won the Nobel Prize for his work on how antibodies protect us against infection. Later in his career, while working at the Rockefeller Institute, he turned his attention to brain development. Edelman's idea was essentially that the architecture of the brain was created in the same way as living beings competed for survival. As animals compete for food and space and mating rights in their bid to survive, so do the synapses within our brain have to compete with one another.

In the womb, at the dawn of life, the foetal brain produces new neurons at a staggering rate of 250,000 a minute. These neurons attach to their neighbours in a frenzy of seemingly random wiring – though, as we shall see later, genetic and environmental factors also play a role at this stage. But in the first year of life, there is a process called apoptosis, a programmed mass suicide of brain cells. Neurons are, in effect, forced to compete for their survival, to show that they are part of a useful circuit in order to maintain their connections with other neurons and their supply of neurotransmitters.

For Edelman, the whole evolution of the human brain can be seen as a tale of ever more efficient competition. The simplest creatures, like worms and jellyfish, he suggests only adapt over many generations. Slightly cleverer creatures, like snails, develop a sort of reflex memory over a lifetime. In humans, this process reaches its apogee, such that we are able to make continuous adjustments to our brain in response to the demands of the environment.

It's that final phrase that stands out as the most important for me. For through an understanding of Neural Darwinism in the human brain, we arrive at a clearer indication of the relationship between nature and nurture. If we say a person is 'nervous', or 'aggressive' or 'a great talker', we do not then have to fall into one of two camps and say that this attribute is *only* something they could have inherited, or conversely *only* something that happened to them because of experiences in early life. It is both. Attributes of mind – traits of personality, habits and skills – emerge as result of the interplay between the brain and the environment.

A classic example is shown in an experiment performed by Michael Merzenich, a pioneer of the 'plastic brain' theory. Merzenich carefully measured the cortical areas associated with one digit of a monkey's hand. He then amputated the finger and watched what happened to the brain. He discovered that the related area shrank in size, while the areas associated with the other fingers grew. In further experiments, he trained monkeys to use their middle fingers in a certain task – and noted that the related brain area grew in size, as if the increased demand resulted in a need for greater cortical space.

We do not need to look at monkeys to see this process at work. Research has shown that musicians have 25 per cent more of their auditory cortex given over to the processing of music than non-musicians. The greatest amount of extra music area is found in those musicians who started to play an instrument earliest. Similarly, non-musicians given the task of performing a simple finger exercise over a number of days have been found to grow extra connections in the relevant area of their brain.

This is the basis of Cognitive Behavioural Therapy – a process that seeks to correct behaviour detrimental to self or others by replacing 'negative' thought patterns with 'positive' ones. Jeffrey Schwartz at the University of California, Los Angeles School of Medicine performed brain scans on patients undergoing CBT and noted a decrease in the neuronal activity associated with the original, negative impulse.

We might also look at the thoroughly invigorating example of the nuns of Mankato, a religious order based at a remote nunnery in Minnesota. The nuns have been the object of study for David Snowdon, a University of Kentucky professor, due to their remarkable longevity – many of them living for over 100 years. They are not just long-living, but they have a lower rate of Alzheimer's disease and other impairment of their ability to think than is found in the general population. Now, of course, nuns may not be an ideal population to study – there are certain aspects of their lifestyle that exclude activities given to other women. But their secret, Dr Snowdon is convinced, lies in their ceaseless activity. For the nuns, idleness is a sin and so even the oldest of the sisters keep their brain occupied with quizzes, puzzles and political debates. Snowdon has examined a hundred brains donated by deceased members of the order and concludes that their intellectual activity stimulates the growth and connection of neurons. Significantly, the nuns who were engaged in intellectual pursuits resisted the debilitating effects of ageing far better than those nuns who were kept occupied with menial work.

The brain deprived

Of course, it therefore follows that brains deprived of stimuli produce corresponding deficiencies in their owners. Only by regular usage does a neuronal pathway remain strong and healthy. If neurons are not being used in a certain pattern – for instance, the patterns associated with doing bar-room maths – they will quickly be co-opted into new groupings for new tasks.

Hence, for example, the sad plight of the Romanian orphans discovered in the late 1980s, practically abandoned by the state in vast, bleak orphanages, devoid of any emotional contact or environmental stimuli. The Harvard Medical School researcher Mary Carlson studied babies abandoned at one such orphanage and found that, in addition to being mentally and physically retarded, they also had abnormally high levels of the stress hormone, cortisol.

A University of Washington study undertaken by Geraldine Dawson found some physical abnormalities in the brains of babies raised by clinically depressed mothers. These babies who were less exposed to smiles or other lively shows of excitement by their ill mothers showed reduced activity in the left frontal region of the brain. At three and a half years of age, these babies were significantly more likely to exhibit behavioural problems. However, thanks to the plasticity of the brain – and the especially plastic infant brain – it is nearly always possible to reverse the negative effects of environment with timely intervention. The brain is remarkably resilient. For those readers following another BBC television series in which I've been involved, *Child of Our Time*, this reversal, and the extent to which it is possible, is poignantly shown by some of the children we have filmed.

The brain can even recover from physical damage. Young children who have had an entire brain hemisphere removed due to severe epilepsy manage to compensate for this loss, often with only minor mental or physical impairment. And stroke victims who lose the capacity of speech may regain it, in a reduced capacity, by using neighbouring areas of their brain circuitry. In general, however, we have to be aware that there are limits to the brain's plasticity. Young children recover from brain injuries significantly faster than teenagers, for example.

A good example of the limits to the brain's plasticity is in language acquisition. As we now know, babies' brains are hard-wired to learn language. Research has shown that newborn babies can recognize and develop a preference for their mother's language by the time they are four days old. Russian babies will suck harder, for instance, when they hear Russian spoken, than when they hear Portuguese. By less than a year old, a baby can only recognize as language the set of sounds particular to its mother tongue. Anything else is rejected as non-meaningful in the language sense. When babies babble at this age, they do it using the set of sounds, or phonemes, particular to their native tongue. The same process occurs in children born with cataracts – if these are not removed before they are six months old, the child will never see. The neural circuits vital to the process will swiftly take on other functions if they are not used.

We can see this most vividly demonstrated in the examples provided by the plight of Genie and Isabelle – children who spent a key period of their developmental years deprived of human contact. Until the age of thirteen, Genie was locked in a Los Angeles attic and beaten if she made a noise. She grew up without human interaction and the language areas of her brain were severely distorted. When she eventually did learn to speak, scientists discovered that she was using the auditory regions of her language faculties – which presumably had been kept working because of background noise, such as birdsong and traffic. With only this narrow facility at her disposal, Genie was able to master basic vocabulary, but never grammar.

In contrast, Isabelle was released from her silent imprisonment at the age of six. Within a year and a half, she had acquired a vocabulary of 1,500 words and could use quite complex grammar. It seems likely therefore that the brain has a window of opportunity when an understanding of grammar can be most easily gained, which becomes more difficult somewhere between the ages of six and thirteen.

In short, what happens to us has a direct and observable effect on the shape of our brain, and this, in turn, informs who we are. But thanks to the brain's plasticity, much can be changed throughout our lives. To deepen our understanding of how we become who we are, we need to know not just that the brain changes in relation to the information it receives, but also how it receives and processes that information. If this feedback relationship between brain and environment makes us who we are, then it follows that differences in the way we perceive the world equate to differences in personality and character.

John Ratey, author of *A User's Guide to the Brain*, and associate Professor of Clinical Psychiatry at Harvard Medical School, suggests psychiatrists stop asking their patients how they feel, but instead ask how they perceive the world, in order to understand the origins of their disorder. In the next chapter, we shall be applying this principle to the broad spectrum of human experience. What goes on inside the brain as we see, feel, hear, touch, run from danger, kick a ball or complete a crossword? And could it be that differences in the way our brains handle these tasks lie at the root of differences in our personalities?

chapter**three**

Coming to your senses

In the twentieth century, there was something of a revolution in opera. An increasing number of composers were aware of new scientific discoveries, public interest in psychology was mounting and the understanding of human consciousness was growing. So composers wrote operatic works focusing strongly on the workings of the mind. Leoš Janáček's opera *Jenufa* (1903), Richard Strauss's *Salome* (1905) and *Elektra* (1909), Alban Berg's *Wozzeck* (1925) and Benjamin Britten's *Peter Grimes* (1945) are all psychodramas that examine the mental state of the main protagonists, whether sane or frankly pathological. Another of these great works is Sergei Prokofiev's *The Fiery Angel*. There can be few more gripping, yet terrifying, operas; it always sends shivers down my spine. In it, the central character, Renata, is possessed of an evil spirit, the Fiery Angel, which haunts and obsesses her to the exclusion of all normal love. Eventually, to escape her spirit she enters a nunnery. In an attempt to exorcise her religious delusions, the Grand Inquisitor visits her, but the Fiery Angel's control over her mind is so powerful that, in an extraordinary hysterical scene at the end, the Abbess and all the nuns become equally possessed, strip naked and have what must be the only full-scale orgy ever put on an operatic stage. I strongly recommend going to see this opera.

Prokofiev's remarkable music, with its vivid orchestration, conjures up a striking and horrifying resemblance to some forms of epileptic attack – initially a sense of foreboding and disaster, altered sensory perception of sounds or smells, a religious illusion that seems totally real to Renata, the rapid repeating phrases in the music reaching a climax, a kind of fit, and then, unconsciousness.

Few disorders have caused more superstition, fear and terror than epilepsy. Its fits, convulsions and loss of consciousness stem from flurries of sudden overactivity within areas of the brain – rather like localized electrical storms. These seizures, sometimes regarded as the work of the devil or the work of God, have been recognized since earliest times. Epilepsy is mentioned in the Ayurvedic literature and in ancient Egyptian tablets. One of the most ancient accounts comes from Babylonian times; the tablets (all forty of them) are preserved in the British Museum in London. Each type of seizure these tablets describe was associated with a different god or, more usually, an evil spirit. Many famous individuals suffered from epilepsy, including Julius Caesar, Peter the Great of Russia, Fyodor Dostoevsky and, probably, Lord Byron. Some of the ancients, and medieval scholars such as Maimonides, the great Jewish philosopher and doctor writing in the twelfth century, attached stigma to the disease and wrongly regarded it as mostly inherited. Basic understanding of the mechanism behind epilepsy did not really occur until the nineteenth century, when the neurologist John Hughlings Jackson suggested that the sudden fits were caused by electrical discharges from specific areas of the brain. Jackson, who has been called the father of English neurology, gave his name to the fits

– Jacksonian seizure. As we see throughout this book, the study of epilepsy and focal stimulation of the brain has had a huge influence on our understanding of how the brain, and therefore the mind, works.

Penfield pioneering brain mapping

One of my great heroes – ever since I first heard of his work when I was a medical student studying brain anatomy forty years ago – is Wilder Penfield. He pioneered brain surgery for epilepsy in the 1930s. Originally from the USA, he went to Oxford to study medicine. On his return to the USA in 1921 he is reported as saying, 'Brain surgery is a terrible profession. If I did not feel it will become different in my lifetime, I should hate it.' He had what could be a unique experience and one that most surgeons would dread. When he was in his thirties, and brain surgery was in its infancy, his own sister, Ruth, was diagnosed as having a brain tumour. Once it had been discovered that the tumour was extensive and undoubtedly cancerous, he operated on her himself and removed more brain tissue than most neuro-surgeons would have considered remotely safe. This must have been an incredibly difficult decision, but it gave Ruth another three good years before the tumour eventually returned.

Ruth's difficulties motivated him to establish the famous Montreal Institute of Neurology, which he went on to direct. The understanding of epilepsy at the time was primitive and, although it was very common, the treatment was frequently ineffective. The main drugs, barbiturates, with their depressant side effects, frequently did not prevent fits recurring. It is not surprising therefore that, as a neurologist, he decided to specialize in, among other nervous diseases, epilepsy. Sometimes the electrical storms paving the way for a fit emanate from scarred areas in the brain, produced as a result of old injury. They tend, in some patients, to be preceded by warnings – referred to as auras – such as a sensation of tingling, or of a strange or unpleasant odour. It was Penfield's theory that if he could get into the brain and stimulate different areas electrically while the patient was conscious, this might provoke the onset of an aura. He would then identify where the faulty brain tissue was and cut it away.

You possibly cannot help a tiny shudder at these last words – the idea of cutting into the brain seems crude. It feels wrong because it is difficult to believe that it is possible to treat this seat of our being as if it was an ordinary organ. Surely it must be more – something that transcends physics and biology and exists beyond time and space? This is the organ of consciousness, of original thought and inspiration, the means by which we feel soaring joy, burning rage and perceive the glory of God. Could it be improved by simply cutting bits away with a knife?

Even though well over sixty years of serious scientific activity have left us in little doubt that the brain is responsible for each and every attribute of the mind, these personal feelings demonstrate that brain research is emotive. It turned out to be just as emotive for Wilder Penfield. He noted that, when certain parts of the surface of the brain were electrically stimulated, his patients – who were lying awake on the operating table with the top of their skulls sawn off like a boiled egg – reported a range of sensations. It may seem odd but the brain has no pain fibres so it is insensitive to touch or injury. Most pain experienced in the head is perceived through other tissues, such as blood vessels. Penfield's stimuli produced no pain at all but the body underwent a series of involuntary tics and muscle spasms, and some patients experienced sensations of light, heat or cold. Others reported very subtle mental states – such as foreboding, and déjà vu. And some even experienced quite detailed hallucinations – including in one case a man who heard the strains of Beethoven so clearly that he seriously thought Penfield must have had a radio concealed within the operating theatre.

Penfield's findings shocked even himself – to the extent that, towards the end of his life, he backtracked somewhat. Although his research provided a very solid foundation for the view that consciousness itself resides in the electrochemical processes of the brain, he argued that it could never account for the sum total of experience. Attributes like free will, Penfield argued, could never be pinpointed within the structures of the brain. But a good deal of the evidence now points to the contrary. And to an extent, some of the most magical aspects of human consciousness have been demystified and almost 'normalized'. This is certainly one of the by-products of some more recent findings in Switzerland.

A more detailed atlas

A Swiss neurologist called Professor Olaf Blanke made an interesting discovery when performing a latter-day version of Penfield's procedure on a woman who suffered severe epilepsy. The surgeon builds his specialized map by opening up the scalp and the skull after it has been treated with local anaesthetic while its owner – the patient – is fully conscious but sedated. Various areas can be stimulated with an electrode, just like Penfield used. The surgeon can ask his patient to perform simple tasks, like counting or reciting a nursery rhyme, or simply to report the sensations he or she undergoes as an area is stimulated. By first painstakingly mapping out areas in this way, any surgery to remove pieces of brain or tumours can be done with minimal risk of damage to other, vital brain faculties that have been clearly identified, such as language or motor function.

Increasingly nowadays, before any surgeon starts cutting into this most precious human organ with his scalpel, he is able to build up a detailed map of his patient's brain, often using methods we have already discussed, like MRI. And fMRI, an increasingly sophisticated technique, has even made it possible for surgeons to watch the brain as it undergoes the 'storms' associated with a seizure and to identify the areas where the activity is most concentrated. But some readers might be surprised to learn that there is no reference book a surgeon can consult at this point, because the geography of each human brain is different. Some might, at this juncture, be hurriedly flipping back to my opening chapter, in which I categorically stated that my brain, your brain, every human brain there ever has been or will be is structurally the same. On the surface, every brain looks more or less the same, even though there are variations in some of the minor contours and divisions. But beneath, at the level of individual neurons and glial cells, there is a vast number of variations.

We can talk about certain brain areas with confidence, because we know they perform the same function for everyone, and when they are damaged, they produce the same symptoms in everyone. For example, if I had a stroke – essentially a disruption to the blood flow – in the left side of my brain, I might well lose the ability to speak. And so would you. If there was a lack of activity in three areas of either your brain or mine – the dorsolateral prefrontal cortex, left basal ganglia and anterior cingulate – then we would display many or all of the symptoms associated with Tourette's syndrome: tics, twitches, a compulsion to utter inappropriate words or make strange sounds.* This,

* Some people who have listened to my lectures might argue that these three areas of my brain are indeed likely to show abnormal activity.

roughly speaking, would be because the dorsolateral prefrontal cortex, among other things, is concerned in all brains with inhibiting inappropriate actions. The basal ganglia are concerned with movement. And the anterior cingulate is associated with control and focused attention.

But, as we shall see on the course of this journey, there are almost as many points of individual difference as there are similarities. We could make an analogy here between a road map and the knowledge that a resident has of his own street. The current state of scientific knowledge about the brain enables us to know the location and purpose of many sectors of the brain – in the same way as a road map can tell us how to get from Leicester to a remote village in the Hebrides. We know, for instance, where the long-term memory is situated, or which parts of the brain become active when we listen to music, recognize a friend, solve a crossword, see something dangerous, become angry or sad or fall in love. But to be more specific, to see precisely where discrete functions of those larger categories are situated, it's necessary to talk to the residents.

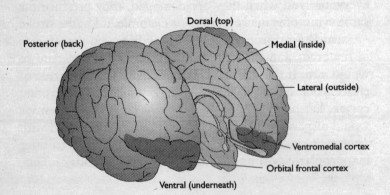

The two hemispheres separated, seen from the right side

Awareness: seeing or feeling

Olaf Blanke and his team noticed that whenever they stimulated a certain region of their epileptic patient's brain – the angular gyrus of the right cortex, in this case – she reported an out-of-body experience. 'I see myself lying in bed, from above,' she told them. When she was asked to lift her arm up and look at it, she believed it was trying to punch her. Rather than having tapped into the 'seat of the soul', Olaf Blanke reported in *Nature* that he had simply identified a region of the brain where two functions overlap. The visual system – which we use to see our own bodies as well as the rest of the world – and the system which uses touch and balance to create our physical sense of self-awareness.[4]

Much as Blanke's findings are interesting – and potentially depressing for people who believe that out-of-body experiences are sterling evidence of an afterlife, or of some 'soul' that exists outside the physical body – they are not new to the scientific world. There are numerous, well-documented accounts of people who have experienced the doppelgänger effect, otherwise known as an autoscopic delusion, in which they are confronted by an exact replica of themselves. And they don't need to be lying on a surgeon's table for it to happen, nor overcome by a spiritual revelation. In one case, published by the psychiatrists Louis R. Franzini and John Grossberg, an engineer dogged by his doppelgänger spent large chunks of his day making faces and aiming punches at it – even though he knew quite well that it was his own image.[5]

As Blanke suggested, it is thought that these phenomena result from a disturbance to the part of the brain where visual information intersects with information about the body. As I write this, I am aware of myself as being seated at a desk in a rather noisy lounge at Frankfurt airport, with one foot resting on the other, and my hands over my laptop computer keyboard. I don't need to see a picture of myself in order to know this. I know it from memory, but also from touch and balance, from areas situated at the top-back of my brain known as the parietal cortex, which tell me where my body is in space and what it is doing. On the other hand, I know that in front of me lies a computer screen and an expanse of desk. For that knowledge I rely on my visual faculties, which sit in an area right at the back of my brain, the occipital cortex.

But if, instead of hearing the call for my next flight to St Petersburg, I was to undergo an autoscopic delusion at this moment, it would be as if the current information in my 'touch and balance' files was mistakenly transferred into my 'what's in front of me' files. For a period of time, then, I would end up 'seeing' a version of the information I should normally be 'feeling'. I might then be confronted by an image of myself, seated and typing. Truly terrifying, if I didn't know something about the workings of the brain – and, come to think of it, even if I do. Unsurprisingly, people over the centuries have seen doppelgängers as the harbingers of imminent tragedy.

Dr Blanke's story is more than an intriguing anecdote, for it expresses an important point about something we take all too often for granted. The way we respond to the world, the way we *are* in the world, and the way the world *is* to us, are constructed by this rubbery organ inside our skulls. If, for whatever reasons – genes, the chemical and hormonal environment inside our mother's wombs, childhood experiences, illness or surgical intervention – some tiny modification results in the inner structure of my brain being different to yours, then I will see a different world to the one you see and, what's more, I will be a different person. Two brains may look identical on the surface, but no two minds can ever be alike.

Think about the ways in which we use the word 'mind' in English. For instance – we use it to imply attention and perception, as in the phrase 'Mind out!' We use it to imply preference for things and, by association, emotion, as in 'I don't mind what music you put on'. We use it to imply memory, as in 'It slipped my mind' or 'Mind my words'. We use it to signify our conscious, calculating selves, in that someone who is 'out of their mind' or who has 'lost their mind' is taken to have lost all ability to behave in a rational way.

Taken together, these four idiomatic expressions provide a neat summary of the chief services that our brains perform for us. Without attention and perception skills, we can't 'mind out' and would hence be in great danger. Without emotion, we would be unable to signal our intentions to other people, or understand theirs. Consequently we would not be able to express whether we 'minded' something, or know whether someone else did. Without memory, we would be in a state of permanent confusion – never knowing where we were, or what we were doing. If our higher cortex could not override our emotions and impulses, we would be uncontrolled, unable to formulate plans and carry them through. So it's clear that the mind, at least in the way we speak of it in a day-to-day sense, is not a single entity so much as a set of skills. It is a toolbox that enables us to function. But to take that analogy further, a hammer might be used by both a carpenter and a sculptor, or to perform a bodged bit of DIY. An artist uses a scalpel and so does a surgeon. The way we use those tools – the way we individually make use of faculties like perception, attention and emotion – informs our individual identities. So let's begin by looking at how we perceive the world.

The senses

Once during a hot summer I was driving over an Alpine pass. At the very bottom of the pass I noticed a simple advertisement on a bend. It was one word, advertising a local brand of mineral water. I looked away and concentrated on the tricky business of negotiating the pass, with its many bends and twists. The process was rendered less enjoyable than I had hoped, in part because of the relentlessly beating sun, and also because I had landed up behind a very slow farm vehicle which, in addition to belching out thick, black fumes, also happened to be carrying a load of gently baking pig's manure. I was hardly enjoying myself. Then, about halfway up the mountain pass, I saw another sign. This time, the name of the brand of mineral water was repeated – alongside a very simple picture of a bottle of the product. From that point onwards, I became aware of a powerful thirst. I grew utterly unconcerned about the tractor in front of me, the pounding sun or the time my journey was taking. The thirst had taken over – and the only thing that concerned me was finding some means to quench it, as soon as possible.

At the top of the pass, there was a small car park set up so that people could enjoy the view. There was also a small café and – like someone staggering in from the desert – I made a beeline for it. The waiter asked me what I'd like. I immediately mentioned the mineral water whose advertisement I'd seen. As he went away, he laughed. And then, as I settled into my seat, I realized why. Everyone else in the café was drinking the same brand of mineral water. When it came, it was expensive. It tasted like any other mineral water, too. But it didn't matter. Two strategically placed little signs had convinced me, and a whole café full of other motorists, that what I most wanted was a glass of fizzy mineral water.

This experience, which is hardly unique or particularly transcendental, may partly be due to a subliminal message, which we shall encounter later. But haven't you ever looked at a painting of a sunset and thought that it

made you feel 'warm'? Seen one of those cleverly manip-
ulative gin and tonic adverts on a hot day and felt
unaccountably thirsty? Heard a piece of music and felt sad
or at peace? I should hope so, because if not you would be
living a considerably compromised sensory existence. In
all of us, there is a crossover between the information that
comes in through our eyes, ears, noses, mouths and skin,
and faculties further down the line, such as emotion and
hunger and memory. We see, hear, smell one thing, and we
feel, taste and remember another. It all happens so automatically, without us consciously making it happen, that
you may not have considered it much before. So what is
happening inside your head?

Why not try this? Professor Vilayanur Ramachandran
from San Diego, California, in his recent opening 2003
Reith Lecture on BBC Radio 4, asked his audience at the
Royal Institution to picture two objects. One he described
as a shattered piece of glass with jagged edges, the other
an object resembling an amoeba with undulating curvy
edges. One he called a 'bouba', and the other a 'kiki'. He
then asked his audience which was which. Ramachandran
pointed out the amazing thing that 98 per cent of people
automatically and spontaneously associated the word
'kiki' with the jagged piece of glass. Try it yourself with
the photograph in the plate section. To be fair, this may
not be quite as impressive as it first seems. There may be
a number of explanations for this association, not least
that the 'k' and 'i' of 'kiki' look jagged and the letters of
'bouba' are rather round.

Synaesthesia – blending the senses

Some individuals have a condition known as synaesthesia:
they experience an odd blending of their sensory perceptions. They can taste sounds, for example, and feel

words. It's a faculty that, as I stated above, exists to a degree in all of us – otherwise we'd be dead to the charms of art or music. But in synaesthetes, the confused perceptions are particularly intense. Vladimir Nabokov, the author of *Lolita*, describes his own synaesthesia as 'coloured hearing'. In his autobiography, *Speak, Memory*, he reveals its peculiarity, and also indicates something of the richness this disorder lends to his experiential life. '. . . I see [the letter] q as browner than k, while s is not the light blue of c, but a curious mixture of azure and mother-of-pearl . . .' Arguably, if it weren't for synaesthesia lending such a vivid range of colours to the way Nabokov heard language, he might not have become a celebrated novelist. In his, as in so many cases, we can see that the specific and unique way our brains are wired makes each of us who we are, for better and for worse.

This disorder, if I may call it that, is surprisingly common – it has been suggested that as many as 10 per cent of the population experience some kind of synaesthesia. It tells us something interesting about the way we take in sensory information and what we do to it. If someone feels they can taste the colour green, for example, then it implies that perhaps there is no real, objective, 'out-there' colour green, merely some raw data which goes in through the eyes to be assigned a meaning within the brain. In a non-synaesthetic person, the electrical pulses stimulated when we see a green wall would go directly to the visual cortex of the brain. But in a synaesthete, they might be travelling to the taste centre as well. So there is no real colour green – merely a route of information-processing that is similar within most, but not all, people's brains.

Ramachandran points out that synaesthesia was described by Francis Galton, the distinguished geneticist, in the nineteenth century. Some people who are otherwise completely normal get their senses mixed. Every time they hear a particular tone they see a particular colour. So, for example, when they hear C sharp, the colour red is evoked, while F sharp is blue. One of my friends at university, who was a particularly good pianist, had a strong sensation of a blue colour when listening to music in A minor; when the music modulated to a different key, the perception of this colour faded. What is of considerable interest is that this trait may be genetically determined, says Ramachandran, and indeed Galton pointed out that it runs in families. Work by Simon Baron-Cohen in Cambridge confirms this.

Another aspect of synaesthesia is that some people when shown numerals printed in black and white experience a particular colour. So, for instance, '5' is always seen as red and '6' is always tinged green; '7' may always be indigo and '8' always yellow.

Ramachandran asks why this mixing of signals occurs. Together with one of his students, Ed Hubbard, he studied brain atlases and they were impressed by the fact that the fusiform gyrus is where the colour information seems to be analysed. And the number area of the brain, which represents visual symbols of numbers, is next to it, almost touching it, in the same region. Ramachandran suggests that in some people there is some accidental cross-wiring. In his Reith lecture, he argued that in some people this may be caused by a gene or a set of genes that causes abnormal connections between adjacent brain regions. In this instance, the connections are between the areas perceiving numbers and colours.

Ramachandran devised a test for discovering synaesthetes of this type. Using a simple display on the computer screen, he had a number of blue 5s scattered on a white screen. Embedded among those were a number of 3s. The figure 3s were placed to form a shape like a triangle, hidden among the other figures. Normal individuals looking at this pattern see a mass of 5s. But a synaesthete who sees numbers in colour immediately sees the 5s as blue and the 3s as red and identifies the fact that they are forming a red triangle on a white background (see the photograph in the plate section).

MRI imaging makes it more possible for us to at least partially understand what is happening. When subjects are asked to look at a colour, or to smell an odour (or even to imagine it – though this is a matter for later discussion), rapid firing of neurons down a well-travelled sensory path occurs, leading from the organ in question to various destinations in the brain. As the information travels through the brain, it is split into streams, which then go to various separate and distinct 'centres' to be processed. For instance, some information will go to the visual cortex. Other parcels of information will travel into the limbic system, to be tagged by the emotion and memory centres. In this way, we may not just think 'green' when we see green, but might also be reminded of the institutional corridors at our first school, for instance, and how fearful they made us feel.

The region in the cortex for each sense is split into further smaller zones, which deal with more specific properties. So for instance, in the visual cortex, there are separate regions that interpret shape, size, depth, colour and movement. Once these areas have done their job, then the information is sent onwards into larger cortical regions known as association areas. Here, the sensory information is married to corresponding cognitive associations – for instance, the sight of a hammer may be paired up with associations like 'heavy', 'cold' and 'banging', i.e. its function.

Research has shown that kittens, if deprived at an early age of certain features in their visual environment – such as horizontal lines or right-angles – are unable to recognize them when they grow to adulthood. This may seem like a strange phenomenon. If their eyes are working properly how can they be blind to what is right in front of them? But because the visual cortex of the brain has been deprived of a certain kind of input, the cats never developed the apparatus to perceive it. In my childhood, one of the common threats made to 'lazy' children was that if they didn't get up and do whatever was expected of them at that moment, they might lose the use of their legs. And to some extent, this threat is not far short of the mark. Hostages kept for months, even years, in conditions of enforced solitude have great difficulty using their voices once freed. The associated areas of their brains, no longer used for the purpose for which they were originally designated, may take on new functions. As we have already seen, the plastic brain neural connections that are not used die away, and with their death, even vital sensory functions like seeing and hearing can be lost. The reverse is also true – the more use a brain region receives, to some extent the bigger it may become. So it follows that, to an extent, our brains will differ purely because we are each exposed to subtly different environments.

Sensory overload

There is a Zen Buddhist tale of a frog who lived in a well and dreamed of escaping to the river. But when its wish was granted and it saw the river for the first time, the frog fell down dead on the spot. The implication was that, in its narrow little world, the hapless frog had been unable to imagine a thing so vast as a river, and its brain and body experienced a sensory overload. A similar, almost certainly apocryphal story, is reported of Captain Cook's voyages in the South Seas – he claims that the inhabitants of one particular island were unable to see his huge galleon anchored a few yards off the shore. Their brains did not have the sensory capacity to encompass an object so huge. Whether this tale is true or not, consider the fact that some societies only count up to ten. Anything beyond that is simply 'many'. Living in small human groups, with small numbers of possessions and accompanying animals, they do not have any need for larger numbers. This expresses the point that brains develop only the faculties they need. The proportional layout is much the same in everyone, but excessive use or disuse results in individual differences.

I was reminded of this frog recently as I was experiencing another kind of sensory overload. Not so long ago, I visited the Jewish Museum in Berlin designed by the remarkable architect Daniel Libeskind. Part of the museum is a memory of the Holocaust. There I spent a short time in the 'Garten des Exils' (Garden of the Exiles). You can only spend a short time there because the sensory overload becomes unpleasant. Here, in a high-walled space, over which you can just perceive the tops of other buildings, are forty-nine square pillars made of smooth concrete, seven metres tall (each placed a metre from each other made out in rows of seven, so they form a square seven by seven). The planting is on the top of the pillars, so the garden is high above and completely out of reach. The pillars, though regularly placed and parallel, are themselves planted in sloping ground, and this sloping ground is uneven and cobbled. So, within the walled space, vision is giving one sense of horizon, balance on the sloping ground another, and (because of the cobbles) proprioception – the feeling of where one's limbs are in space – another neural message. With the senses so completely confused, the effect is extraordinary and extremely disorientating – indeed, after about five minutes I had an overpowering need to leave because I was feeling dizzy and sick.

Sight

Information derived from our sight has a long way to travel from its initial entry point at the lens of the eye. Firstly it goes to the retina, at the back of the eye, where it is converted into electrical impulses. These travel along the optic nerve and half of them cross over (left visual field information to the right brain, and vice versa). The majority of the fibres connect with an area of the thalamus called the lateral geniculate body and a few with the pulvinar nucleus. As we saw, the thalamus acts as the connection site between the sensory components and the higher functions of the brain. Visual information then proceeds to the visual cortex, situated at the back of the brain, above the cerebellum.

Like all other areas of the brain, the visual cortex needs regular use to grow and remain healthy. A particularly sad example of this is found in the children who received the first lens transplant operations. Born blind, and typically having remained so for ten or more years, the children and their families welcomed this new surgical innovation with great enthusiasm. But after the operations were completed, it became obvious that none of the recipients was able to see without significant problems. They experienced visual stimuli as glaringly bright and painful. They were unable to use visual information to orientate themselves in space. They were, in short, unable to learn to see, because their brain had passed the point of plasticity for that skill – the years of disuse had robbed their brain of the potential. Following the surgery, these children (who had had such high hopes) felt severe depression, and some even committed suicide.

But the future may not be so gloomy. Eventually, it may become possible to train the brain to form a 'new' visual cortex. Dr Susan Blackmore[6] of the University of the West of England, Bristol, points to a preliminary report from

Dr Paul Bach-y-Rita of Strasbourg.[7] One of the patients he describes is a woman who was congenitally blind. He showed that, using specially devised electrodes on her tongue, she could learn to see and manipulate objects within a few hours of training. And Dr Blackmore also comments on Dr Mriganka Sur's experiments at MIT, Boston,[8] involving rewiring the cortex of newborn ferrets. They have suggested that there is nothing intrinsically 'visual' about the visual cortex. Visual input was rerouted by these researchers to what should have developed as the part of the cortex that is involved in hearing. But later examination showed that it had, with time, become organized into maps and columns in a very similar pattern to that normally found in the visual cortex. In other words, the characteristic organization of different parts of the brain's cortical areas might be driven by the input rather than being purely pre-wired and therefore determined by genes.

Smell

Smell is detected high up in the nose by two small patches containing around six million or so cells. While our sense of smell is not as acute as it is in most animals, normal humans can detect some chemicals even when they are diluted to one part in several billion of air. Our sense of smell is most acute when we are children and drops off as we age. There is some evidence that premature ageing – such as Alzheimer's disease – results in a more rapid decline in our ability to smell. It is also true that a number of conditions such as schizophrenia, depression and anorexia are associated with decreased ability to smell and at least one study suggests that migraine sufferers may also have a lowered sensitivity to some smells, at any rate during if not before an attack. What is not clear is whether this lowered sensitivity is caused by the migraine.

Smell is perhaps the oldest sense available to us, a remnant of our reptilian past. Whereas all other sensory information takes a lengthy route through the thalamus, into the cortex and thence onwards to the limbic system to be tagged with emotions and memories, the smell route goes directly to the limbic system and then up to the cortex. This might explain why we find smells so instantly evocative of moods and memories. In one study, students learnt new words while sniffing an unusual smell and then sniffed it again when trying to recall the words. They showed a 20 per cent boost in their memory recall.

It is easy to see that, in our evolutionary past, having a quick response to smell information could be vital to our survival. That's why there is so swift a route from nose to limbic centre. It seems that smell was a major factor in the development of our brains, as the original, reptilian brain consisted of little more than some olfactory tissue sitting on top of a nerve cord. The entire limbic system is thought to have evolved from this point – originally interpreting odours and sending out corresponding chemical messages. The olfactory cortex is no smaller in modern humans than in our ancestors, though arguably modern life – with its sensory onslaught of exhaust fumes, perfumes, deodorants and the like – does not help our smell system to guide us as it once did.

Nevertheless, there is evidence that we still unconsciously respond to smells, and sometimes to things that we cannot even smell but perceive through our nostrils. Women may have a sense of smell up to 1,000 times more sensitive than men's, and various studies have shown these powers to be eerily sharp. A study by researchers at the Institute of Anthropology in the Ludwig-Boltzmann Institute at the University of Vienna involved showing a number of female subjects a frightening film while having them wear underarm pads to collect their perspiration. For those of you that are interested, the film was *Candyman* – said in its blurb to be 'the most frightening film since *Silence of the Lambs*'. A further group watched a neutral film while undergoing the same process. And a third group of women were able to distinguish by smell alone those pads that were carrying the 'scent of fear'.

Many animals produce pheromones and there is good evidence for their role in communicating alarm and fear. Fish, insects (such as bees), and some worms even, communicate alarm, which encourages members of the same species to avoid a threat coming from a particular region where the alarm was triggered. Mammals certainly communicate by smell, and it is known that cats can seek out stressed mice by the scent caused by their stress. Until recently, the evidence for the existence of human pheromones has been more controversial.

Pheromones are very volatile substances that affect either the behaviour or the hormones (or both) of an individual who perceives them. There are two different types. 'Signalling' pheromones result in a direct and observable change of behaviour. 'Primer pheromones' have a long-term effect on the receiver, presumably by changing his or her hormonal state. Pheromones act directly on the brain and are generally not recognized consciously by their recipient. They are frequently present in sweat, and in recent years research has shown that they may often play an important part in human sexual behaviour. Norma McCoy and Lisa Pitino, from San Francisco State University, recruited volunteer female undergraduates by asking whether they might like to participate in testing whether a synthesized female hormone, when added to their regular perfume, could 'increase the romance in their lives'.[9] To qualify for inclusion in the study, these young women had to be menstruating regularly, heterosexual, not using oral contraceptives and not cohabiting with a regular partner.

During the experiment, some of the volunteers used their regular perfume which had been spiked with a totally odourless pheromone. Others, selected at random, had their perfume spiked with a neutral substance. None of the volunteers was told whether they were using a pheromone or not, and the control group was only recruited with the promise of a free vial of the real stuff for everyone at the end of the experiment.

Each volunteer was asked to use their perfume in a standard fashion and at set times. The women using pheromone-spiked perfume turned out to be three times more likely to have sex, or just share their bed with a male friend, than those in the control group. The researchers demonstrated that, rather than make the women more responsive to sexual advances, this particular pheromone made the women more attractive to men. So men, too, may have a keen sense of 'smell' even when a substance seems odourless. I regret that I am unable to divulge the name of the pheromone – the researchers report that Athena, the American company marketing this product, are in the process of patent application – but no doubt you could try buying it on the internet for a price.

Women are undoubtedly more sensitive to smell at certain times during their menstrual cycle. Their olfactory powers are usually at their sharpest when they are most ready to conceive, at the time of ovulation. Women produce pheromones that not only influence men, but also other females, particularly as regards their menstrual cycles. Thus in a relatively closed environment it is not at all unusual for women, after being together for several months, to have their periods at the same time. One of my senior colleagues at work tells me that in our laboratory, in which 85 per cent of the staff are female, after several months of working together the women tend to have synchronized menstrual cycles.

Dr Laurence Katz and his team at Duke University have

discovered that animals have a pheromone-processing machinery in the brain which seems to produce a specific 'pheromonal image' of another animal.[10] Such an 'image' of the other animal's identity, sex and reproductive status governs a range of mating, fighting, maternal–infant bonding and other behaviour. The neurons that process pheromonal signals are similar in many respects to the 'face neurons' in the part of the visual cortex of the primate brain that are specifically triggered by facial features of other animals. A wide range of mammals, from mice to elephants, possess such a 'sixth sense' for detecting pheromones. These animals have a specialized sense organ, the vomeronasal organ, in the nose cavity. This organ pumps samples of pheromones into a sensory cavity, where they are detected by chemical receptors similar to those used in taste and smell – even though they cannot be detected by either smell or taste. These sensory neurons send connections to a structure called the accessory olfactory bulb, where Katz's team made electrical recordings of individual cells. To measure the signals, Dr Katz used a method first developed to measure the activity of neurons in the brains of birds as they sing. Three tiny micromotors attached to hair-thin electrodes are insinuated into the pheromone-processing region of the animal's brain. These electrodes can be retracted or extended by remote control by minute amounts, seeking out individual neurons to record their activity during different behaviour. The electrodes are so tiny that the animal can move freely, interacting with other animals, while their brain is sending signals over a fine wire to a recorder.

But animals like mice live largely in the dark, and generally have rather poor vision. So, as Dr Katz points out, communicating by pheromones would be critical to their survival. Whether a structure like the vomeronasal organ persists in humans remains controversial but it must be likely.

When smell information goes into the limbic system, it passes to both the amygdala and the hypothalamus. Some epileptics report a sensation of unusual smells just prior to having a fit, and brain imaging shows that in these individuals, the electrical activity is concentrated in the limbic area. The amygdala, as we shall see later on, is vital for the expression and interpretation of emotions. The hypothalamus, meanwhile, is the brain's hormonal centre – among other things responsible for the 'fight-or-flight response' when we are startled. That is why certain odours have the power to increase heart rate and blood pressure, and others to induce calm. Aromatherapy is not simply a New Age fad; it may really play a therapeutic role by relieving tension.

Taste

Taste and smell are probably the most linked sensory faculties. Most taste is actually smell – we can discriminate only five tastes with our tongue. These are salt, sweet, bitter, sour and umami. Umami is a kind of savoury taste given by foods that are rich in amino acids – glutamates. This is why monosodium glutamate is such an effective taste enhancer. But a mere whiff of roast chicken or banana and we can imagine the taste so vividly it might as well be on our tongue. Unsurprisingly then, the taste area of the cortex is situated right above the olfactory area in the right hemisphere. Damage to this area can result in some surprising phenomena – a Swiss study identified thirty-four brain-injured patients who had suddenly developed an obsession with food. In all these cases of 'gourmandism', lesions were found in the right frontal lobes. Interestingly enough, though these people developed an obsession with food, buying it, cooking it and eating it, none of them became overweight. Having tasted the dairy-heavy delights of Swiss cuisine on many a ski trip, I find this hard to explain.

It is easy to see how our taste faculties would have been vital to us in our ancestral past. Something that tastes 'bad' to us might well be a source of poison. We are, incidentally, thousands of times more sensitive to bitter tastes than to salt, sweet or sour ones. 'Bitter' is likely to be more dangerous. For that reason, survival manuals advise anyone trying to forage for food in the wild to be guided by their senses – as a rule of thumb, avoid anything that tastes or smells bitter, rancid or otherwise disgusting. The area of our brain that becomes active when we taste something we don't like – the anterior insula – has been reported to light up in the scanner when we see someone else registering disgust. From an evolutionary perspective this could be very useful – if another of our species tries to eat something and registers disgust, we don't need to waste time trying to eat it ourselves. We can concentrate on chasing after that tasty antelope instead.

Hearing

The last sense lost when slipping into unconsciousness, and the first to return on waking, seems to be hearing. For this reason it is important to be cautious when speaking in front of a very ill, apparently unconscious person, or in the presence of a dying relative. Many years ago when I was rather more callow, I was doing a simple Caesarean section operation for Julie, who had already had two previous deliveries by Caesarean section. Consequently, a third delivery by this route was inevitable. In such circumstances, once the baby is delivered safely, the rest of the operation procedures – removing the placenta, stopping the bleeding, repairing the uterus, cleaning out the abdomen and closing the wound – are pretty hum-drum. Twenty years ago it was quite normal to give a fairly light general anaesthetic for this kind of surgery, the reason being that the baby would be born fitter and more ready to breathe. My anaesthetist, well known in the hospital to be urbane and totally unflappable, was con-gratulating himself on the excellent condition of the baby boy as I delivered him – moving all four limbs vigorously and crying with gusto. Some devil got into me and made me want to prick his complacency. 'It's all right for you,' I declared, 'your anaesthetic may be good for the baby but it's resulted in a lot of dilated vessels – we've a brisk haemorrhage here.' Moments later, I complained about spurting from the inferior branch of the left uterine artery, and shortly afterwards asked my assistant who, I regret to say, had just as inappropriate a sense of humour, how many pints of blood had been cross-matched before the operation. On hearing his false-apologetic answer, 'Only two, I'm afraid,' I said with as much alarm in my voice as I could muster that we might need another five. It soon became clear to my anaesthetist we were just trying to shake him and the operation finished without hitch.

Next morning I went to see Julie with her little boy. Both were fine. Then she turned to me and apologized for giving me so much trouble. Uncomprehending, I said that the operation had been easy and routine. 'But what about the haemorrhage?' she asked. 'And the trouble with, what did you call it, the left inferior branch of the uterine artery – and the five pints of blood you needed to order?' She then told me that she had heard – and could recall – everything that had been said up to the time when we started to close her abdomen. She felt no pain, and just at times the remotest sense of touch. After an initial anxiety, Julie said she felt no fear – 'Everything seemed to be under control and there was nothing I could do, any-way.' But we learnt at least one important lesson from

Julie's case. We were very lucky that we hadn't frightened her, but this taught us to ensure that sufficiently deep anaesthesia is always maintained.

As in all other faculties except smell, information crosses over from one side to the other: information presented to the left ear travels to the right side of the brain to be processed. Both left and right hemispheres play a distinct role in sound processing, so a person with hearing problems in one ear may perceive auditory information very differently from someone with perfect hearing. For example, the left side of the brain seems to deal mainly with the identification and naming of sounds, rather than their quality. So somebody who can only hear in their right ear may have a blunted perception of aspects of music like rhythm and melody.

It is worth noting that although there is a direct link between the auditory cortex and the language region of the left hemisphere, you do not need to hear to develop language. MRI studies of deaf children born to deaf parents who grew up using sign language reveal a language region just as developed as that of hearing children. Moreover, non-deaf children born to deaf parents never attain the same degree of fluency in sign language as their deaf counterparts – they use a 'broken' version of the language, similar to a foreigner's attempts to master English. Other senses are somewhat similar. Blind people who learn to read using Braille show activity in the visual and language areas of the brain, as well as in the sites responsible for receiving touch sensation through the fingers.

Touch

A sense that develops early is touch – and it is more acute in newborn babies than sight or hearing. The 'rooting reflex' can be seen from day one, when the baby turns its head towards anything that touches its face, the reflex that helps it to find the nipple.

Touch is vital to a baby's development. It seems curious that the practice of giving newborns to their mothers seconds after birth is a development only of the last twenty years or so. Previously, they were often whisked away to be washed and weighed by the nursing staff – hospital etiquette that seems to fly in the face of human need. But, arguably, while this practice might have been injurious to the mother–child relationship, it did not necessarily do any harm to the baby. All the while it was with the nurses, receiving the touch of human hands. But in some countries, of course, babies are routinely wrapped tightly in swaddling and restricted in movement – whether this is harmful is an interesting question. Brain scans of infants deprived of touch show that vital areas of their brains are underactive – like the Romanian orphans I described earlier. Dr Tiffany Field from the University of Miami has reported that premature babies massaged three times a day for ten days gained nearly 50 per cent more weight than premature babies who didn't receive this extra bit of simple tender loving care.[11] Moreover, they were able to leave the hospital an average of six days earlier. She claims that massage stimulates the production of human growth hormone, and hormones such as insulin that aid growth. Although I have not seen hard evidence of this, she maintains massage increases levels of the neurotransmitter serotonin, an action not dissimilar to that of some anti-depression drugs.

Touch and movement information are processed at the top of the brain. Pain, as distinct from ordinary touch, is received at the site of its production by units called nociceptors. These send signals to the spinal cord, which then pass up through the brainstem to the thalamus. From there, the information travels into the cortex to our sensation regions, telling us what type of pain is being experienced and where it is located. A region in the frontal lobes called the anterior cingulate cortex – one particularly associated, as we shall see, with attention, monitoring and concentration – plays a particular role in pain perception. Many painkillers – particularly the ones that have a similar structure to opium: for example, codeine, found in many cough medicines – act by blocking the receptor sites in the brain's neurons that would normally be filled by the brain's own painkilling chemicals, encephalins. Opioids also decrease activity in the anterior cingulate cortex – and, accordingly, people find that they are paying less attention to their pain. When I was a medical student, after repairing a thick waterproof anorak while lying on my bed, I dropped the massive bodkin I had been using onto the floor. Some minutes later, foolishly forgetting what I had done, I jumped off the bed and rammed the thick iron darning needle into the sole of my foot, the needle breaking into two pieces deep in the muscles of my foot. It could only be extracted after a substantial incision under general anaesthetic. The pain after the surgery was intense, but an injection of pethidine – an opiate-like drug – relieved the unpleasant aspects of the pain completely. I could still perceive a throbbing feeling in my foot but it wasn't unpleasant and it didn't feel as if the rest of me was there at all. I think I can understand why some people become addicted to such drugs.

Have you noticed that if you rub the site of a pain – provided it is not severe – it begins to feel better? Mothers will instinctively rub or kiss the limb of a child who has hurt itself. This is not superstition – it works, because of a process called competitive inhibition. By rubbing the site of the pain we are sending a second set of signals to the brain in addition to the pain signals. The brain now pays attention to two streams of signalling at once, with the result that there is a reduction in the perceived severity of the first, 'pain-type' signal.

This raises some interesting questions. If pain and touch are things perceived inside the brain, do we need a body to feel? The obvious answer would be yes, of course we do. But there are a number of people who have had whole limbs amputated but who feel a ghostly sensation of the limb still being there as part of them, to the extent that they can even feel pain. The Phantom Limb Syndrome is still not fully explained, but neuroscientists know that the areas in the brain that are involved are the thalamus and the touch areas of the cortex. Vilayanur Ramachandran, who has done much research in this area, noted that when pressure was applied to other, intact parts of the body, his phantom-limb patients, missing, say, their left leg, would often say they felt touch in specific, and consistent, regions of the non-existent leg. Ramachandran suggests that, when an area of the brain is not in use, it becomes coopted by neighbouring neurons for new purposes. So the area of neurons once responsible for an arm, now amputated, might be taken over by the face area of the brain. These coopted pathways still carry traces of their original function – and hence the feeling of pain in a leg or arm that no longer exists. There may be other reasons for phantom-limb pain, too. If there has been intense pain immediately before amputation in the limb to be removed, this pain can persist afterwards.

There is a curious, opposite phenomenon, known as apotemnophilia, which might possibly be related to 'phantom limb'. A few people report feeling that normal limbs which are very much a working part of their bodies should not be so. They may go to a doctor asking for amputation of the offending limb. They frequently find it very difficult to explain how they feel except that, for some inexpressible reason, they feel their leg or arm is 'not theirs'. Rather understandably, medical professionals have been apt to treat apotemnophilia as a form of hysteria, a psychotic state such as schizophrenia or associated with severe depression, or sometimes a sexual fixation with limblessness. Some apotemnophiliacs have succeeded in persuading surgeons to perform amputations on them in order to alleviate their mental suffering. The most famous recent cases in the UK resulted in questions being asked in the House of Commons by Mr Dennis Canavan, MP, in February 2000. Mr Robert Smith, a surgeon operating privately at Falkirk and District Royal Infirmary, had cut off the lower legs of two patients who were requesting amputations. One of these patients was British, the other German. The men had previously been turned away by other surgeons from different hospitals in Europe before Mr Smith agreed to operate. Mr Smith said that he saw nothing wrong with his actions but that he did 'not want to specialize in the procedure'. 'The last thing I want to be is a world centre for cutting off arms and legs,' he said. The Health Minister at the time told Mr Canavan that the government did not think a public inquiry was necessary.

Our advancing understanding of brain science may make it possible to alleviate this bizarre syndrome without recourse to bone-saw and ligature, by treating the related area of the cortex. The desperate measures taken by some people with rare disorders of the awareness of their own body are a reminder that it can be hard to communicate your perception of yourself, and the world around you, to anyone else. It's a problem that stems, once again, from that hoary old chestnut of philosophy and language. The only device we have for interpreting and communicating about our brain is the brain itself. Indeed, while microscopy, electroencephalography, MRI and PET scans and X-rays are essential tools in understanding how the brain works, the investigator's own brain – with all its flaws and subjectivity – will always remain the best tool of all. As each person's brain is different, I cannot possibly know what it feels like to be in possession of a working limb and yet want it amputated. If you say you see the colour green, I can only think of my brain's version of green, not your green. When talking about your brain and my brain, we have to accept a certain, necessary amount of indistinctness.

The thingummy-whatsit-doo-dah

Our senses are not really separate and discrete skills, because we tend to use them in combination with one another. Certain linguistic visual cues – the things we see, such as the shape of someone's lips – activate the hearing areas of the brain, while non-linguistic movements of the mouth do not. It's as if our brains are primed for the task of communicating with others. A study at Brandeis University in the United States also found that we some-times 'see' with our hearing. Volunteers shown an image of two blocks moving towards one another on a screen reported that they had seen the two blocks collide when in fact all that had happened was that they had heard a sharp click. Similarly, subjects playing ping-pong with their ears muffled perform less well than those who can use their hearing, as well as their sight, to judge the trajectory of the incoming ball. Have you noticed, for example, that when you are trying to park your car in a tricky space, it helps to turn down the music on the car radio? This might be purely because you need to concen-trate. But equally, although you are not immediately conscious of it, you are relying on hearing, as much as visual and motor skills, to get your car safely into that tight spot.

One of the chief areas in which all our sensory faculties work together is in recognition. Recognition is one of the mind's most vital skills – we call upon it daily in the most simple and complex of tasks. When I go into my laboratory office in the morning there is a tiny, un-conscious 'Aha!' inside my brain as I pick out the familiar surroundings from the other laboratory offices immediately next door. The familiar smell of old coffee and the odour of damp paper underneath the leak from the top window, the unique view of the high-security wing of Wormwood Scrubs prison just over the hospital wall, and the depressing sight of mounds of correspondence which my secretary has thoughtfully left on my untidy desk (still with a soggy packet of biscuits) all inform me that this is indeed my office. As I look downwards, I can pick out from the huge mound of post the really important letters in the pile. All this happens so swiftly that my conscious brain barely registers the process at all. Indeed, I am already thinking about a whole host of other things.

In the BBC TV series of *The Human Mind*, I wear a 'magic hairnet', a series of hundreds of electrodes covering the whole of my scalp, while watching a TV screen on which there is an optical illusion. At first difficult to identify, there is a still photograph of a Dalmatian dog hidden against a background of a wider pattern of large and small blots of black. Only when the dog starts to move do I recognize the image. When I recognize it, Gaynor Evans, a doctoral student of the University of Warwick, is on hand with an EEG machine to measure the electrical activity inside my brain as the 'aha' moment dawns.

In order to get clear results, it is necessary to perform the test many times. Each time I was asked to recognize the dog, there were two significant bursts of gamma waves. The first occurred as my brain registered that I had seen 'something'. The second, very shortly afterwards, occurred at exactly the point where I experienced a little internal 'aha' – I had 'got' the trick. This gap between the two bursts of activity is likely to be the time it takes for the brain to match what it sees with a meaning. In other words, the gap is where my brain matched up the 'something' with my own internal images of dogs.

Our faculty for recognition does sometimes break down. This American patient described a very familiar object in the following way: 'I can see a lot of lines. Now I see some stars. When I see things like this, I see a lot of parts . . .' The man was suffering from a brain disorder known as agnosia and was attempting to identify the American flag. His speech demonstrates that his problem was not merely that he didn't know the correct word. He quite genuinely had no idea what the item in front of him was. He was having difficulty because agnosia is, in essence, an inability to recognize objects. The disorder tends to be very specific – for instance, sufferers may be unable to recognize faces, or other body parts, or items of food, or man-made objects. The discovery that such disorders can result from a single bit of damage to a specific area in the brain suggests that there are compart-mentalized sections of our memory in which we store items of particular kinds. Biological objects are generally stored by how they look in parts of the brain that do perceptual analysis – while inanimate objects are commonly represented in the brain by their function, so they are stored in regions associated with actions that might use those objects.

One of the problems for me in writing about the working of the brain is that it is easy to descend into a description of what may be seen as a 'freak show'. But the truth is that in research into the way the brain functions it is so often the study of the disordered state that enables us to understand what's happening in a normally functioning brain. Studies into agnosia have found that the brain seems to make a strong distinction between living and non-living objects. Rather more bizarrely, people who are unable to recognize living things are also often unable to recognize food and musical instruments. Yet these same people have no problem identifying other man-made objects or human body parts.

These pairings may seem weird but they led scientists to conclude that the recognition faculties of the brain are both rigorous and at the same time quite generalized. From the examples above, for instance, we can say that the brain seems to put food, animals and musical instruments in one compartment, and man-made objects and body parts into another. A study at the University of Iowa, led by husband-and-wife team Antonio and Hannah Damasio, tried to pinpoint the exact location of the brain's various recognition categories. They found a whole range of possible 'storage sites'. For instance, lesions in the right temporal lobe seem to be related to difficulties in facial recognition. An inability to recognize tools and instruments was associated with lesions in an area of the left hemisphere where the temporal and parietal lobes join together.

Recognition units

One explanation for this is that the brain stores and categorizes things, not according to how they look or what they do, but according to our individual relationship with them. Take a hammer: it could be used for work or it could be an instrument of violence. A musical instrument could be seen, felt, played or heard. It seems that each quality of an item is stored in a separate region of the brain – in her book *Mapping the Mind*, the medical writer Rita Carter terms these 'recognition units' or RUs. So a musical instrument like a trombone would have not one 'unit', but several: a shape recognition unit in our visual areas, a word RU in our vocabulary area, a touch RU in the touch area, a sound RU in our hearing area. Each region of the brain may be stacked up with the recognition units of objects that might, on the face of it, not seem to have anything in common. They are together because they share the particular quality that concerns that region of the brain. So our trombone might sit in the same little corner of the brain's library as a cigarette or a straw – because all are non-food items that we put in our mouths. When we think of a trombone, all of our separate trombone RUs are drawn together from their separate storage regions and united to give us what we recognize as a long, sliding brass instrument that makes wonderful sounds in Tchaikovsky's Fourth Symphony.

So once again we see that our own particular experiences of the world create the geography of our brains, and these in turn inform our behaviour. A life-long vegetarian, for example, would not have a taste recognition unit for a chicken or a cow. A musician would have different RUs for a trombone than a casual jazz listener. I have it on good authority that a colleague's mother is unable to bear the taste of bananas because, as a child during the second world war, she was given mashed turnip doused with vanilla essence and told that this was the rationed fruit. Now, when she eats a genuine banana, she still tastes its distinctly unappealing substitute. Clearly some trick of the recognition facilities is at work here – for her, the word RU for 'banana' still has a connection to the original turnip-and-vanilla-taste RU, a connection so strong – and probably reinforced by the emotional memories associated with her disgust – that more recently formed RUs, such as what a real banana looks and feels and tastes like, cannot break it. Seemingly some qualities of both are filed next to each other in her brain's recognition site.

Nearly all of us are familiar with the dreadful experience of arriving at a party and being greeted by someone who seems to know us intimately, but whom we cannot remember. Some people suffer permanently from this, in a condition known as prosopagnosia, and cannot recognize people they have known all their lives and see every day. Still others suffer from its equally alarming opposite, known as Fregoli's Delusion (or Syndrome), in which they believe total strangers to be completely familiar. Leopoldo Fregoli was a famous actor from Rome, who died in 1936. The syndrome is named after him because of his mastery of disguise. It is said that, in one performance, he played no fewer than sixty different characters – a sort of forerunner to that great comic satirist Rory Bremner. Sufferers from Fregoli's Syndrome, on encountering people they have never previously met, may come up with very convoluted reasons to justify their belief that this really is an old friend. They may argue, for example, that this friend or relative has adopted a clever disguise.

But we cannot blame our lapses on another person's delusion. At the cocktail party, it is not that the other person is suffering from Fregoli's Syndrome, but much more likely that we have forgotten their face. Why does this happen? Why do we forget one face and yet remember another – even though both might be similarly unstriking faces?

Partially the answer boils down to the 'use it or lose it' maxim. Humans are social animals. We live as we have always lived, in groups, and we depend upon the ability to communicate with others of our species for our survival. For that reason, we have evolved with special brain wiring for face recognition – such that even a newborn child will orientate itself towards objects that resemble faces, like a balloon with eyes, nose and mouth drawn on it, and also show a preference for symmetrical faces over non-symmetrical ones. Faces occupy a special category of our recognition faculties – and brain scans show heightened activity in these areas not only when we see faces, but even when we imagine seeing them. If a face is seen or imagined frequently, its RU stays active, the neural connections remain strong. If on the other hand we do not see or think about a person for a long time, the facial RU for that person falls into disuse. Try remembering someone you have not seen for a very long time – the likelihood is that you can remember many of their attributes, but not accurately recall their face.

Of course, some other faces from long ago remain very vivid – and we will be examining the relationship between intense experiences and memory at a later point. For now it is enough to say that facial RUs do not necessarily need a strong stimulus to become activated, or to remain active. Imagine that the car-park attendant near your office has a strong resemblance to a cousin who emigrated to Australia. In seeing the car-park attendant five mornings and evenings a week, you are also keeping the facial RU for your Antipodean cousin stimulated. We can also see this at work in the grief process: someone in mourning is imagining the lost person more or less constantly, and the process is keeping that person's facial RU permanently switched on. Hence it is not uncommon for bereaved people, or people who have recently separated from a lover, to keep thinking they have seen their deceased or lost loved ones in the street. In reality, the 'heated' state of the facial recognition unit merely predisposes it to be triggered by the weakest of stimuli – so the sight of a face only vaguely similar to the face of our loved one, or the object of our grief, is enough to convince us, momentarily at least, that we have really seen them.

Facial recognition of all kinds involves a strong emotional content. This is why we are never entirely neutral in our dealings with other humans – however much we would like to be. Some people 'look' trust-worthy or kind to us. (In fact, research has shown that men with softer, more feminine features are more likely to be found 'not guilty' in court.) Other people may seem sad, or threatening – without us being able to find any solid evidence for our verdict. This is because facial recog-nition takes place along two pathways within the brain –

one conscious, and involving the higher cortex, and one unconscious, taking place within deeper structures of the brain. The unconscious pathway runs through the limbic system – and includes the amygdala – which you will by now recognize as one of the centres for our emotions.

The conscious pathway works at a slower speed – it is here that we work out, for instance, whose face we are looking at, and how we should behave towards them. This is why we often have an almost instantaneous 'first impression' of a person, which is chiefly an emotional feeling – of unease, for example, or affection – which we might find hard to put into words, and which might, upon further inspection, turn out to be quite wrong. As we shall see time and again throughout this book, the amygdala does not merely generate emotions, but seems to encode very basic emotional memories. So meeting someone whose face has vague similarities to the feared and hated maths teacher at school, for example, triggers a swift, brief and shadowy reliving of our feelings towards the original figure.

And now we can go some way towards accounting for the bizarre recognition disorders I mentioned earlier on. Prosopagnosia – a generalized failure to recognize faces – results from damage to the higher cortex – in particular the temporal and occipital lobes. In other words, it affects the higher, conscious recognition pathway. Patients with this problem, when shown photographs which include familiar faces, will typically say they recognize no one. But brain scans show that there is activation in their lower, sub-cortical pathway. Electrical changes in their skin also suggest that their brain is registering more than they are capable of saying. It remains, in effect, possible for them to recognize someone without being conscious that they are doing so.

This also helps to explain why people suffering from Fregoli's Delusion – in which they believe strangers to be familiar – feel the need to come up with complex and creative justifications for their beliefs. The same is true for people with the opposite disorder: Capgras Delusion – in which familiar faces are perceived to be those of strangers. People with Capgras suggest, for instance, that their friends or loved ones have been replaced by aliens or androids. They no longer have any emotional attribution to familiar faces. In both cases, their explanations are a noble attempt by the conscious brain to make sense of information coming up from the unconscious.

Walk a mile in my moccasins

So differences in the way our brains receive and assess incoming sense information create differences in the way we each see the world. In fact, it might be more accurate to speak of individual sense-worlds rather than some objective world. A Native American Indian proverb states that in order truly to understand a man, one must walk a mile in his moccasins. And we can see how this wisdom applies to our enquiry into the mind. The way we see, hear, touch, taste, in all respects sense the environment we're in, makes us who we are. It is possible to gain insight into someone else's character, if not our own, by understanding how they sense the world.

Harvard psychiatry professor John Ratey cites the example of Buckminster Fuller, architect of the geodesic dome, who used to walk around with specially developed eyewear. Here was a man who felt overloaded by the strength of his visual perception. The glasses shut out all but a small part of his view, and left him able to walk around while thinking. Ratey suggests that many of the disorders we now label as autistic, or corresponding to Attention Deficit Hyperactive Disorder, may have their root causes in abnormalities, or rather simple differences in perception. I am reminded strongly here of the scene in the film *Rain Man*, where Raymond, the autistic man played by Dustin Hoffman, screams, as if in pain, when Tom Cruise attempts to give him a hug. Many autistic people are highly sensitive to touch, to noise, to quickly moving visual stimuli. In a real and observable way, their faculties seem too finely tuned to the sensory input of the world around them – rather like the fairy-tale princess who could sense a pea hidden underneath the tower of mattresses upon which she slept. Unsurprisingly, for many autistic people, the world out there is a bewildering, painful experience and they withdraw from it into a private reality, governed by their own rituals and routines.

A similar principle may be at work in the brains of people with the reading and writing disorder dyslexia. This is a widespread phenomenon, and not, as generations of teachers and parents once believed, related in any way to intelligence. Many dyslexic people have made sterling contributions to the fields of medicine, architecture, art and literature. Many more lead normal, intellectually fulfilled lives.

While most dyslexic people have normal hearing and visual capacities, many others report hearing and seeing things that people without dyslexia do not. Tests have indicated that dyslexic people have a more developed talent for hearing slower, deeper sounds, like vowels in speech, for example, or the bass 'riff' of a piece of music. But it seems that it is the relation between sight and perception that is most variable in cases of dyslexia.

There are two principal pathways leading from the eyes into the visual cortex, known as the geniculostriate pathway and the tectopulvinar pathway. The latter pathway seems to be what we use to orientate our eyes towards a particular element of our view. It provides the mechanism whereby I might find my pen amid the clutter of my desk, for example, or notice the brightly glowing lights of a

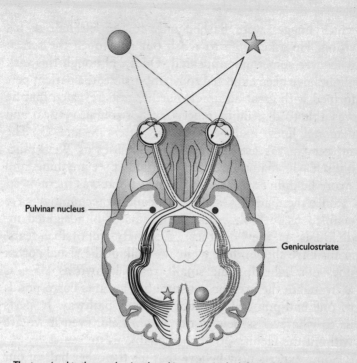

The two visual pathways, showing that objects seen in the left visual field are processed in the right hemisphere and objects seen in the right are processed in the left

Pulvinar nucleus

Geniculostriate

service station on a dull horizon. The pathway to the pulvinar nucleus senses the interesting object and then the geniculostriate pathway kicks in to help me see it.

Investigations into a phenomenon known as 'blindsight' have demonstrated the importance of the tectopulvinar pathway. Research into this most peculiar aspect of the brain began with the Oxford psychologist Larry Weiskrantz. In the 1970s, Weiskrantz was studying a patient who had damage to an area of the visual cortex known as the right calcarine fissure. The damage resulted in him being unable to see anything just to the left of where he looked – he had a reduced field of vision.

Larry Weiskrantz sat his patient in front of a screen, on

which images were flashed only to the left half of his vision. He asked him to reach out and touch the screen whenever something appeared. Oddly, although this task might have been expected to be impossible, the patient performed with great accuracy. Further tests revealed that he was able to discriminate between horizontal, vertical and diagonal lines, and between the letters X and O. The subject was as astonished by his abilities as Weiskrantz himself – he maintained that he couldn't see anything, and swore he didn't know how he knew what was there.

Since then, advances in brain-scanning technology have enabled us to look inside the brains of blindsighted individuals as they do these seemingly impossible feats. Scans have shown that, while the bulk of the visual cortex does not light up, one small area – known as V5 – is activated in the presence of stimuli. This area corresponds to one end-point of the tectopulvinar pathway. It alerts us to whatever is new in our visual field, even if we are without the faculties that then enable us to 'see' it.

Some experts now believe that it's the other half of our visual mechanism, the geniculostriate pathway, that is faulty in people suffering from dyslexia. This pathway begins in an area of the thalamus that is subdivided into two regions – the first is a slow processing system that appears to deal with colour, and the second a faster system that provides information about movement and location. Many dyslexic people report that when they try to read, the words seem to move about on the page. Various studies have shown that dyslexic individuals process visual information more slowly than non-dyslexics. For instance, they have problems identifying the order of two sequentially flashed visual stimuli – like two numbers or words flashed one after another upon a screen – but they can perform the task perfectly well when the sequence is slowed down.

Research performed by Margaret Livingstone and Al Galaburda of Harvard Medical School showed that neurons in the fast layers of the thalamus, dealing with movement and location, were more disorganized in dyslexics.[12] This pattern would, presumably, result in the brain having a limited ability to clear out one image before it moved on to the next. And this might explain why the words seem to shimmer or move around on the page.

Paula Tallal of Rutgers University and Michael Merzenich of the University of California, San Francisco, used findings from research that Dr Merzenich had done in monkeys showing how well the brain can adapt to new sound stimuli.[13] They devised a CD-ROM game for dyslexic children called FastForward which is played fifteen hours each week. It trains their brains to register the quick changes between one sound and another in normal speech. After a four-week period, exercised brains showed a leap of as much as two years' worth of language ability.

In his book, John Ratey provides a fascinating glimpse of how such a minuscule difference of cellular organiz- ation could give rise to a whole facet of a person's character. One of his researchers reported to him her surprise on realizing that her mother had always been dyslexic, but had covered it up. She had also always been wary of crowds, or any busy, unfamiliar place. This aspect of her behaviour might have arisen from the same abnormality that created her dyslexia – being unable to process fast-moving information, being overwhelmed by it. Over time, she developed behavioural strategies to avoid being beleaguered – and these caused her shyness and lack of spirit of adventure.

Ratey suggests that even milder differences of person- ality might, in the final analysis, be the result of faulty processing at the sensory or motor level. The archetypal 'nerd', for instance, may really be someone who has with- drawn from a society they find hostile and bewildering.

This is perhaps because they don't see people the way others do or because they cannot accurately judge when to chime in with an opinion. The underlying principle that follows is clear. Health professionals, teachers, court officials and others must stop labelling people as 'stupid', 'lazy' or 'bad' and consider the reasons why they might be the way they are. Such understanding is the hallmark of a healthy society and we are still a long way from that goal.

One key observation cautions against taking this theory too far. No two brains are alike in every way. So every single human brain, in that sense, could be said to have some 'abnormality' or other. If we notice that someone is shy and withdrawn, for example, and we then notice that the area of their auditory cortex responsible for processing language is smaller than other areas, we could say that they are withdrawn because they find communication with others difficult. Fine, as far as it goes, but we could also say that someone with an especially large language processing area was shy because they were overwhelmed by their extreme sensitivity to spoken sounds. No physician would have to look very far to find some curiosity in each and every human brain, which could then be seen as the cause of a disorder.

chapter**four**

Paying attention

You might be vaguely interested in learning how I got hooked on making television programmes. In the 1960s, when I was a registrar – a stripling, trainee obstetrician in a hospital in rural Essex – one patient elected to have her baby delivered using hypnosis as her sole pain relief. Pamela had had three babies delivered conventionally so she was a deal more experienced than her trainee doctor. She was convinced that the next labour might be more comfortable if she employed a hypnotist to attend and anaesthetize those parts that other painkillers did not easily reach.

It happened that at that time the BBC were making a programme on hypnotism – possibly one of the first programmes made about neuroscience. They heard of Pamela through her hypnotist, and got permission from the hospital to bring a lightweight crew to film her delivery. Pamela's due date was the Thursday before Easter, so learning by phone that she had been admitted, the crew battled to the countryside, hotfoot from Shepherd's Bush. In those days of good budgets – how TV funds have dwindled – a 'lightweight' crew included a cameraman with assistant, a soundman and assistant, an electrician and, of course, the producer and his assistant. They stood eagerly in the maternity wing, more expectant than any patient, and ready to shoot. None of them had seen a baby delivered, and only the blasé electrician had been in an operating theatre before, so they were trembling with excitement at the prospect in store. But Pamela was not in labour; she was reading the newspaper in a sitting room.

Crestfallen, they hung around all day awaiting the onset of labour. It seems incredible now that a media crew could be so naive as to believe that a pregnant woman delivered on her due date. By evening nothing had happened, but they decided to continue to wait as Pamela was having mild contractions. I took pity on them and suggested they came for a meal I offered to cook in my bachelor flat. Over the next few days, Pamela teetered on the verge of labour, the crew dithered about staying over the entire Easter weekend, and I became concerned about my patient because her blood pressure had risen – an unwelcome sign. By Easter Monday, the crew had camped four nights in my three-bedroom flat using every available space – it was not a pretty sight. That morning it became obvious to me that Pamela needed a Caesarean section. I phoned the consultant, who agreed and told me to get on with it. The BBC had invested a fearful amount in this adventure – massive overtime on a Bank Holiday weekend. The producer asked if I would ask Pamela if she was willing to have the Caesarean done under hypnosis. When I flatly refused he asked if they could at least film the operation (a Caesarean had never been previously shown on TV, even in black and white) and have the anaesthetist inject the drugs painlessly to induce sleep through a hypnotized arm beforehand.

I phoned my boss again. He was not impressed. I was well capable of doing a Caesarean section, he said, and he certainly did not intend to come in through the foul weather and spoil his Bank Holiday. If I wanted to let them film, he said, that was my affair. Sadly, I have always been a thespian and temptation proved too much. Pamela was wheeled into the anaesthetic room, the camera was switched on, the hypnotist pompously came in and, at great length, demonstrated to the cameras how good he was at making her right arm completely numb, and then went home.

Unhappily, it being a Bank Holiday weekend, we only had a locum anaesthetist, a pleasant foreign lady whose sole shortcoming was her extremely imperfect command of the English language and her consequent failure to understand why she, and her anaesthetic, were the focus of a BBC film crew's attention. Entering the anaesthetic room, she picked up a syringe full of prepared drugs and, without waiting, banged the lot into the patient's un-hypnotized left arm. Pamela was wheeled into the theatre and there was that curious scurrying sound I associate with the media. It is the sound of an entire film crew scuttling behind their object of interest, desperate to keep quiet and not lose focus while adjusting lights, camera, microphone. I stood proudly at the operating table with the entire circus behind me, filming my every move. Unable to handle anything at all outside the operating field because of sterility, I cleaned the skin, draped the patient, opened the abdomen and incised the uterus without mishap. But as I was pulling the baby from the abdomen, I experienced the sensation that every surgeon, experienced or novice, deeply dreads because he cannot touch anything. It was the sensation of the pyjama cord around my theatre trousers inexorably loosening and then, with dread inevitability, the terrible sensation of my trousers inch by inch creeping down to my knees and then ankles. But not until the baby was safely handed to the waiting midwife did I recall the chaos in my flat and think, 'Did I remember to put clean knickers on this morning?'

In spite of all the distractions, at the key moments during that delivery I was using my attention, firing up specific parts of my brain. Attention directs our senses like a conductor, guiding them in different directions, disengaging them from one task and engaging them upon others. We all make use of a common mechanism within our brains to make sense of our sensations. Just imagine for a moment that you were able to see, hear and smell everything in your immediate surroundings all at once. The humming of your computer printer. The smell of your carpet. The car alarm outside and the dog barking. Your neighbour's stereo system turned up a bit loud. Somebody peering over your shoulder when you are trying to write. The unfolding events of the TV news in the adjoining room. A child's constant fidgeting. Imagine trying to think clearly for a moment in the midst of this sensory tsunami. Imagine trying to conduct a telephone conversation or look for your car keys in those surroundings. It ought to be well nigh impossible. So how do we manage? What within our brain enables us to choose what we are going to see, hear, smell? How can we tune into a conversation with someone else and ignore other sounds, find that missing set of keys amid the clutter of the office?

Paying attention unconsciously

In the 1940s, a group of American psychologists conducted some groundbreaking research into the factors that influence attention. The experimenters used a tachistoscope – basically a laboratory device to flash images for a usually short and very specific moment of time. The research team flashed words to a group of volunteers over varying periods of time and asked them to identify the words they had just seen. The words were of two types: neutral – including words like hammer, sheet, pin; and taboo – words with a high emotional content, like whore and penis.

At the very briefest duration – tiny fractions of a second – people were unable to identify what they had seen at all, registering only that they'd seen a burst of light. But as the time period became longer, something very interesting emerged. People took longer to recognize taboo words than the neutral words. The more socially unacceptable the word, the longer it took to register it.

This experiment was largely done to support the psychoanalytic, Freudian view that we employ certain unconscious 'defence mechanisms' in order to repress information that is upsetting us. But it is of interest here because this early research suggests that some agent inside our brains is making decisions about what we pay attention to, and what we choose to ignore. Psychologists sometimes call this the 'cocktail party effect' – based upon the observation that, even in environments that demand so much of our senses simultaneously, we are often able to follow conversations and at the same time respond if someone on the other side of the room mentions our name.

In a later experiment, conducted by Drs Corteen and Wood in the 1970s, participants were conditioned to expect a small electric shock when they heard the name of a city. Their fear was measured by means of a sweat-sensitive pad placed on the skin. Possibly these hapless participants were offered financial sweeteners in recompense for their ordeal, but I doubt it. Neuroscientists and psychologists are sufficiently knowing about the human mind usually to be able to find a stream of willing research-fodder close at hand – possibly PhD students who refuse at their peril. Sometimes, of course, small cash sums, record vouchers, cans of beer and even humbugs have been used as a reward, but never in my experience for the most deserving – the PhD students.

Having tortured them in this way for some time, the

experimenters then got the subjects to listen to two streams of words, one in each ear. They were instructed to repeat aloud the words they heard in one ear, but to ignore the words they heard in the other ear. Even when they were not paying conscious attention to it, the sound of a city name in the ignored ear provoked the same fear response. It became clear that we can pay attention to complex streams of information when we have sufficient motivation to, and that only some of this occurs at the conscious level. This is the cocktail party effect: we can hear our own names mentioned on the other side of a noisy room, even when deep in conversation, because the information has emotional significance.

Much more recently, Daniel Simons and Christopher F. Chabris at Harvard University asked subjects to watch a short film of some basketball players, one team in white and the other in black T-shirts.[14] They were asked to note down the number of passes that the white-shirted players made in the game. During the film, a woman entered the field of vision, and opened a dark umbrella. Of twenty-eight observers, only six noticed this event. The experimenters then expanded the experiment to include a person in a dark gorilla suit performing a nine-second routine in a stretch of footage that only lasted sixty-two seconds. Roughly half of the viewers failed to notice the gorilla.

I mention these experiments chiefly to demonstrate that paying attention is not the conscious, all-encompassing process we might consider it to be. We notice some things without knowing we notice them, while other things pass us by, even though we would consider ourselves to be on full alert. This highly selective aspect of human attention has given rise to considerable controversy over the years.

In the year 2000, during the Bush–Gore election campaign in the United States, the Republican party was accused of using foul means to influence voters. The

Democrats alleged that, during a Republican party political broadcast, the word 'RATS' appeared on the screen for exactly one-thirtieth of a second during a sequence in which other elements of the word 'DEMOCRATS' moved in and out of vision. They were saying, in effect, that their opponents had used a subliminal message to influence viewers. The Republicans defended themselves by maintaining that this was pure chance – even though the word 'RATS' was also in a larger typeface than 'DEMOC' and the ads had cost them a mere $2,576,000. This particular battle became forgotten amid the wider, and much more publicized, furore over the issue of the miscounted votes, favouring Bush.

This was not the first time that there had been a scandal over the use of this technique. In 1957, there was a massive outcry when James Vicary, a market researcher, announced that he had successfully inserted messages – for periods of time so brief that they escaped conscious detection – into a reel of cinema film. The messages, targeting no fewer than 45,699 patrons of a particular cinema in Fort Lee, New Jersey, were inserted during successive screenings of the film *Picnic*. The letters that were briefly flashed on screen (3/1000th of a second every five seconds) urged cinemagoers to 'Eat Popcorn' and 'Drink Coca-Cola'. Over a six-week period, Vicary boasted that this had resulted in a 58 per cent increase in food sales and an 18 per cent increase in drink sales. People in the 'Land of the Free' took great umbrage at the idea that it was possible for people to tamper with their thoughts without their consent. Their outrage was particularly strong in the tense milieu of the 1950s, in which it was suspected that all manner of unsavoury agencies, from aliens to communists, might be able to gain control over people's brains. When congressmen got involved in the protests – Representative Dawson said, 'Put to propaganda purposes [it] would be made to order for the establishment and maintenance of a totalitarian government' – Vicary's experiments were called to an abrupt halt. His team later announced that there hadn't been any subliminal messaging at all, and the whole affair had been an elaborate hoax. It remains unclear to this day what really occurred but it seems likely that Vicary was not exactly a reliable witness.

The modern view of subliminal manipulation of attention is a little less paranoid. So-called subliminal cuts are banned in advertising, both in the US and in the UK. But at the same time, experts on both sides of the Atlantic consider it to be of only very limited power to influence people's thoughts and actions. Some studies have shown that the influence of subliminal messages may last for as little as one-tenth of a second – which is probably not long enough to go out and buy some popcorn or vote for the Republicans.

It also seems that these messages only work when the circumstances are right. It is believed that subliminal messaging works – when it does – through a process called priming, the same mechanism we examined when looking at facial recognition. If a word like 'Coke' is flashed subliminally to my brain, then this message will have the effect of priming, or stimulating, other recognition units connected with my experience of Coke: quenching thirst, sweet taste, bubbles on the tongue, good with a shot of rum . . . But as priming has an effect on the brain for quite a long time afterwards, this liking for Coke may last – at least for longer than one-tenth of a second.

Arousal

What parts of the brain are we using when we pay attention? Neurologists have argued that, in fact, attention is not one process but four – arousal, orientation, reward and novelty detection, and executive organization – each with its own related set of brain regions.

Arousal is the faculty whereby our awareness increases. It is controlled by what we call the reticular activating system in the brainstem, the frontal lobes, the limbic system and sense organs. Information from the sense organs, or even from our own thoughts, can arouse us. The hippocampus, the seat of long-term memory, also communicates with the reticular activating system, comparing what we see in the present with the past.

The arousal system plays a big part in what we call the fight-or-flight response – that combination of brain and body activity that comes into action when we are startled by something. Imagine you are lying in bed, just drifting off to sleep. Suddenly you hear a loud crash outside. Immediately, you are fully awake, searching for the origin of the sound. Information from your auditory cortex is being hurriedly assessed and parcelled – some of these parcels travel to the amygdala, triggering the sensation of fear. The amygdala in turn communicates with the anterior cingulate in the frontal lobes, and then the hypothalamus and spinal cord.

The hypothalamus, Grand Vizier of the body's hormonal system, starts to fire up. Briefly, it suppresses bodily activity – heart rate is calm, blood pressure stays constant, breathing is quiet. This creates a bodily delay so that the upper cortex can work out what's going on outside our window. Then it swings into reverse – heart rate and blood pressure soar, the secretion of adrenaline heightens awareness and primes your muscles for action.

Meanwhile, the frontal lobes are at work, assessing the situation. If you determine that some neighbourhood cats have knocked a dustbin lid off, then your frontal lobes send a message to the amygdala to calm down. This sets a chain in action which concludes with the heart rate slowing down, and blood pressure and adrenaline levels dropping. What's happening here is that the lower, unconscious brain is calming down and handing control of the situation over to the higher brain. You begin to 'think' about what's happening rather than simply reacting to it.

If, on the other hand, you determine that someone is trying to kick your front door in, signals travel swiftly to the hypothalamus. This triggers the release of corticotropin releasing factor, or CRF, the brain's stress hormone, and of the neurotransmitters adrenaline and cortisol to prepare body and brain for swift action. It also talks to the pituitary gland, which secretes hormones affecting every major gland of the body. This rush of activity stimulates the amygdala and the brainstem – you become prepared to run or fight.

The intensity and speed of these processes differ from one individual to another. Calmer people will generally spend more time in the 'cool' mode, their lower brain activity dampened down as their higher brains gather and assess incoming information. More panicky types of people may do the opposite, immediately entering fight-or-flight mode, even if the information out there indicates that they do not need to.

Nowadays, I am not afraid of public speaking, but I can remember enough of my early days of stage fright to know that it is very real. In this instance, what's happening is a loop between the upper and lower brains. In our upper brains we see the expectant faces before us, feel our dry mouths and our shuddering limbs and conclude, 'Oh dear, I'm afraid.' This sends a message to the amygdala and thence to the whole nervous system, sending heart rate and blood pressure sky high, letting the stress hormones such as adrenaline flow. Our upper brain observes these fresh symptoms of impending disaster and says, 'Oh no – now I'm *really* terrified.' And it sends that message back to the amygdala – a vicious circle *ad infinitum*. People with stage fright who have taken beta-blocker drugs – which suppress the production of adrenaline – report that, while their brains still feel alert, and in some cases more than a little nervous, their bodies do not take up the baton, as it were, and thereby the cycle of panic thought–panic symptom–panic thought is broken. I have to say that for me, as for all public speakers I know, certain environments are more intimidating than others. I can relate from experience that standing to interrupt a minister in a debate in the House of Lords, speaking in front of the 24-carat-gold throne and with the Lord Chancellor on the Woolsack, knowing that there are real experts in the audience who are bound to have more knowledge than oneself, is pretty frightening. The pulse races, the mouth goes dry and breathing feels tight. Not surprising that more than one member has collapsed with a heart attack either during, or immediately after, making a speech in that formidable chamber.

Orientation

Once we are aroused, the orientation system allows us to fix our senses upon the item that has aroused us. This is essentially unconscious – tiny babies turn their heads towards a novel sound. So do we adults, without thinking about it first. It has been shown that this apparently seamless procedure is made up of three components. First of all we use the posterior parietal cortex to disengage from whatever we were doing and pay attention to the new stimulus. If the stimulus is visual, the frontal eye fields are engaged. Then our basal ganglia and parietal attention areas shift the focus of our attention onto the new stimulus. Thirdly, the lateral pulvinar – the area within the thalamus we discussed above in relation to dyslexia – acts as a filter, focusing our attention on

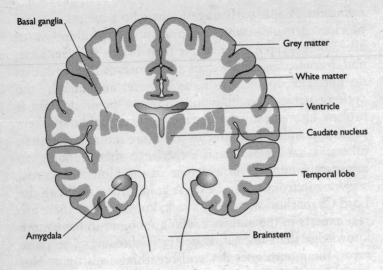

Basal ganglia

Grey matter

White matter

Ventricle

Caudate nucleus

Temporal lobe

Amygdala

Brainstem

Brain in cross section

the new stimulus and blocking out other information.

If the posterior parietal cortex should be damaged for any reason, a phenomenon known as Balint's Syndrome can arise, in which people are unable to pay attention to multiple objects. They are unable to shift their attention from one focus to another. Eric Courchesne at the University of California at San Diego has argued that a similar process may be at work in people with autism, leading to an inability to shift the focus of attention in response to social cues. John Ratey gives what may seem a rather surprising example of this in his book: if a mother points at a tree and says to her child, 'Look at the tree,' the child may be unable to detach its attention from her face and look at the object. This is surprising because, in general, autistic children have difficulty in attending to faces. But here, the child cannot take a cue from its mother's face or voice. The result is a difficulty in conducting social relationships – which is certainly one of the hallmarks of autism. People with that condition can also be curiously indiscriminating about where they direct their attention. An autistic child might, for example, remain fixated upon a bunch of keys, even though a seemingly enticing toy is being dangled under his or her nose.

The normally functioning brain has something like a hair-trigger mechanism, tuning it to notice and lock onto anything that is novel or surprising. When a subject is read a list of words: 'book', 'paper', 'volume', 'chapter', for example, and then a surprising word like 'treetop' is thrown in, an 'oddball' response can be measured in the brain. It has been known for a long time, from EEG measurements, that our brain activity shows a rapid surge in electrical potential about 300 milliseconds after we first report being aware of an oddball. This is known as the P300 response. Brian Strange, working in Ray Dolan's group at the Institute of Neurology, London, has used fMRI scanning to identify which part of the brain is activated.[15] When we experience various kinds of oddball stimuli – be it an odd noun, an odd font in which a noun is printed, or an emotional connection – the same parts of the cortex, the inferior prefrontal and the posterior part of the temporal cortex, show activity. Curious that anybody by the name of Strange should be working on oddballs. A number of researchers – such as Eric Halgren at the University of California at Los Angeles – have concluded that as the P300 surge occurs, there is a corresponding dampening down of activity in other areas of the brain. In other words, the peak stands out because it stands alone.

Research shows that it takes the brain 100–200 milliseconds to fire up in response to novel stimuli. Then comes the period of dampening down in other areas – then the P300 response. It's as if the brain is taking in everything, all the time, but mechanisms within it act to turn the 'contrast knob' up or down – focusing our attention on some features, shutting off our attention to others.

We can see how this would be a vital part of our memorizing procedure. If we were to record every aspect of every moment, our memory would swiftly become overloaded. Indeed the renowned Russian 'Memory Man' S.V. Shereshevski seemed to suffer from exactly this kind of filtering problem. Shereshevski could reproduce a table of fifty numbers in any order, vertically, horizontally or diagonally. And according to reliable reports he could recall these numbers months or even years later. When memorizing tables of numbers, Shereshevski could apparently see the table in his mind. Any mistakes he might make were invariably because a number was not properly legible, or because he had been distracted perhaps by a noise when he was studying the table. While he could remember the minutest details of a long story, he lacked an overarching awareness of its point or message.

It is interesting that the amygdala has a role here. It seems that the amygdala biases people to recall the gist, rather than the detail, of pictures with an emotional content. Normal people tend to remember the gist of an emotional story or image while people with a damaged amygdala don't show this switch.

Other brain structures seem to play a role in this filtering procedure. David LaBerge, working at the University of California, Irvine, performed a visual detection study on eight students. They were asked to look at a screen upon which a display of characters was flashed, and to search for a letter O among characters visually similar, such as C and G. With these letter displays flashing up for just one-fifth of a second, it was no easy task. The subjects had to keep the target letter O in mind, and also blank out the red herrings. LaBerge discovered that the harder the task, the more a specific region of the thalamus – the pulvinar again – had to work.[16]

The basal ganglia seem also to play a part in this process. Dick Passingham at University College, London, began his enquiries by asking human subjects to perform simple motor tasks like moving an arm while undergoing PET scans.[17] He then asked them to perform more complex tasks, such as rapping out a rhythm or tapping an intricate sequence on a keypad. His brain mapping studies showed that, when a task was new, there was a widespread and highly active network of neuronal activity. But as the sequence was mastered, became 'second nature' to the subjects, the circuits involved shrank back to the barest levels of activity. In other words, there was a visible difference between an activity that requires our conscious attention, and one that we have learnt to the point of it being automatic.

Passingham's studies were boosted by the arrival of cheaper, more repeatable fMRI technology. With this, he was able to see precisely which areas of the brain were involved in attending. He discovered that when subjects were having to think carefully about a task, most of their frontal cortex was active. The prefrontal areas leading to the brain's motor system were active, as were the cingulate and parts of the thalamus and parietal cortex. But crucially, once the task was mastered to the point where the subjects performed it without conscious attention, most of the higher cortex dropped out. It was now the lower rungs of the brain's motor system – the premotor cortex and the cerebellum, and the basal ganglia – that were doing the work.

It seems that, where attention is concerned, the basal ganglia act as an output filter in the same way as the thalamus acts as an input filter. Ann Graybiel, working at the Massachusetts Institute of Technology, studied what happened in the brain when a monkey learnt a simple association task – the sound of a click with a sip of juice. She found that, at first, as the monkey was still learning, the activity in the basal ganglia simply mirrored what was taking place elsewhere, in the higher cortex. But over time, a special class of basal ganglia neuron began to react on their own. They activated as the click sounded. In other words, as the skill was learnt, they took on the role that had been in the hands of the higher cortex. The task became an un-thought habit. We can conclude that, just as the thalamus enables us to focus on particular aspects of what we experience, the basal ganglia enable us to focus on the appropriate actions – in both cases, freeing up the higher cortex for other tasks.

Reward and novelty detection

Once we are duly aroused and orientated towards the source of the arousal, our reward and novelty detection faculty comes into play. This depends upon the action of the mesolimbic pathway, a grouping of dopamine-containing neurons that controls the limbic system. When this is at work, we take note of new stimuli, and our reward system produces minute sensations of pleasure. It also tags the stimuli with an emotional value – pleasure, disgust, fear, etc. – which then stays with them in their transit into the memory. If the same stimulus recurs later, the memory of the emotions provides a response, which we use to devise a strategy to deal with the situation. So we know to avoid the cocktail party bore – the second we hear his loud, hectoring voice and see his pompous, overconfident body language, we remember similar situations in the past, and we know to make a beeline for the kitchen.

A key region in this process is the nucleus accumbens – situated in the forebrain and connected to the amygdala and limbic system. It contains the highest store of dopamine in the whole brain and is also acutely sensitive to other neuro-transmitters such as serotonin and endorphins. These 'pleasure chemicals' result in feelings of reward and satisfaction, and are thus very important motivators of action.

People with Attention Deficit Hyperactive Disorder are believed to have abnormalities in this region – in particular a low density of dopamine receptor sites. Unable to experience pleasure beyond a certain low threshold, they become predisposed to seek constant, immediate gratification, and ignore anything that might offer a more long-term reward. So, a child with the disorder would find it impossible to spend an afternoon practising his violin in order to win a competition. He'd be out of the door and into the nearest amusement arcade, but might not even arrive there. Monkeys with lesions in the region of the nucleus accumbens show similar traits – they would rather eat a peeled nut now than hoard unpeeled ones for the future. And Rudolph Cardinal, working in Cambridge, found that rats with a damaged nucleus accumbens showed similar tendencies to be impulsive.[18]

You may be puzzled that this disorder can be successfully treated with drugs like Dexedrine – which is a form of amphetamine known on the streets as 'speed'. It seems paradoxical to be giving anyone with attention and hyperactivity problems a compound that seems to accelerate everything – from pulse to thought processes. But such drugs have quite the opposite effect on people with ADHD. They act upon the synapses to increase the output of dopamine and block the action of certain enzymes that cause dopamine to be flushed away. Accordingly, the nucleus accumbens is able to maintain a more stable feeling of 'reward' and there is a reduced drive to go in search of excitement.

Executive organization

The final component of attention is the executive organization facility. This allows us to sustain our attention in the new direction, by blocking out irrelevant information. A key player in this process is the anterior cingulate gyrus (ACG), situated within the frontal lobes but sometimes considered a part of the limbic system. This acts as a filter for incoming data, prioritizing and throwing out what we do not need. The ACG shows increased activity whenever we concentrate on a task. It is a bizarre coincidence that we humans often scratch our head over roughly this area when we are thinking hard, almost as if we have some unconscious sense of where the activity is.

Mostly, the mechanism of attention involves using this four-step process to lock onto sensory information coming from the outside world. But it may not have escaped you that we also use it internally. Sometimes we pay attention to our own thoughts, for example, or to our own feelings. Sometimes we deliberately set a chain of thoughts in action, and keep the chain going, to the exclusion of other thoughts or incoming sensory stimuli. We might cope with the frustration of a traffic jam, for instance, by very meticulously working out exactly how we are going to spend our lottery jackpot, when it finally comes. We might work out how much paint we need to brighten up that spare room as we queue in the super-market. The process of thinking requires attention.

When we think, our brains show activity in the very same areas that become active when we are actually doing something. If we imagine a colourful object visually, that part of the visual cortex is activated just as it would be if we were seeing it. And a different part of the visual cortex would respond if we imagined a moving object. If we feel fear, an MRI scan will show activity in the amygdala. The same activity will be observable if we see a picture of a fearful face, or if we are asked to imagine being afraid. The same would be true for picking up an object, or travelling a familiar route. When we imagine something, we build on patterns built into our circuitry through previous imaginings, and previous instances of actually doing.

A crucial region in the thinking process is the dorsolateral prefrontal cortex (DPC). Planning takes place in this area, as does decision-making. Areas within this region have more specific functions, rather like the different aisles of a supermarket – for example, there is an area that becomes active when we think about objects that we cannot see. Another area of the DPC sparks into life when we think about how many times we have done something before.

Intention, planning and initiating action are faculties particularly restricted in people with frontal lobe damage – as we will see later. This can result in them forming dozens of plans every day, yet failing to carry any of them out. The importance of the frontal lobes has been demonstrated by Chris Frith and his team at the Wellcome Department of Cognitive Neurology in London.[19] They asked normal subjects to perform a simple task – in this case, lifting a specified finger at a given auditory cue. The execution of this task caused activity in the auditory cortex and the motor cortex, as we might expect. But then Frith added an extra element to the task by asking the subjects to pick for themselves which finger they wished to lift. When they performed this, self-willed activity, a specific area within the frontal lobes – the prefrontal cortex, just behind the forehead – sprang into action.

So the 'will' is a brain faculty like any other – one that can be strengthened with use, or lost through damage. Neurologist Antonio Damasio described one patient who, after a stroke, had damage to the frontal lobes and lay for several months without speaking or moving. When she eventually recovered, she said that during her period of immobility she had not felt that she couldn't do anything, but that she hadn't wanted to.

As well as being the place where the will is generated,

the frontal lobes play a role in the way we link our perceptions and our awareness of ourselves into a coherent experience. In other words, this is the area where we attach meaning to life. People suffering from depression show a low level of neuronal activity in this area – and correspondingly they often report that their lives seem devoid of meaning or purpose, fragmented and pointless. Conversely, the area is very active in people suffering from mania – for whom every outside stimulus and internal process seem linked in a sort of gloriously meaningful but deluded grand design.

In our exploration of the mechanisms of attention and intention, we have started to touch upon a question that, as yet, has no answer – though philosophers and scientists are set to fight it out for many years to come. This book is intended to be a journey through the mysteries of the human mind, and no enquiry into that subject would be complete without considering one of its most unique and perplexing faculties – the self.

Losing consciousness

A recent television series, in which ordinary people underwent rigorous SAS training, highlighted the deleterious effects of sleep deprivation on the workings of the mind. The subjects were forced to stand against a wall while listening to static 'white noise' and were shaken by guards whenever they showed signs of nodding off. This has been shown to be highly effective in 'breaking' the will of people before they face interrogation. The longer they go without sleep, the more disorientated they become. Their mental agility and memory become impaired. Such is their craving to enter this natural brain state that they will do or say anything to be allowed to curl up in a corner of their cell – even betray their comrades and their countries.

Sleep is an essential process. But, remarkably, its biological function is mysterious, in spite of a vast amount of research. It is not just humans that eventually crave sleep. All species that have been studied in detail appear to sleep, whether reptiles, birds or mammals. There seems no obvious evolutionary advantage in sleep as a sleeping animal is defenceless, cannot feed, cannot take care of its offspring, cannot avoid the risks of being eaten by a predator and cannot have sex and so procreate. The very fact that it is costly suggests that it must be a vital process. In some animals, for example whales and birds, it seems that just one hemisphere of the brain sleeps in turn. Rats chronically deprived of sleep will die within about two weeks – which is roughly how long they can last without food. But surprisingly, the cause of death is far from clear.

It is curious that humans vary so much in how much sleep they need. The average is around eight hours a night – but age has an impact on this figure. Newborn infants sleep most of the day, whereas the elderly may require only a few hours a night. For years I have required only about five hours' sleep a night, although after several days and nights with just five hours, I find I need a binge – a longer period of sleep. As I get older, I have found I need slightly more, not less, sleep and I also find that I feel I need more sleep if I have been on holiday or if I have regularly been sleeping longer on previous nights.

The stages of sleep

What exactly is sleep? It seems to be a set of four separate brain stages. We all know the difference between nodding off gently on the sofa to the news and that full-blown state of leaden unconsciousness that causes us to wake up hours later without having seemed to move in our beds. We also know that one is infinitely more refreshing than the other. It also seems more refreshing if we have been more or less horizontal – perhaps this is why sleeping in a first-class seat in an aircraft leaves one feeling better than sleeping half upright in a slightly reclining chair in tourist class.

When we are awake, our neurons are extremely active. Even neurons not being specifically used in a given task will fire off approximately one 'blip' of electrical current every second. We can measure this electrical activity with an EEG machine – the device that gives a pictorial representation of electrical activity in the brain, the brain waves. When we are awake, our brain waves are a series of short, tightly spaced teeth – like the serrated edges of a bread knife. These are so-called beta waves.

As we start to relax and close our eyes, there is a difference in the brain waves. The little 'peaks' of activity become slightly larger, and more spaced out, more regular. The 'Chinese firecracker' effect of alert wakefulness turns into a gentler, more rhythmic sequence of pulsed activity, like the hum of distant machinery. This is the shift from beta to alpha wave activity. When we are hypnotized, or when we are meditating, this is the wave activity going on inside our brain.

As sleep deepens – the first stage of actual sleep – we see theta wave formation. Theta waves are less regular, more like bad handwriting on the paper roll of the EEG machine. We can be easily aroused from this stage of sleep and may not even be aware that we were asleep. While this is going on, we may see vivid images before our eyes – sometimes a version of what we would see in the room we are in if we were awake. We may experience sudden jolts in our arms and legs – which can sometimes wake us up with the sensation that we have fallen to earth with a sharp bump.

As we enter the second stage of real sleep, breathing becomes more regular and we are less likely to be woken up by external intrusions. EEG scans still show theta wave activity, but there are occasional bursts of activity, known as 'sleep spindles', and larger spikes and dips, called 'k-complexes'. It may be that these little patches of activity represent the brain operating to shut out external stimuli, so that we can get on with the 'work' of sleeping.

In the remaining two stages of sleep, associated with delta wave activity, we become increasingly oblivious to our surroundings and are quite likely to feel startled and disorientated if we are woken suddenly. Delta waves are characterized by a 'high and low' pattern of regular peaks and troughs on the EEG machine – becoming more pronounced as we move from stage three to stage four of sleep. Interestingly, though, we can still take in information. Parents may snooze away, oblivious to traffic noise outside, but be instantly aroused if they hear the sound of their child crying.

During normal sleep, there are periods back and forth across this scale of brain-wave activity. After roughly ninety minutes of sleep, for instance, the cycle goes into reverse – going rapidly back to stages three and two – and then still further back to the beta-wave activity we see in the state of alert wakefulness. At this point, we enter rapid eye movement (REM) sleep – a term coined in 1953 by its discoverers, Nathaniel Kleitman and Eugene Aserinsky at the University of Chicago. Sleep researchers call this 'paradoxical sleep' – because of the contrast between our sleeping body and our seemingly alert mind. Some areas of the brain – including the brainstem – are more active during REM sleep than in our normal waking state.

Perchance to dream . . .

Around 80 per cent of people, if woken during REM sleep, will say that they have been dreaming. But if awoken during any non-REM stage of sleep, no more than half will claim to have been dreaming. And the nature of the dreams experienced is radically different. Non-REM dreams have a 'procedural' nature – for example, performing dull, day-to-day activities, like typing, or dialling numbers on the telephone. REM dreams, by contrast, are a more fantastical mixture of reality and the outright bizarre. They can involve intense emotions, illogical events and a suspension of the normal rules of time and space. Familiar people may turn into strangers, and vice versa.

Sophie Schwartz and Pierre Maquet at University College, London, noted a similarity between many of the aspects of REM dreaming and the symptoms of people with disordered brains.[20] For instance, the misidentification of faces and the disproportion of objects can be consequences of lesions to the temporal and frontal lobes. They are also very typical elements of REM-type dreams. Lewis Carroll's *Alice in Wonderland* gives a good account of the dreams associated with REM sleep. When people are scanned by PET and fMRI, they show similarities in their brain chemistry to what happens in some brain disorders.

There is increased activity in a region of the brainstem just above the spinal cord, called the pons, in REM sleep. Levels of the neurotransmitter acetylcholine increase sharply just before REM sleep, as if preparing for a burst of communicative activity. This then takes place, as the pons communicates not only with the spinal cord, but also with the thalamus and occipital lobes, site of the primary visual cortex. This is what seems to cause the rapid eye movements. There is also a heightening of activity in the amygdala, partnered with a corresponding dampening down within the dorsolateral prefrontal cortex. Since we know that this part of the cortex is responsible for rational thought, and the amygdala produces strong emotions, there could be some explanation for the power-ful, yet irrational content of our dreams.

The sleep cycle repeats itself throughout the night (or whenever we are sleeping) – becoming shorter as we move towards the appointed hour of the alarm clock. As the cycle shortens, relatively more time is spent in REM sleep. We may pass through this cycle to the point of actually waking up several times a night, but if our brain is healthy and there is an absence of distracting stimuli, we will go back to sleep and not remember these intervals. People suffering from depression or anxiety, however, may experience trouble getting back to sleep after they have first woken up.

Disturbed sleep

Sleepwalking probably only occurs during non-REM sleep. Sleepwalking is close to wakefulness, and a number of brain functions are observable in sleepwalkers which wouldn't be seen in the deeper stages of sleep. Movement is the most obvious, but there is also goal-orientated action. Sleepwalkers always have a goal. Lady Macbeth spent the night washing her hands of the blood she had spilt. More prosaically the action might be to visit the toilet – something which had unpleasant consequences (for me) in the case of my sleepwalking climbing com-panion when I found that the foot of my sleeping bag was mysteriously damp in the morning. Sleepwalkers can even sometimes conduct limited conversations. In spite of old wives' tales, there is no evidence to suggest that it can be fatal to wake up a sleepwalker.

Various brain structures and neurotransmitters are involved in the sleeping-waking cycle. Humans tend to wake up in daylight and go to bed when it is dark. This depends upon a particular part of the hypothalamus, which responds to light – even in blind people – and individual neurons within it show a peak in firing activity once a day. The hypothalamus also communicates with the pineal gland – telling it to secrete the hormone melatonin. Melatonin concentrations are lowest in daylight, and highest during the hours of darkness.

Waking is partially controlled by areas in the brainstem called the locus coeruleus and the nearby raphe nuclei. These pump out noradrenaline and serotonin to various parts of the brain, increasing alertness. They are most active when we are awake, and almost completely inactive during REM sleep. The preoptic area of the frontal cortex also seems to be involved in sleep. Neurons in this part of the brain are particularly active during non-REM sleep and people with damage to this area often suffer from insomnia.

A few people – between 3 and 6 per cent of the population – can get to sleep and stay asleep without any trouble but then they experience the horrors. Sleep paralysis is frightening and tends to happen just as people are entering or leaving sleep, and can last for minutes. SP sufferers find themselves awake but dreaming at the same time – undergoing hallucinations that can be deeply disturbing. At the same time, their body is paralysed. They cannot move to hide from the imagined 'danger', nor can they rouse themselves from their disordered state.

One explanation for this might be that the shunting backwards and forwards across the sleep cycle sometimes brings us close to the point of waking. In certain cases, we might slip right over into wakefulness, but there is a time lag, ensuring that the spinal column remains in its 'off' mode, and the brain continues to behave as it does in REM sleep – pumping out emotional signals at high intensity. As a result, SP sufferers find themselves pinned to their beds, beleaguered by all manner of fear-inducing sensations. Some antidepressants can relieve this because they inhibit REM sleep. Many sufferers learn to develop their own 'escape routes', such as forcing a tiny movement by a massive effort of will. Research suggests that some form of sleep paralysis is experienced on occasions by as much as 60 per cent of the population. I know just how unpleasant it can be, because when I was a boy I used to be a regular sufferer and I eventually learnt how to deal with the paralysis by moving in a particular manner to my left. Following this I used to often wake in a cold sweat. Some researchers have suggested that reported 'paranormal' experiences may stem from an incident of SP.

Why sleep at all?

The risks associated with losing consciousness for eight hours a night are minimal nowadays – with our secure houses and high-tech burglar alarms. But they must have been pronounced when our ancestors lived on the savannah – prey to all manner of wandering hazards. So why do we sleep? At least three purposes of sleep have been proposed: restoration, regulation and learning.

We all know the restorative properties of a good night's sleep. We feel refreshed – full of energy, our brains alert. Human growth hormone is released during the deepest stages of sleep and, apart from stimulating growth, one of its functions is to trigger the repair of damaged tissue. It is also significant that when we are ill, we tend to sleep more.

What happens when the brain is deprived of its nightly fix of sleep? Two or even three nights without sleep can be tolerated by many people; it may not significantly impair our strength or even our ability to perform complex tasks. But our performance on routine or repetitive tasks is impaired. This, of course, is why driving when you are sleep-deprived is particularly dangerous. The brain's apparent ability to cope with more complex, more challenging activities might be because we coopt other areas to help. Brain scans of sleep-deprived subjects undertaking a verbal fluency task showed that the parietal lobe – an area primarily devoted to touch and spatial awareness – was being drafted in to assist.

Sleep deprivation can have a positive side. It may seem bizarre to deprive depressed people of their only means of escape from the daily 'pain' of consciousness, but various studies have shown that it can result in improved mood.[21] This is thought to be because sleep deprivation leads to an increased activity of serotonin receptor sites within the neurons – mimicking the effects of antidepressant drugs like Prozac and Seroxat.

The second function of sleep – regulation – would have played a key role for our ancestors. Our brain and body processes are organized into regular cycles known as circadian rhythms. And these cycles are present in species distant from us such as the fruit fly, where there is good evidence that these rhythms are regulated by genes switching on and off. The human body clock, which tells us when we should be feeling sleepy, or when we should be eating, and which can be thrown off balance by flying across time zones, may seem a little less relevant in modern life. But in our distant past, circadian rhythms may have existed to ensure that we were most quiet and inactive during times when there was the greatest danger. In other words, at night.

According to this theory, all animals need only be awake for the time it takes them to find and eat food and to mate. At all other times it benefits them to be asleep, and hidden away from danger. Humans, so the argument goes, have developed a specific circadian cycle that keeps us out of the way at night-time. However, there are some obvious problems with this theory. It is certainly true that all the while we are out looking for food and mates we are also at risk from predators. But it is also true that, unless we are very clever at hiding, we are even more vulnerable to predatory attack when asleep.

A third function of sleep is also likely: sleep acts as oil for the cogwheels of the brain. Earlier on, I mentioned the importance of REM sleep in memorizing and learning. It does indeed seem that the brain 'goes over' itself during sleep, reinforcing new firing patterns and etching them into memory. Numerous studies have shown that human and animal subjects master tasks more quickly and more thoroughly if sleep follows the learning procedure. It is also interesting to note that infants and small children – whose brains have the greatest amount of learning to do – need far more sleep than adults.

The meaning of dreams . . .

What place do our dreams have? Our ancient ancestors seemed to believe dreams were messages from another dimension, means by which spirits and gods could communicate with living humans. Jacob's dream in Genesis is of great symbolic importance. Near Beer-Sheba he lays his head on a stone pillow and dreams of the ladder ascending to heaven whereupon God is revealed and reiterates his promise to protect Jacob and the generations to follow. Where biblical dreams are not a form of divine communication they tend to reveal omens. Joseph, Jacob's favourite son, dreamed repeatedly. His dreams persuaded him of his own subsequent greatness (and his precocious attempts to boast to his brothers resulted in his being incarcerated). Later, when he had got out of the pit and then into and out of prison, he was able to forecast the fate of Pharaoh's unfortunate baker and his luckier butler, as well as the famine which eventually hit Egypt. In Roman times, soothsayers, of course, would have needed a fee to interpret dreams in this way.

Early psychoanalysts took this many steps further – merging the ancient belief that dreams 'told us something' with their scientific forays into the human psyche. Freud believed that dreams symbolically express the conflicts, desires and terrors of the individual unconscious mind. The dreams of each person would have their own, unique symbolic language, but it would be possible, through cross-referral, to assume that two different dreams could purport to the same hope or fear. Such a 'science' of interpretation is highly subjective and ultimately inconclusive. Freud's own powerful preoccupation with the libido led him to see sexuality as the raging torrent from which all mental phenomena, dreams and otherwise, would spring.

A leading theoretician in this field is J. Allan Hobson of Harvard University. Hobson was the first person, in 1977, to note that REM sleep shares many properties with a state of wakefulness and that this may tell us something about the nature of our dreams. More specifically, he noted that neural stimulation proceeding from the pons activates brain mechanisms that would normally be processing visual information if we were awake. Finding no 'parade' to interpret, the sleeping brain effectively summons up its own – using both memory and emotional faculties – and these are our dreams. Meanwhile, the slumbering state of the frontal lobes ensures that dreams are significantly different to what we experience with our eyes open – our dreamed parade might thus be full of all sorts of mythical beasts and bizarre events. In short, Hobson's theory suggests that dreams are just an entertaining by-product of brainstem activity.

Antti Revonsuo, of Turku University in Finland, suggests an evolutionary explanation for dreams.[22] He argues that dreams allow us to rehearse behaviour that we may rely upon in daily life – particularly in life-threatening situations. Certain features do support this theory –

notably the high level of activity in the amygdala during REM sleep. There is also the fact that fear dreams – in which we are falling, running, fighting, hiding, etc. – are the most commonly cited type of dream, and fear is central to the sensations experienced by SP sufferers (see above). On the other hand, these dreams may be rare, but more memorable. In fact, if we perform a repetitive activity for a long period of a day – typing, for instance, or operating a cash register – we will often report sensations of repeating this activity during non-REM sleep. This is backed up by studies of rats, whose brains show a certain pattern of neuronal firing after they have been trained to run through a maze. When they sleep, their brain activity repeats the original firing pattern. Neither can Dr Revonsuo's explanation, compelling though it is, account or provide any rationale for the simply bizarre content of our dreams. Perhaps, as Dr Hobson suggested, this is just a by-product of the fact that the rational part of our brains goes to sleep, even if other parts do not.

Another interesting theory is that dreaming helps the human memory to clear itself of what is not needed. The neurotransmitter acetylcholine is produced by the pons in increased quantity just prior to REM sleep. We know from James Bower's work that this neuro-transmitter may play a role in 'cleaning up' memories, preserving the relevant and junking the extraneous. And Allan Hobson has argued that the difference between waking and sleeping can be reduced to a difference in the balance of one type of neurotransmitter over another. In waking, the neural activity is dominated by amines, like serotonin and noradrenaline. In deep sleep, there is a balance between aminergic neurotransmitter systems and cholinergic systems, which would include acetylcholine. Hobson has suggested that a range of brain states are dependent upon the balance between these two systems in the brain – including meditation, day dreaming, REM dreaming, psychosis, depression and mania.

In 1983, Francis Crick, the Nobel Prize winner for his discovery of the structure of DNA, and Graeme Mitchison adopted the controversial stance of saying that 'we dream in order to forget'. Since the human brain takes in 100 billion billion bits of information in a lifetime, but can only store 100 thousand billion bits at any one time, there must be some procedure whereby memories are dumped. Their argument that dreams are the brain's waste-paper basket seems quite compelling. But this does not account for the fact that the act of dreaming itself produces memories, some of which stay with us.

When I was reading around the subject to write this book, I came across an account by Carl Jung of a dream experience he had in 1913: 'Before me was the entrance to a dark cave, before which stood a dwarf with leathery skin . . . I squeezed past him through the narrow entrance . . . waded knee deep through icy water to the other end of the cave, where I saw a glowing red crystal.'[23] More than one commentator has noted that Jung's dreams sound incredibly close to what we might see in fantasy novels by people such as Tolkien, and horror films like *Alien*. My first thought on reading such novels was that they didn't seem real. Which then made me think about that idea more closely. If they were dreams, then of course they wouldn't be real. What I actually meant was that Jung's dreams didn't sound like anything I'd ever dreamed, or would be ever likely to. And of course they wouldn't, because my brain isn't Jung's brain. Dreams are very personal and peculiar to the dreamer – which is one of the reasons why listening to someone else describing their dreams can be a boring experience. I cannot imagine what it must be like to be a psychoanalyst. And just as my brain creates my own very personal dreams, so it is responsible for many of the defining aspects of my identity. And I have heard of one architect who dreams of cement every night!

Where am I in all of this?

Think about yourself for a moment. What does that mean, exactly? Do you have a visual image of your body – now or in the past? Do you think in terms of situations you have been in, things you have done? Do you envisage some bodiless 'I AM' sitting just behind your eyeballs?

The sense of self is a crucial concept. If we define mind as the experience of having a human brain, then it follows that some conscious 'I' is experiencing that experience. So what are its attributes? Can we find it inside the brain? Can it be switched off?

Many tribal societies believe in not one, but a plurality of souls. And philosophers sometimes similarly distinguish between three levels of the self. There is the minimal self – which means the conscious experience of self in the here and now, reacting to but distinct from the immediate environment. Insects, birds, most animals can be said to have this form of self, and this alone. We have it too, but in addition to others. I am no touch typist so my minimal self at the moment is experiencing the touch of the plastic keys as my right forefinger prods away at my computer, the sensation of the chair under me and behind my back, the warmth of the room, the birdsong and the roar of traffic outside. I am also worrying about a crucial lecture I have to give tomorrow. And I am also experiencing levels of tiredness and hunger. Maybe it's time for a break . . .

We also have an objectified self. This is the capacity to serve as the object of one's own attention. For instance, I can ask myself – what sort of a mood am I in today? And further to that, we have a symbolic (sometimes called narrative) self. This is the capacity to form mental representations of ourselves through language. It is in evidence when I talk about how I have changed since I was twenty-one, for example. It provides an awareness of myself as existing over time.

It has been argued that this third level of the self is a relatively recent evolutionary development, an adaptation that helped us survive.[24] We are able to process personal and environmental information more successfully and this heightens our chance of continued existence and, hence, of reproducing our genes. Thus the symbolic self could include the capacity to reflect and to be introspective, to make long-term plans, to have a sense of one's own mortality, to develop moral arguments and even form governments.

The ability to think of yourself and your actions in the abstract enables reflection, and thus forward planning. For instance, if I were a wolf, and did not possess this level of self-awareness, I would hunt in a pack, the rules of which were determined by the relative killing and reproductive powers of its members. The strongest leads, has rights to the greatest share of the kill and reproduces the most.

But as a conscious human, it is possible there are more cooperative strategies that allow for a greater equality and greater specialization of roles, and hence a greater overall chance of survival. I can conclude, for instance, that I do not want to die and want to do as much as I can to avoid it. I can conclude that my neighbour, Ug, is a ruthless killer of bison but that I am better at seeing and tracking prey. I can suggest a transaction to him – I find the prey, he kills it and we share the spoils equally. Instead of only one of us surviving and passing on our genes, two can survive this way. And we can coopt others into our scheme, people who are nimble and agile tree-climbers and flint-knappers, ensuring a greater number of more efficient hunting expeditions, to the extent where we can even acquire a surplus of food that allows some people to stay at home getting better and better at making weapons or building shelters. It all begins from an awareness of myself.

But this still does not say precisely what the self is. The sociologist George Herbert distinguished between the self that knows (I) and the self that is known (me). The two have separate functions. The I-self does the thinking, the feeling, the acting. The me-self performs more objectified tasks – planning, choosing, exercising control. It's this second me-self that we mean when we talk about self-awareness.

Myself and others

Babies develop a me-self some time between eighteen months and two years of age. A classic experiment to prove this self-awareness involves painting a red dot on a baby's nose and then placing him or her in front of a mirror. Babies who are self-aware understand that the dot resides on their nose and will thus touch it with their finger. They may even cry because they think they have a blemish; I have seen some chuckle and keep rubbing it. But babies who have not yet developed this sense, or who have some impairment, will reach out and touch the mirror. Studies performed on chimpanzees in the 1960s seemed to show that they too may have this degree of me-self awareness, though, if they do – and it is some-what controversial – they develop it substantially later than human infants, at around six years of age. Other animals such as dogs, which are undoubtedly intelligent, do not understand what they are looking at. I did this experiment with a spaniel recently who grew very excited at seeing 'another dog', and who soon was looking behind the mirror to search for his new friend.

The next stage is to understand a self as existing over time. Beyond the age of two, most babies can identify themselves in photographs, even if not taken recently. By the age of three, a child understands that its shadow is created by its body. I think I can recall when this happened to me, when I would deliberately try to tread on my younger sister's shadow to assert my authority over her. But children also acquire notions such as shame, guilt and embarrassment. These ideas depend on a deepened awareness of self, and on self as something that others can evaluate. Accordingly, the children have to have developed an ability to understand others as separate and distinct from themselves.

At this stage, at about four years of age, something we call 'Theory of Mind' develops. A famous test of the Theory of Mind is the so-called Sally-Anne experiment. In this task the subject watches two girls play with a marble, a basket and a box. Sally puts the marble in the basket, and goes out of the room. While Sally is away, 'naughty' Anne puts the marble into the box. Sally returns to the room, and the watching child is asked where Sally will look for the marble. Children who have developed a Theory of Mind will correctly select the basket. They understand the state of knowledge that exists in another's mind, rather than relying solely on what they themselves know. Research has shown that children with autism show marked difficulty developing a Theory of Mind and will say that Sally will look in the box, where the marble actually is.[25]

A sense of self also allows us to misrepresent information, to deceive others for our own personal advantage, in a word, to lie. In one classic experiment, small children were given a set of stickers and asked to pick a favourite. They were then told that they would be visited by two puppets, who would ask them which sticker was their favourite. The first puppet, they were advised, would be a friendly sort who, after asking his question, would pick another sticker for himself. But the second puppet was a mean character who would almost certainly grab the child's favourite. Children up to the age of three were unable to deceive the mean puppet, thereby losing their favoured sticker. But from the age of four onwards, they were able to lie with ease, while still being truthful to the nice puppet.[26]

Some years ago, when making the film *The Human Body*, I watched four- and five-year-old children through a one-way mirror. They had been led singly into a room full of nice toys but with one irresistible toy – such as a wonderful painted engine. The adult bringing the child into the room told them they could play with whatever they wanted, only they must not even touch the really special toy. Once the adult left, for even just a few minutes, nearly all the children succumbed to temptation. And on the adult's return, when asked directly, nearly all the five-year-old children swore really touchingly that they didn't handle the forbidden toy, though one little boy paused and then burst into tears. Apparently, the more intelligent children tended to lie best in this experiment. And the ability to deceive represents an understanding of other people's state of mind – and how they can be manipulated. Four-year-olds are too young generally to lie like this.

As much as self-awareness may give us the capacity for deceit, it is also a vital prerequisite of self-policing. Only when we are aware of ourselves can we measure our own behaviour against that of everyone else. We have already looked at the role of the frontal lobes in regulating the impulses and drives of the lower brain. Unsurprisingly, the frontal lobes develop slowly in the human, only reaching maturity when we hit the age of twenty, or possibly (according to recent research) even later. A sense of self and the ability to self-regulate develops in tandem with the neuronal networks of the brain.

The role of the frontal lobes in self-awareness can be clearly seen in those who have suffered damage to that area. They often have difficulty, for instance, in remembering where they learnt specific information, even though they know the information perfectly. It is as if they are unable to conceive of themselves as an objectified self – someone who exists in time as well as space. Similarly, they tend to show a low degree of self-interest – they can identify disorders and abnormalities in other people, but not themselves. And quite often they do not show much interest in their appearance.

A classic example of the relationship between the frontal lobes and self-awareness comes from the neurologist Donald Stuss at the Rotman Research Institute in Toronto. He treated a highly intelligent man who had a tumour removed from his frontal lobes. The patient retained all his intelligence and knowledge afterwards, but was highly unproductive at work. Despite his poor performance, he was unable to recognize that he had a problem and was mystified by the attitudes of his boss and work colleagues. Stuss asked the patient to perform a role-play exercise, in which he, the patient, played the part of his boss. With this 'hat' on, the patient was able to swiftly and clearly identify the problems at work. But when asked to then evaluate himself from his own subjective viewpoint, the patient said he could not agree with any of the recommendations he had just made. He was unable to be objective in this way.[27]

Problems with understanding the self, and its relation to others, also seem to be related to the traffic between the two halves of the brain. David J. Turk, Michael Gazzaniga and others, of the Center for Cognitive Neuroscience at Dartmouth College, New Hampshire, showed photos of faces to J.W., a subject whose corpus callosum (see Chapter 2), the bridge joining the two halves of the brain, had been severed during surgery for epilepsy. Using computer-assisted manipulation techniques, the faces were morphed to form a sequence – one extreme of which was J.W.'s own face, and the other a picture of one of the researchers (in fact, it was the face of Michael Gazzaniga, whose face is well-known to J.W. and whose research features frequently in this book). The sequence was shown to J.W.'s left and right hemispheres, and he was asked to state the identity of the faces as they were shown to him.

J.W.'s right hemisphere was biased towards identifying faces it saw as 'others' and, conversely, his left hemisphere tended to identify faces as 'self'. The same thing happened when he looked at morphed sequences featuring his own face and that of two US presidents. These results suggest that two separate brain systems are used to formulate our understanding of ourselves and others. When there is a smooth flow of neuronal traffic between the left and right hemispheres, our brains can create a harmonious view of the world, as a person composed of a self among other people.

So the self appears to be created by an interplay between specific parts of the brain. However, our enquiry into this most philosophical element of the mind does not end here. This is because self-awareness is itself a product of an underlying state, that of consciousness.

Consciousness: a tool like any other

Consciousness has always proved extraordinarily difficult to define. One possible approach is to try to confine ourselves to the five observable properties of consciousness. Let us first examine the property of *sentience*. Sentience refers to the unique perspective that we each have about our own experiences. As we discussed earlier, there is no objective colour green, merely a route by which it is recognized in most people's brains. And differences in wiring create differences in perception – so I can never truly convey to you what the colour green means to me, nor can you to me.

Secondly, there is the property of *variable access*. When I am conscious of seeing the colour green, I am also relying upon mental processes of which I am unconscious. I see green, but I do not see or in any way sense the neuronal activity inside my brain. I can learn a list of the properties of serotonin and I can repeat it to you, but I cannot consciously sense the way my neurons and neurotransmitters store and retrieve this information. So we can talk about processes that are conscious and unconscious.

Next there is the property of *unified experience*. Consciousness, we can say, is like a movie, made up of the individual camera-shots that form our sensory and cognitive activity. It is the process that unifies our sensory and cognitive activities into a continuous experience.

Fourthly, there is *self-objectivity*. I am not just an experiencing I but an experienced Me, who can report on my mental experiences. As cognitive neuroscientist Steven Pinker puts it, 'I can not only feel pain and see red, but think to myself, "Hey, here I am, Steve Pinker, feeling pain and seeing red!"'[28]

Last, there is what we call *intentionality*. Intentionality implies a difference between what our sensory faculties tell us and what we do with the information, how we interpret it. For instance, I am always struck, when first arriving in a foreign country, by the unique smell. I do not necessarily know its components – it may be the tobaccos smoked, the food eaten, the trees that grow, the fuel used – but it always strikes me for the first few hours in any new country. Should I return to that country, I am still aware of the difference, but it is a *different difference* to the one I first perceived. And if I stay in the country for more than a few hours, my perception of the smell changes once again, to one that is barely observable.

Clearly nothing has happened to either the molecules creating the smell or to my sensory apparatus. The difference lies in the way I treat the incoming information.

So we have arrived at some useful definitions of consciousness. But can we, with the techniques we have used to discover the brain activity underlying sensory perception, or attention and self-willed actions, subject consciousness to a scientific study? Can we find the brain units involved in consciousness and see them in action? The answer is yes, but there are some surprises.

Half a second later

Think of a blue elephant. And now you've done that, here is a second exercise. Don't think about a green elephant.

How did the second exercise go? I assume it went less well, since in order to think about *not* thinking about a green elephant, you first had to think of one! But let me ask another question. How long did it take you to do each of these little exercises? A nanosecond? Half a minute? Or do you think it took no time at all? It just happened.

We talk, especially in an enquiry into consciousness, about the present, the 'here and now'. Furthermore, we tend to assume that what takes place in the brain is instantaneous; that there is no point asking how long it takes me to think of a chair or to react to a flashing light because it is instant, it just happens.

But your brain does indeed take time over all its processes. You are just not aware of it. Experiments conducted back in the 1840s by Wilhelm Wundt at the University of Heidelberg showed that when two stimuli – noises, for instance, or flashing lights – occur within a short time of each other, the human brain tends to fuse them into one. The same thing happens with touch – if your arm is touched at four points in rapid succession (and I am talking of a maximum time lag of one-tenth of a second, so it's not one you can try out at home), you will perceive this as a single object trailing down your arm. This is, of course, the feature of the brain that makes movies and TV possible – we don't see twenty-four frames a second flashing across the screen, but experience them as a harmonious whole. Consciousness is not an instantaneous experience – it takes time to build itself, and the brain covers up for its own time lag.

Some 120 years later, Benjamin Libet at the University of Southern California became interested in this time lag.[29] He experimented on people undergoing stereotactic neurosurgery for intractable pain. Having first given a local anaesthetic to the scalp, he sent a pulse of electrical current into various parts of their body, such as the skin or the cortex, asking them to report on their sensations. Using EEG to measure the responses, he noticed that it took a surprisingly long time of pulsing before people reported that they felt anything. His conclusion was that it took around 500 milliseconds before consciousness caught up with the sensory reaction to the electrical current and manufactured it into a reportable experience.

Meanwhile, a German scientist called Hans Kornhuber was doing his own work on the consciousness of self-willed activity. He, like Libet, noticed that there was a time lag – a period of mounting neurological action inside the motor cortex – before people make a voluntary action. Libet took this research a stage further in an attempt to isolate the moment at which subjects first decided to make a movement. The volunteers reported that they first became aware of an intention to act about 200 milliseconds before they then produced the act – in this case, lifting a finger when they chose to do so, or in response to a buzzer. But the EEG data showed that, although they reported their first awareness of intention at 200 milliseconds, there was a stirring of motor activity even earlier

– at 500 milliseconds. In a nutshell – the decision to take action occurs first subconsciously and takes around 300 milliseconds to transfer into consciousness.

Although the raw material of these experiments – buzzers and lifting fingers – might not be quite so appealing to readers as, say, human sexual behaviour, or the brain processes involved in dreaming, Libet's findings are quite shocking in their way. For they imply that we are not entirely conscious of the workings of our own brain. By the time we know that we want to do something, another part of ourselves has known it for roughly a third of a second. The purpose of this seems to be economy – by having two levels of processing, one conscious, one unconscious, we keep our higher brains free for the tasks that are the most demanding or unique.

Some years ago, a while after he had retired from first-class cricket, I happened to meet Tony Greig, the England ex-captain and all-rounder, in the green room of a television studio. How was it, I asked, that any batsman could face a fast bowler and repeatedly score runs when he might be consistently bowling at around 90 miles per hour? How could he face the four great West Indian bowlers – Holding, Roberts, Holder and Daniel – in the 1976 series, hour after hour, without being decapitated? Given that the distance between the stumps is only 22 yards, the ball leaves the bowler's hand and arrives at the bat in well under half a second. Allowing for the body's natural reaction time it seemed that it ought to be highly unlikely that any batsman, no matter how good, could consistently make high scores, particularly if the ball was moving erratically off the ground or swinging through the air. Greig thought for a brief moment and then told me that he didn't feel he could explain how he had personally done it but he thought that all good batsmen must anticipate what the bowler was about to attempt.

It seems that research done with sportsmen supports

Libet's work. In the 1980s, Peter MacLeod at the Applied Psychology Unit in Cambridge studied world-class batsmen to see how they prepared themselves for balls that only take 440 milliseconds to leave the bowler's hand and arrive at the bat. MacLeod's results showed that the batsmen did not react immediately to variations in the speed and trajectory of an incoming ball. They modified their swing at a point 200 milliseconds before it hit. This is almost certainly faster than you or I could do it in a pub with one of those reaction-time machines, pressing the red button – even before the first drink.

More research with both cricketers and tennis players indicates that top-class performers are particularly skilled at anticipation. The best players make the earliest and most accurate predictions about what is about to happen. Tests were performed, for instance, on professional and novice tennis players, involving film footage of a person serving the ball. A hundred and twenty milliseconds into the flight of the ball, the film was stopped, and novices and professional players could predict accurately at this stage whether the ball was going to arrive on the forehand, backhand or right in the middle. But the professional players could do the same some 40 milliseconds *before* the ball had left the server's hand. Similar studies of batsmen have shown that they are making appropriate preparatory movements 100 milliseconds before the ball leaves the bowler's hand. Significantly, few of the subjects when questioned could accurately self-report on what they were doing. They were taking in information, forming plans of action based upon it and acting on their plans up to one-tenth of a second before something happened, but they were not aware of it. This kind of analysis is now being taken up at Sheffield Hallam University, where I am currently university Chancellor. There they have a particularly well-equipped sports science centre with sophisticated motion-analysis equipment which helps break down exactly what happens and how sportsmen react. They hope that this approach may also lead to improved performance in people who are training to play sport at the highest level.

So the research suggests that, either way, our consciousness is never exactly 'in the moment' but always occurring a significant degree of time before or after it. But what areas of the brain are involved in creating this extraordinary pre-awareness? Michael Posner at the University of Oregon investigated this by asking volunteers to look at a screen upon which were flashed lists of words printed in different colours. Their task was to report on the colour of the word, not the word. A cinch when the word is, say, 'pill', in yellow type. But when the word itself is 'yellow' and the colour blue, the task becomes harder and requires more conscious effort. In such instances, the anterior cingulate was highly active. Whenever the brain is being mindful of what we are doing or experiencing, the anterior cingulate plays a part. As well as fulfilling other functions, we might call it the brain's concentration centre.

But rather than assume that this structure is the site of consciousness, we would do better to look at the other areas with which it works. Jeffrey Gray at London's Institute of Psychiatry did just that. He noted that the hippocampus, site of our long-term memory, has a particularly strong connection with the anterior cingulate, through a thick knot of fibres called the cingulum. Gray suggested a sequence from this physical connection: the cingulate would ask the hippocampus to expect a certain outcome, and the hippocampus would then report back on what had just happened and what was different or new about it by comparing it to its amassed memories of everything that had ever happened. The interaction between the two would create a sort of arc of consciousness – concentration combined with significance. A flowchart of further activity can then proceed from this point: how do we respond – with action or inaction? If with action, then what action?

In that sense, we can see consciousness as simply the high point of the brain's evolution. At some distant time in our pasts, we were only able to adapt to our environment over progressive generations. Then we became able to learn certain reflexes over a lifetime. Later on, like chimps, we could respond by trial and error to novel events in our environment. The simple reflexes grew fatter, heavier, more enmeshed, because, as we know, the more a pattern of firing occurs within the brain, the more deeply established it becomes. To be the best at the business of survival, we had to perfect the ability to respond instantly (or within a few hundred milliseconds at least) to changes in our environment. Was consciousness what we developed to do this, a tool that allowed for an immediate adaptation of our plastic brains to the needs of each moment?

The term homeostasis is widely used in medicine. Homeostasis is how different parts of the body achieve a steady state, despite changes around the body. There are many examples of homeostatic systems and the brain is one such organ which could be said to operate homeostatically. It 'seeks' to maintain a state of constancy in spite of the ever-changing landscape outside to which it responds. We can compare it simplistically to the heating and air-conditioning system of a large building – when the air temperature outside is cold, the system detects the low temperature, the thermostat kicks in, and the furnace responds by giving out heat. When the weather is hot, the system gives us cold air until a comfortable level is reached. To perform the same self-regulating function, the brain makes ceaseless adaptations within, changing levels of neurotransmitters and hormones, for example, or growing and pruning back synaptic connections. Consciousness is thus an ultimate refinement of a homeostatic system, enabling adaptations, tiny and massive, that are geared to each and every moment.

I am always aware, when venturing into such territory, that it can seem far removed from the day-to-day experience of having a mind. That tends to be the result of analysing something so much a feature of our existence that we rarely consider it. However, it is also clear that there is a lot more to us humans than simply being aware and paying attention. To be human is to be able to escape from the moment, as much as respond to it – to dream, to create, to remember, to feel. In the following chapters, therefore, we shall be looking at some of the more remarkable aspects of the human mind – beginning with emotions.

chapter**five**

The emotional mind

A famous accident over 150 years ago in a minor railway cutting near a tiny village in New England, USA, was eventually to have a huge influence on our knowledge and understanding of the working of the brain. Some might observe that had it had just a bit more influence, at least one Nobel Prize would not have been won for the wrong reason just over one hundred years later, and many thousands of lives may possibly have not been irretrievably ruined as a result of one of the biggest scandals in the history of medicine.

On a sunny afternoon on 13 September 1848, a gang of men were preparing the ground before laying railway track in a wooded cutting about three-quarters of a mile from Cavendish (population 1,300), Vermont. Their foreman, Phineas P. Gage, an able, hard-working and universally popular man, was in charge and they were making excellent progress. The gang was using explosives to level the shale rock on the floor of the forest. There are no totally reliable eyewitness accounts of what happened, but it seems that their meticulous foreman was uncharacteristically careless. All day they had been drilling holes in the rock, partially filling them with dynamite, inserting a fuse and then pouring a large amount of sand into the hole over the top. A tamping iron, 3 feet 7 inches long, which Gage had designed, was used to compact the mixture before triggering the fuse. But on one occasion that afternoon, Phineas Gage rammed the tamping iron into a hole without checking that his assistant had first poured in the protective sand. The resulting explosion drove the tamping iron out of the hole with huge force and it pierced Phineas Gage's left cheek. Although it weighed nearly seven kilos, it went straight through his face, removed his left eye from behind by piercing the base of the eye socket, and shot out through the top of his skull, leaving a massive hole extending back to almost the middle of his head. The bone loss in the cranium and the broken bone around the hole left a gap big enough for a man to insert a fist.

The force of this solid iron thunderbolt threw Gage several yards distant onto his back. The tamping iron lodged in the ground some sixty feet behind him. It isn't clear whether Gage lost consciousness, but he lay there twitching, apparently mortally wounded, until Christopher Goodrich came from the township with his oxcart and deposited him at Joseph Adam's inn in the centre of town. Clearly he was expected to die because Thomas Winslow, the neighbourhood cabinetmaker who lived just across the street (and who doubled as an undertaker in his spare time), came over to measure him for his coffin.

The story becomes a little confused about who first treated Gage for his terrible injuries. It appears that a Dr Edward Williams from Proctorsville, Vermont, was the first physician to examine him; by this time Gage was certainly fully conscious and able to speak because he is reported as saying: 'Doctor, here is business enough for you.' But the local general practitioner, Dr John Harlow, is given the credit for nursing him back to health, suturing the wounds and using clean dressings and a liberal supply of antiseptics. Remarkably, there seems to have been no infection and some weeks later his patient was well enough to return to his family in Lebanon, New Hampshire, to rehabilitate. Dr Harlow gets credit for the quality of his care but also for, importantly as it turned out, publishing the case later. He wrote it up for the *Boston Medical and Surgical Journal*: 'Passage of an iron rod through the head', where it appeared at the end of that year. What was remarkable to Harlow was that, at first, Gage appeared to recover unharmed and intact. He walked and talked normally, had normal sensation, and moved all his limbs without restraint.

It has been said that the story of Phineas Gage, and the insight that his injuries give into the workings of the brain, has resulted in his being mentioned or written

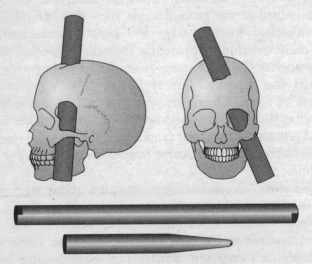

Phineas Gage and the famous tamping iron

about in no fewer than 60 per cent of all books on neuro-science. Because after this injury, Phineas was, in the words of one of his colleagues, 'no longer Gage'. He now had a totally different temperament and character. Whereas before he was patient and amenable, now he was irascible and fitful. Before the accident he was well organized and sociable, getting on well with everybody with whom he came into contact; now he was aggressive, demanding and irreverent. He showed no deference to anybody, made shocking sexual suggestions, and was obstinate, erratic, impetuous and unreliable. Indeed, though he wanted his job back as a foreman, he was undoubtedly unemployable in such a responsible capacity, particularly as it involved work with explosives. He became a bit of a drifter and took a variety of jobs as a coachman, ostler and cleaner. There being less and less chance of employment locally, it seems he spent several years in Chile caring for horses. At some point he ended up exhibiting himself at Barnum's Circus in New York. Over the next ten years, his health deteriorated, and he died, a chronic epileptic, in 1860. In 1867, his body was exhumed, some kind of post-mortem done and Dr Harlow, who had continued his keen interest in Phineas Gage's plight, gave his skull – together with the famous tamping iron – to Harvard's Medical School as an exhibit. They can now be seen in the Countway Library of Medicine, Harvard.

Professor Moniz and the 'white cut'

Almost a hundred years after Gage had his world-famous accident, a doctor in Portugal was coolly and deliberately trying to reproduce a very similar injury in severely disturbed, mentally ill patients. Admittedly, he was attempting this in a much more controlled fashion. Professor Egas Moniz was then at the University of Lisbon, working as Professor of Neurology. Moniz's proper name was Antonio Caetano de Abreu Freire, but his godfather named him after the hero of the Portuguese resistance who fought against the Moors. His christening set the tone of what might be thought a rather macho existence. He became a Portuguese member of parliament at the age of twenty-nine and Ambassador to Spain during the latter part of the first world war before becoming Minister for Foreign Affairs. In consequence, he became one of the signatories to the Treaty of Versailles at the end of the war, but had to resign from politics in 1919 after he was challenged to a duel as a result of a political disagreement.

By the time Moniz was ready to return to full-time medicine in Lisbon, he had already written a book on sexual function and the pathological aspects of sex life, two on hypnotism and its history, and, in 1917, a book on neurological injuries in war. The next few years in Lisbon were a stunning success personally and he gained a large international reputation as a neurologist as he developed new methods for looking at the human brain, while also continuing experimental work on animals. Until Moniz's day, the conventional way of examining the structure of the brain in an intact patient was using X-rays, with the risks that repeated exposure might bring the patient – although, of course, it was only much later that people began to be much more concerned at the risks of diagnostic radiation. That is, incidentally, a relatively recent concern. When I was a child of ten, it was normal practice in the more fashionable shoe shops to have your new shoes fitted under an X-ray machine. It seems incredible that the first X-ray I saw was my own foot in my local shoe shop, with the bones of my toes wiggling in the eerie greenish light of what was undoubtedly a powerful source of highly risky gamma radiation.

But the problem in Moniz's day was not a perceived risk of radiation; it was the fact that doctors could only obtain a very poor-quality image from X-rays of the nervous system. While my toes and other bones stick out well and cast a strong X-ray shadow, soft tissues such as brain, for example, hardly leave any image at all. A major problem was to find better ways of diagnosing such diseases as tumours inside the brain. One method which somewhat improved the outline of the brain was to increase the contrast of the picture by injecting air into the head. But this technique, so-called air encephalography, still gave a blurry, unspecific image, was not without hazard and often left patients with crippling headaches for some time afterwards. Moniz began injecting human corpses. He used, among other compounds, sodium iodide in solution, and injected this into the cerebral arteries. Iodine salts are relatively opaque to X-rays, and so this way he was able to map the position of the arteries inside the brain. And when the location of the various arteries was abnormal this suggested displacement from their normal position by a tumour, for example.

One cannot help but suspect that the international acclaim and adulation that Moniz was accorded after his important studies on cerebral angiography were published led him a step too far. He started to experiment with psychosurgery. Moniz set out to try to treat what then (and even now) were very difficult and distressing conditions, such as severe paranoid states, extreme depression and schizophrenia. By the 1930s he had come across earlier research performed on the brains of monkeys and chimps. These particular animals, hitherto aggressive and anxious, became friendly and passive after certain areas at the front of their brains were cut away. Moniz, working in his mental hospital in Lisbon, started to think about trying this out on humans who showed somewhat similar symptoms. He was pleased with the results.

Prefrontal leucotomy, or lobotomy, was first performed on a human patient on 12 November 1935. Together with his assistant, Dr Lima, Moniz drilled a number of holes in a paranoid female patient's skull under local anaesthetic. He then injected pure alcohol into the area of the brain connecting the frontal lobes to the rest of the central nervous system in the expectation of pickling the nerve fibres. Immediately after the operation, the woman was less agitated and much calmer. As it turned out, other patients in this asylum were rather more fortunate because the Medical Director – either out of professional jealousy or because of genuine concerns about the ethics of Moniz's approach – refused to allow any more of the patients for whom he was ultimately responsible to undergo this surgery.

Moniz cast his net further afield, and there is a strong suspicion that he reported his clinical results with some considerable spin. He, together with Lima, now conducted twenty operations on patients from various places. Some suffered from schizophrenia, others from psychotic depression. Moniz admitted that many of the patients were somewhat sluggish after the surgery, but understated the extent of the negative effects of his work. Apart from being apathetic, they suffered incontinence, nausea and disorientation. Moniz appears to have committed the cardinal sin of not evaluating the state of his patients at long-term follow-up, and also disregarding the very negative symptoms that most of them would have reported – had they been able. But Moniz was now borne along by enthusiasm. Because Moniz felt that alcohol did not have a strong enough effect, he replaced this approach with a special knife, called a leucotome. A few blind strokes with this knife quickly inserted into the brain, and the famous 'white cut' was developed. Psychosis, particularly when patients were uncontrollable or dangerous, was a common and untreatable disease and so there was extraordinary interest in any therapy that might make these very ill patients more tractable.

Dr Freeman's 'improvements'

One neurologist who developed this interest was Walter Freeman from George Washington University. He saw an opportunity in the 'white cut' revolution and was soon prepared to try it for himself on human subjects. His first patient, Mrs Hammatt, was a highly agitated and depressed woman from Kansas and she and her family were faced with her internment in a mental institution, or the knife. In those days internment in a lunatic asylum was virtual death; they chose the knife.

But just before being laid on the operating table, Mrs Hammatt became very agitated. She realized that in order to have the surgery, the doctors needed to have her head shaved. To calm her down, Freeman promised her this would not be done and that the curls of which she was so proud would not be harmed. Six cuts in the skull and several blunt cuts in the brain later, Freeman reported 'she no longer cared'.

A week after surgery, Mrs Hammatt became unable to speak properly. She constantly mumbled repeated incoherent syllables and could not recall simple facts like the days of the week, or the current date. Writing was completely impossible for her. But gradually she seemed to get better, her speech returned and she was eventually discharged home, apparently free of the symptoms of extreme anxiety that had previously plagued her. In his autobiography, Freeman wrote of his first lobotomized patient that, after the operation, 'She survived five years, according to Mr Hammatt the happiest days of her life . . . She could go to the theatre and really enjoy the play without thinking what her back hair looked like or whether her shoes pinched.'

Dr Freeman, together with his assistant from George Washington University, Dr Watts, now recklessly embarked on an indiscriminate career in psychosurgery. Within a relatively short time, he was able to describe a series of 623 surgical operations, of which he reported 52 per cent of the results as 'good', 32 per cent as 'fair' and 13 per cent as 'poor'. Eighteen of the patients died. The documentation of what constituted a 'good' case was dubious and inadequate, and many of the nurses who cared for these patients afterwards recognized that they were frequently unable to use the toilet, often made sexual advances to nurses, sat unmoving for hours and showed apathy, dullness, loss of memory and delusions.

By now, Freeman and Watts were operating on a variety of disorders diagnosed as psychiatric, including drunks found in downtown Washington. Once they had established their lucrative private practices in offices there, Freeman started to experiment with a variety of 'improvements', the precise details of which he even kept from his closest colleague and accomplice Dr Watts. He also commenced a campaign of pure self-advertisement. He set up stalls at medical conventions and conferences, and used the press to make claims about his ability to treat what had previously been thought beyond treatment. Over subsequent years he set out to improve his surgical approach, to make it swifter and simpler. He travelled the USA, operating on a variety of often poorly diagnosed patients in some twenty-three states.

Among his improvements was the so-called 'transorbital technique', often done as an 'office procedure' with the patient being sent home in a taxi afterwards. His son, who sometimes used to assist him, described his method of anaesthesia, which was to stun the patient with a large electric shock. An assistant held the patient down on the operating table and, once the convulsions had subsided, Walter Freeman lifted an eyelid of the patient –

the transorbital approach is graphically described by a bystander: '... lifting the eyelid, he inserted an ice-pick-like instrument through a tear duct. A few taps with a surgical hammer [sometimes Freeman is reported as flamboyantly using a carpenter's mallet when he was showing off his prowess] breached the bone. Freeman took position behind the patient's head, pushed the leucotome about an inch and a half in the frontal lobe of the patient's brain, and moved the tip back and forth. Then he repeated the process with the other eye socket.' One professor of neurology is recorded as fainting when he watched Freeman demonstrating his work.

There is a chilling, slightly blurry photograph of Freeman taken in about 1945. An apparently unconscious woman wearing striped pyjamas is lying on a metal operating table. Two nurses are holding her arms; another appears to be leaning on her legs. Freeman is facing the camera, demonstrating his surgical skills. Wearing a white, sleeveless smock, Freeman with bare, gloveless hands (without surgical gown, hat or mask) is tapping home his knife into the patient's brain through an eye socket. Closely surrounding this operating table are eight other men, mostly wearing respectable outdoor clothes, business suits and ties. Nobody is wearing anything vaguely resembling surgical dress, a mask or even a theatre gown, despite the fact that infection in the brain is one of the most deadly.

The wrong Nobel Prize

Once Egas Moniz, and then Freeman, had published their results, there was widespread international enthusiasm. Moniz's work was initially applauded everywhere; he became a commander of the Légion d'Honneur, gaining a number of honorary doctorates and being appointed as an honorary member of many national medical and scientific academies. The Nobel Prize for Physiology and Medicine was awarded in 1949 because, in the words of the Nobel citation, 'It occurred to Moniz that psychic morbid states accompanied by affective tension might be relieved by destroying the frontal lobes or their connections to other parts of the brain . . .' I wrote at the beginning of this chapter that a Nobel Prize was won for the wrong reason – because while there was little justification for the acclaim given to psychosurgery, the earlier work that Moniz had done on cerebral radiology was truly ground-breaking and influences many of the techniques still in wide use today, over seventy-five years later.

I think it very likely that Freeman was consumed by thoughts of the fame increasingly heaped on Moniz. He persisted with a largely unaudited series of treatments and in doing so fell foul of many from the medical establishment. But the criticism was mostly due to other considerations – it was partly because of his advertising, partly because he was not a qualified surgeon but doing operations, and perhaps because a number of his patients died suddenly, sometimes of brain haemorrhage. Freeman may have been regarded with notoriety by the more serious members of the medical profession, but the plaudits from among the less discriminating and informed members of the profession, the press and the public were considerable. It brought him many prominent patients from the upper rungs of society, among them the Hollywood actress Frances Farmer, and John F. Kennedy's sister Rosemary Kennedy, who was crippled by the procedure, requiring constant care for the next sixty years.

But severe psychiatric disorders were, and still are, extremely common, and drug treatments for violent, disturbed patients were then very limited. Largactyl, the famous tranquillizer, was yet to be invented and barbiturates had only a very crude effect. The fate of the troublesome Randall P. McMurphy, hero of Ken Kesey's *One Flew Over the Cuckoo's Nest*, was a fate handed out to thousands of people throughout Europe and the USA as enthusiasm for the full-frontal lobotomy spread through the medical community. The problem was that while people lost their aggressive and destructive tendencies after the operation, so did they – like poor Phineas Gage – also lose vital elements of their personalities. They might have lost their capacity to become angry, for instance, but so had they lost their capacity to feel joy, or sympathy, or enthusiasm. They might have ceased to present a problem to those around them, but they had also ceased, in some cases, to be fully functioning humans. Moniz, and those others following him, had removed their soul. It was such drastic measures as the lobotomy that led the radical 1960s psychiatrist R.D. Laing to conclude that society merely condemned as mentally ill those people it wished to contain and silence.

Understanding other people

There is a rare disorder of the brain, Urbach-Wiethe disease, which can result in calcification in the amygdala. Ralph Adolphs described a patient who had impaired perception of fear after such bilateral amygdala damage in 1994.[30] The patient suffered no impairment of intelligence or motor function – but she found it difficult to recognize emotions in people's faces and voices. Fear and anger were impossible for her to detect. Fear and anger may be understood as concepts, but patients with amygdala damage may not spot them or react to them as they appear in their daily context.

This may appear to be merely interesting for the observer and frustrating for the sufferer. But the role of the amygdala is crucial for survival. It's one thing not to be able to read people's faces in a modern context. But it is quite another not to feel fear as a massive truck is bearing down upon us. During the filming of *The Human Mind* I stood wearing breathing apparatus in a flimsy iron cage in the sea off Cape Town while a two-ton, 18-foot-long great white shark bore down on me, rattling my cage and showing its rows of four-inch teeth. No matter how much my cerebral cortex was telling me I could keep calm and that I was not going to get hurt, the alarm bells in my amygdala were ringing unbearably.

In an age when survival was more precarious and humans bonded together in groups in order to ensure the best chances, group communication was vital. Among bushmen and other hunter-gatherers, the most severe sanction for breaking the rules of the group is to be excluded from the group – punishment that meant death in harsh environments. We can see the traces of this in the custom of 'shunning' – a terrible curse for the ancient Israelites, prevalent in the Welsh valleys until the early decades of the last century and still practised among Romany traveller groups. Perhaps this is why they still use the black ball at the Garrick Club in London. Group communication, at its most basic, entails an ability to read and express emotion. If another member of the group is angry, for instance, or fearful, then it pays to be able to interpret that and modify our behaviour accordingly. Monkeys released into the wild after having their amygdalae removed do not survive very long. They neglect their young, and prove unable to pick up signals of approaching danger from other members of the group.

If asked to describe 'emotion' I might be tempted to see it as a set of feelings – fear, anger, sadness, joy, etc. But looking at the above evidence, it seems less about feelings than a set of survival mechanisms that helps us to avoid danger and direct ourselves towards whatever is most beneficial. According to Joseph LeDoux, a prominent researcher into human emotions – and author of *The Emotional Brain* and *Synaptic Self* [31,32] – the emotional process is a body–brain interaction, the bulk of which remains buried below conscious view. The conscious 'felt' element is merely the tip of the iceberg, or as LeDoux puts it, 'the icing on the cake'.

There is a sad instance mentioned by Antonio Damasio.[33] A patient of his, after an operation to remove a brain tumour, became indecisive and irresponsible. Unable to make correct choices, he threw himself into one doomed venture after another. Dr Damasio's tests suggest that the patient was unable to 'feel'. Shown pictures of horrific scenes, he said he recognized that they were horrible, but he didn't feel it. In his daily life, he was unable to pick up the cues from others that help normal people to make choices. In the case of this particular man, the problem seemed to be caused by damage to the connections between his frontal 'conscious' cortex and his limbic system.

In the opinion of many scientists, emotions would seem to be rather like colours – more complex ones are made up of a combination of a limited set of more basic primaries. The primary emotions, most brain scientists agree, are fear, anger, sadness and joy. Others add disgust and surprise, and some say contempt as well. So the more complex emotion of guilt might be said to be composed of fear and sadness. Other scientists argue that emotions like surprise, disgust and guilt exist alone. And some interesting results from brain surgery patients would tend to back up this more complex, categorized view of emotions. When certain parts of their brain have been stimulated, they have been able to report some very specific and subtle emotional experiences, such as the shame one feels on making a social faux pas.

How does it work?

Information from the body itself plays some part in feeling emotions. There is documented evidence of spinal injury patients, paralysed from the neck down, who report some emotional blunting with a detachment similar to Damasio's patient. Damasio argues that one reason for their limited and incomplete detachment is that they still get direct brain stimuli from the cranial nerves – which directly connect the face, for example, with the brain. Perhaps a 'gut reaction' is not such a meaningless cliché after all.

Normally what happens is that an emotional stimulus – say, the sight of an angry face – passes from our visual cortex to the limbic system. At this stage, the information on that electrochemical email might read something like: 'Human face seen, now, at five feet – eyes flashing and teeth bared.' As part of a loop between the amygdala and the hypothalamus, signals are sent on to the body. The 'email' from the hypothalamus might say: 'Close threat, how are you going to deal with it?' But the hypothalamus also sends signals to the pituitary, the master gland of the whole body. Secretions from there, informed by the hypothalamus, rapidly direct heart rate, blood pressure, levels of tension in the muscles, alertness. This level of alertness is fed back into the hypothalamus – an email that might say, 'We are on alert, ready to fight, check with HQ' – and thence onwards to the higher cortex. The higher cortex takes the bodily information received and concludes, 'This means anger. So let's start responding.' We would then have what we consciously call an emotional feeling.

But without the body acting as a vital part of this loop of communication, the information we receive in our higher cortex is patchy. Think of a teletext page when there is poor TV reception – letters and numbers mixed up together, whole areas blank. If I was suffering from a spinal injury, my hypothalamus – not itself 'knowing' about the spinal injury – would send its information down to the body nonetheless. It would then 'check' to see what was happening, and receive the information that nothing was happening down there – constant heart rate, no muscular movement, etc. It would 'conclude' that, after all, things were in a state of calm. The higher cortex, processing this information, would hence register less.

We started out by highlighting the importance of the amygdala in all of this. So what is its precise role? It seems that it acts as a 'tagging' system for information coming in directly through the senses, a process that also involves consultation with the memory banks of the hippocampus. It can operate entirely outside of consciousness – as we might see if we heard a loud bang in the night and immediately hid under the sheets, even before we had assessed the reasons for the noise.

Essentially, the amygdala may assign an emotional value to information – i.e., this is good, this is bad, this is disgusting, this is frightening – which allows us to then assess the world and react to it accordingly. Dopamine plays an important role in its functioning. The effects of this neurotransmitter are relayed to the amygdala, and there are changes in its supply in response to pleasurable or painful stimuli. Disorders in the way the amygdala 'tags' information could be responsible for many mental conditions. The amygdala of phobic people, for instance, may tag more information as 'scary' or 'disgusting'. A happy or just normally functioning person might look at a tree in winter and see it as a beautiful structure, or not be moved by it at all. A depressed person might look at the same tree and see only its dead, lifeless condition. Their amygdala is tagging all incoming information in the same way – as sad or horrible.

Some people are able to feel emotion, but are unable to express it. This condition is known as alexithymia. In their case, the disruption lies not between body and brain, but between the higher cortex where emotion is consciously processed and the regions of the brain that control facial expression and speech. Imagine trying to convince your lover to marry you using a flat voice without any guise of sincerity in your eyes. Imagine trying to lead your troops into battle. Clearly, we use our emotional faculties not just to read other people, but to influence them.

In expressing our own internal emotional states to others, and in reading their internal states, body language is very important. As a medical student in the 1960s I remember being taught by one of Britain's really great clinicians, Dr Bomford, the importance of good emotional contact with a patient. Shake hands at the start of a consultation, make eye contact, lean forward towards the patient, do not cross your arms or fiddle with your pen, do not answer the telephone or turn to engage another person in conversation. Sit close to them if they are lying or sitting, and try to keep your head at their level. Show interest in what they are saying. Surprisingly, this teaching was much before its time. A fellow consultant in the same hospital, an excellent chest physician technically, was much more typical of doctors after the war. His ward rounds were like a military inspection. Standing erect at the foot of the bed – never moving to get close to the supine patient – he would address each male patient by his surname. 'Ah, Watson, how are we this week?' Without really hearing the answer, there would be a brief, muttered conversation with the registrar and the ward sister, a long glance at the record and temperature charts, and terse recommendations to the staff for investigations. Meantime, the patient, particularly if male, would be lying rigidly to attention in bed. 'Jolly good, Watson, old fellow, splendid progress' – and on to the next patient. And this consultant, whom I cannot name, was absolute kindness, but never understood the devastating emotional effect he had during his weekly rounds. We students picked up the pieces by a second round after he had left.

Gestures, how we stand or sit, the use of our hands are all very important in conveying deference, indifference, trustworthiness. Tony Blair, our Prime Minister, for example, like many good politicians has a particular habit of opening out his hands as he speaks. It tells his audience that he is trying to be open and trying to gain trust. In his case I have no doubt that it is genuine, but many politicians use these gestures in a false way – and quite often, our emotional intelligence is capable of discerning this. This, perhaps, is the success of Rory Bremner, who is brilliant at imitating Tony Blair. He makes the gestures just slightly differently and they take on a different meaning to the beholder. But more importantly, Rory Bremner uses his face. When we communicate emotion, facial expression is by far the most important mechanism.

'Smile and the whole world smiles with you'

We humans have some 7,000 different facial expressions in our repertoire. And they tend to mean the same things wherever you are in the world. (This is not the case for gestures performed with the body – the Greeks, for instance, shake their heads to mean 'yes', and older people might take considerable offence at a thumbs-up sign.) There are experts who make a living out of facial analysis; organizations like the FBI take their work seriously. But for now a simple example may suffice. We use two muscles (the left and right zygomaticus) to produce a polite smile – the sort I might keep on my face at a cocktail party, for example. A genuine smile, on the other hand, such as I might produce when I see my wife after having finally freed myself of the cocktail party, uses four muscles – two extra around the eyes, the left and right orbicularis occuli. When this smile fades, it fades in a more even pattern than my false smile and more slowly. I do not consciously produce this. Facial expressions are a vital part of our emotional vocabulary, and while it is possible to become conscious of them, they largely take place in the region of unthought, habitual action.

Paul Ekman of the University of California at San Francisco pointed out that certain facial expressions are universal in humans, irrespective of their cultural background. These facial expressions do not substantially vary and are recognizable by other members of our species whether on the north-east coast of the USA, in South America, Japan, China or New Guinea. Ekman defined six such basic facial expressions: anger, disgust, sadness, fear, surprise and happiness. He described how we have automatic reactions that unfold within microseconds when we see these facial signals in others.

When we see disgust, for instance, a particular part of the amygdala works to identify it in the face of the person at whom we are looking. Another part of the brain will be working to pick up cues in their voice. It seems that the left side may read voice information, while the right responds to facial information, consistent with the fact that in 95 per cent of people, the language centre of the brain is in the left hemisphere. One of the problems, though, in studying these and other human facial states is that there are wide ranges of intensity in these emotions, such as happiness. I regret to emphasize that I support a North London football club, but the season is much on my mind as I write this chapter. I might feel happy if Arsenal won a match early in the championship season, but I would be ecstatic if it was at the end of the season and Manchester United had just lost. Picking up a pound coin lost in the football ground would give me a very different intensity of feeling from being informed that I had just won the million-pound Premium Bond draw. Cognitive neuroscientists sometimes try to get around this problem by measuring the degree of our arousal in response to a particular emotion. Another strategy is to attempt to measure the depth of a given emotion by assessing the actions it induces. Thus happiness may induce in us a tendency to engage with a situation; fear or disgust, conversely, depending on the depth of feeling, may lead us to turn away, shun contact or withdraw.

We humans are deeply social creatures and when we see an emotion in another person we tend to experience it ourselves. For instance, as I showed earlier, it makes evolutionary sense for us, if we see someone eating a piece of food and registering disgust, not to go and repeat their mistake. As far as we know, the amygdala responds to four basic emotions: disgust, sadness, happiness and fear. It is uncertain whether a specific part of the amygdala is activated with each emotion – probably not. It may just tell the cortical brain, 'Hey, watch out – there's an emotional response needed.' So 'smile and the whole world smiles with you' doesn't always work. As an ad hoc experiment, I asked one of the producers of *The Human Mind* programme to try to smile at everyone when going out to fetch some milk. A few people did indeed smile back but one person asked him, quite aggressively: 'Who are you smiling at? Shall I wipe it off your face?' But even that response is quite telling. The individual in question did just what we are supposed to do when registering emotions coming off another person – he looked to see what the other's face was communicating about the environment. Finding nothing in the congested streets of White City that could possibly produce a smile, he concluded that the information was faulty and therefore coming from a faulty source, hence his threat. Admittedly, this response could have been due to the fact that Queens Park Rangers, the football club down the road, had just lost to Vauxhall Motors after a penalty shootout in a replay at home in the FA Cup. Expressions might be common to everyone, but the understanding of them differs from one individual to another. If your experience of smiles has been that people are mocking you, or are about to do something unpleasant, you won't be happy to see a smile. You may be afraid and defensive.

Dr Ulf Dimberg, of the University of Uppsala, Finland, has recently published a study in which volunteers were asked to respond to images of happy, sad or neutral faces. He discovered that when the volunteers mirrored the facial expressions they saw, the action was smooth and effortless. When, in contrast, they were asked to provide the opposite of the expression they were looking at – for example, to frown at a smiling face – their muscles continued to try to produce a smile. In other words, we are hard-wired to 'copy' our fellow humans.[34] In fact, as we shall see later, there are various neurons within the brain that appear to fire only in response to the actions of other humans. The presence of these so-called 'mirror neurons' would tend to confirm the notion that modern man evolved living in groups, and that imitating our fellow group members was a valuable strategy to survive.

It follows that the expressions we generate produce feelings within ourselves, as well as communicating our feelings to others. If I make a scowling face, the neurons in charge of that expression communicate to my higher cortex that a) I am frowning and b) there must be some reason why I am frowning. This goes to the amygdala, triggering, shall we say, a mixture of fear and low-level anger, which eventually gets back to my facial muscles with the message: something worrying here, get frowning! So my frown intensifies, and this new information goes to the amygdala and so on. A feedback loop may thus be set up – rather like our brief analysis of stage fright earlier. Cognitive Behavioural Therapy (CBT) takes its practice from a belief in this loop. CBT gets people to identify the triggers that make them sad, fearful or angry, for example, and to consciously substitute another emotion in their place. By doing this over and over again, the practice becomes habitual. The frown is replaced by a smile, and the amygdala reads this information as 'happy' and tells us to smile more.

Beyond a joke

A curious incident happened on the morning of 30 January 1962. It was business as usual in the girls' school in the Boboka province of Tanganyika; now, of course, Tanzania. The school hummed quietly with the activities of learning – the teachers' voices, the shuffling of feet, the usual whispering at the back of the class. And then it all went crazy. A particularly banal remark by one teacher set a couple of girls laughing. The teacher, realizing her faux pas, joined in for a moment or two. Like all the best teachers, she understood that getting irate can sometimes make a situation worse. The rest of her class joined in. After a few minutes, the teacher started to call for quiet. It was time to be getting back to the lesson. But it was useless. The girls were rolling in the aisles, tears streaming down their cheeks. Neighbouring classrooms, separated only by flimsy clapboard partitions, began to be affected too – teachers and girls alike, of all ages, became gripped with an epidemic of laughter.

Nearly all of us can remember similar incidents at school, probably equally banal, which gave an uncontrollable fit of the giggles. That combination of being young and being in a situation in which one is supposed to be behaving sensibly can be quite deadly – there is, I think, a case for this reaction being more pronounced among females.

On this occasion in Tanganyika, the situation became

more serious. The laughing continued all day. The girls went home to their neighbouring villages, taking their laughter with them – and setting off their parents. Over the next fortnight, the laughter spread to villages throughout the province. Eventually, it was no longer any joke. The Red Cross was called in, because people were beginning to suffer from exhaustion, dehydration and malnutrition. The army was called in and forced the school to close.[35]

This story is, in a way, nothing more than a demonstration of the maxim that laughter is infectious. The sight and sound of our fellow humans having a good laugh makes us want to laugh, too. But why should that be so? Is this a mere trick of nature, or does it have some purpose? Laughter is a display of emotion – and some work has suggested the amygdala is partly responsible, although the orbitofrontal cortex has also been pinpointed by some recent research work. I recently saw footage of the remarkable 'Giggling Guru' – Doctor Madan Kataria, in Mumbai, India, who uses daily group laughter as a therapeutic tool for a range of physical and mental ailments. Like everyone in the room watching the footage with me, I was overcome by laughter at the glorious sight of these people laughing – indeed I am smiling now as I recall it. At the time of viewing, I felt a sense of great happiness for some moments afterwards. Perhaps now we can see why the Tanganyikan laughter epidemic became so hard to control. The sight of laughter in others makes us laugh. Our brain interprets this laughter as a sign of happiness – bringing on more laughter. Before we know it, we are stuck in the loop.

Laughter functions primarily as a social lubricant – and many scientists believe the smile, and its extension into a full-blown guffaw, is essentially a means of showing others that we are not a threat to them, and a means of promoting group bonding. Other primates exhibit forms of behaviour physiologically similar to the human laugh. Chimpanzees, for instance, will open their mouths wide, expose their teeth and make loud, repetitive vocalizations in situations that tend to evoke human laughter, such as rough-and-tumble play, or when they are being tickled. Even dogs and rats make noises of a certain pattern when playing. These are different to the noises they make during other interactions more related to survival and repro-duction. So laughter may be a form of social activity in both animals and humans. This view is supported by the observation that, in humans, laughter is up to thirty times more likely to occur in a group than when we are alone.

It also seems that human laughter produces a phenomenon called *allomimesis* – in which the behaviour of individuals synchronizes, and results in them all reporting the same physiological and emotional sensations. Karl Grammer and Petra Weixler of Austria's Ludwig-Boltzmann Institute of Urban Ethology tested the allomimetic effects of laughter by secretly filming pairs of male and female strangers. Subsequent to the filming, the subjects were asked to complete a questionnaire, which required them to rate the experience they had just undergone, such as the pleasantness of the situation, their perceived risk of rejection by, and their interest in, their partner. The researchers discovered a pattern whereby if the participants were interested in one another, the frequency and emphasis of the male's body movements directly mirrored the amplitude and frequency of the female's laughter. In turn, the more mirroring there was between their body move-ments and the female's laughter, the more the males rated the experience as pleasurable. In other words, laughter gets us 'on the same wavelength' and thereby promotes bonding. This may help explain how the Tanganyikan laughter epidemic managed to spread so far.

We have yet to isolate the genes possibly responsible for a sense of humour, although we have already seen that there seems to be a link between a dominant left brain and a tendency to experience a more constant, positive outlook on life. However, the fact that laughter is a global phenomenon and that it depends on specific brain regions suggests that, in the past, there was an evolutionary advantage in being able to share a joke. Perhaps the logic is that laughter can defuse violence, promote group bonding and therefore assist in the business of survival. This is something that hostages kept by gangs in various lawless parts of the world seem to believe and report after being freed successfully.

The areas of the brain involved in laughter are at least three-fold: the frontal lobes assess the situation and detect 'what's funny', the supplementary motor area produces the necessary facial and vocal movements and the nucleus accumbens gives rise to the attendant feelings of pleasure. The curious event in Tanganyika might well have arisen from a disturbance to one or more of these brain regions. Diseases like encephalitis – common in that part of Africa – can have bizarre effects on the brain.

In 1998, neurosurgeons at the University of California, San Francisco, were operating on a sixteen-year-old patient with severe epilepsy. They noted that whenever they stimulated the supplementary motor area of the left side of the brain she erupted in laughter. Furthermore, because she was conscious throughout the procedure, they were able to ask the girl what she found so funny. Each time they stimulated the area, she laughed, and provided a different explanation for her laughter – she saw a picture of a horse as being inexplicably hilarious, and on another occasion told the surgeons that it was they themselves who were so funny.

It is interesting to note that the stimulated area was within the girl's left brain. This suggests that the left brain produced the emotion, while her right brain searched for reasons in the immediate environment. We can see a similar interplay at work when stroke patients are asked by neuroscientists to provide the punch line to jokes. For example: 'A horse walks into a bar and asks for a pint of bitter. The barman says . . .'

a) Why the long face?

b) Two pounds fifty, please.

c) Nothing – but he takes out a gun and shoots the horse.

Patients with right-brain damage tend to choose the most literal interpretation – which in this case would be punch line b) – while those with left-brain damage choose the most random, seemingly inexplicable outcome – which might be punch line c). Sometimes I wonder about neuroscientists and their sense of humour; possibly you, too, are reluctant to choose a).

The normally functioning brain depends upon both hemispheres working in tandem – left producing the emotion, and right producing the meaning – for humour to work. Just as we produce the most laughter in interaction with other humans, so must the separate components of the brain interact in order for laughter to take place.

Asperger's and autism

Not all of us possess this in-built capacity to interact flaw-lessly with our fellow men. Some people have great difficulty in reading the facial expressions of others and in working out their mental state by observing their gestures and expressions. It is a particular feature of the disorder found more commonly in males known as Asperger's Syndrome, a condition which is a type of autism and was first described by an Austrian doctor, Hans Asperger, in the 1940s. Simon Baron-Cohen at the University of Cambridge conducted a test in which Asperger's Syndrome sufferers and normal subjects were shown pictures of faces depicting ten simple emotions – such as fear or joy – and ten more complex ones – such as guilt. They were shown both whole faces and separate parts, eyes or mouth. People with Asperger's Syndrome were able to read the basic emotions just as well as their counterparts, whether they saw whole faces or just eyes and mouths. But when it came to the more complex emotions, they had more trouble. People without the syndrome performed just as well at these more complex identifications even when only shown separate facial parts. But for people with Asperger's Syndrome, telling a complex expression from the eyes alone – a key way we communicate how we really feel – proved impossible.

Simon Baron-Cohen has suggested that autism and associated conditions like Asperger's may be variants of the 'male' brain. He quotes Hans Asperger: 'The autistic personality is an extreme variation of male intelligence. Even within normal variation, we find typical sex differences in intelligence . . . in the autistic individual, the male pattern is exaggerated . . .' Baron-Cohen points out there is essentially a type of brain which is systematizing (which he calls a Type S brain) and a more 'female' brain which tends to be more empathizing (the Type E brain). Perhaps what is most important about this is that men and women tend to express their emotions differently – for example, women are more alert to facial signals and pick up social nuances in a very intuitive way. They are more likely to be accommodating and less rigid in their view. Men, on the other hand, tend to be more interested in analysing details. Just possibly this is part of the reason why there are more male chess grandmasters. Men may relish railway timetables, train spotting, aspects of computing or technical details of various kinds, which fascination many women find almost incomprehensible. Perhaps all this is inherited from an aspect of human evolutionary past when men were focusing on hunting and women were more concerned with the social cohesion of the group.

John Ratey of Harvard cites an example of an autistic patient who reported that, until she was forty, she hadn't known that people told each other how they felt about things with their eyes. It seems that the eyes – not for nothing dubbed the 'windows of the soul' – are a particularly important indicator of human emotional states. Some key evolutionary differences have made this function paramount. For instance, the sclera – or white of the eye – is bigger in humans than in all other primates, and the very fact that it is white, not pigmented, would seem to lend it an advantage in signalling messages to other humans. Research has indicated that there is a considerable degree of overlap between the brain systems we use to follow the direction of someone's gaze, and those we use to infer the mental states of others.[36] It has also been argued that domesticated dogs, having lived so long with humans, have evolved with an ability to follow and read the gaze of their human owners – a trait not found in wild animals such as wolves. I find it interesting that a study, recently published in *Science*, suggests that this is a trick which dogs perform better than most primates. Dogs can follow these signals of a man's intentions more efficiently than chimpanzees.

The raging mind

Anger, and its expression in violent, aggressive behaviour, is also a key part of our emotional make-up. In both Western and Eastern cultures, we set great store by the facility to control and even deny our emotions – particularly anger and grief. We could say that we attach a hierarchy of value judgements to what are only chains of chemical activity. Why, for instance, should anger be frowned upon, but joy not? Probably because, certainly in the modern age, anger is more threatening and is likely to cause more social disruption. Much psychotherapy involves getting patients to acknowledge how they really feel and stop judging their emotions, as a way of ultimately preventing them from bubbling up out of conscious control. The phrase 'own your anger' might be a psychobabble cliché of the 1980s, but it expresses an underlying truth – some of our emotions are destructive, and only by acknowledging them and understanding their origins can we hope to submit them to conscious control.

In our evolutionary past, anger must surely have played an important role; indeed, in some human situations it still does. Shows of aggression, particularly between males in the animal world, serve to warn off predators. Within the basic survival group, males engage in combat to gain preferential rights to females and to any food supplies that might be going. We can see the descendants of this at work today in any crowded urban pub on a Saturday night: young males competing in various ways in order to attract females, a process that often spills over into aggression and then violence. If you have managed to grit your teeth and watch programmes like *Ibiza Reps Uncut* on Sky television, or even *I'm a Celebrity, Get Me Out of Here*, you will have seen that these elements are only just below the surface of modern behaviour. Games that rely on strength and physical prowess, such as football or boxing, represent a stylized version of the violent combative instinct, and the response of some of the more vocal spectators demonstrates that such impulses are alive.

I admit to being very edgy at airport check-in queues. At times, I have to exert considerable self-control not to show belligerent behaviour to perfectly innocent members of the airport ground staff. But nearly all of us feel aggressive sometimes. The recently coined terms 'road rage', 'trolley rage' and 'air rage' are merely labels for an essential trait that comes out in everyone when we feel frustrated by the traffic jam, when others gain what seems to be an unfair advantage over us in the supermarket queue, or when we are too closely confined with other humans in the sardine-tin conditions of a long economy-class flight. A psychoanalyst might even argue that the gloating over such relatively small-scale events in the media enables us to cope with our own rage. We are able to see it as something 'out there', not within us – thank heaven it's them doing it, and thank heaven they are in court for it, not us.

So what's happening when we lose our temper? The frontal lobe area of the brain plays a big part. People with damage to their frontal lobes – such as, of course, Phineas Gage – often have difficulty in controlling their anger. MRI scans of people with antisocial personality disorder, characterized often by aggressive, destructive behaviour, also show low activity in the frontal lobe region.

In normal individuals, the frontal lobes act as the policemen of the emotional mind – in particular, the area called the ventromedial cortex. They receive information from the lower cortex pertaining to urges, impulses and responses, but they inhibit it, and form careful plans of action to address it. For instance, if I feel hunger as I walk through a restaurant to the table I have been allotted by the waiter, I do not grab a few chips from the other diners' tables as I pass by. My frontal lobes are effectively saying to my lower brain: 'I see that there is hunger. But we wait. I won't cause a scene.'

In the autumn of 2001, teams from the Toyota Motor Corporation and Sony unveiled a new car with the 'pod' concept. It effectively performs the function of the frontal lobes for its drivers – thereby, hopefully, reducing accidents and dangerous driving. The car uses sensors within the steering wheel to measure perspiration and pulse – both of which increase when we are stressed or angry – and then puts strategies in place to correct our behaviour. When the onboard computer of this car registers an increase in stress levels, or when drivers go too fast or too close to other vehicles, an alert message is flashed on the dashboard and calming music comes out of the speakers. The prototype car – known as Pod – can even convey the driver's mood to other motorists or pedestrians using lights attached to the front and rear bumpers. In other words, somewhat like functioning frontal lobes, the Pod car can attempt to override the possible dangers caused by an angry or stressed person at the wheel. It sounds like an infuriating motorcar.

But, of course, some degree of anger is not only inevitable, but a healthy feature of development. Children, particularly teenagers, go through a period when mood swings and flashes of temper are frequent. As we have seen, research has shown that the frontal lobes take a long time to develop, not reaching maturity until we are in our early twenties. Recent studies have even shown that while the frontal lobes go through a rapid spurt of growth up to about the age of ten or eleven, they may actually be in reverse during puberty. It seems that an economic decision is made to put certain wiring circuits on hold so that more energy and attention can be given to the crucial purpose of creating a human adult ready for survival and reproduction. Suddenly parents have a neurological explanation for why a teenager, if asked to do the washing up, might storm off in a foul temper, while their ten-year-old sibling might comply sweetly.

In the BBC television series accompanying this book, we followed the very courageous efforts made by a man to address his anger problem. We saw, in essence, that triggers for anger created a loop between his amygdala, his body and his memory. Something that 'set him off', like another motorist cutting him up in traffic, created bodily symptoms of rage – quickened heart rate, tensed muscles. His memory fed in instances where he had been angry before – in a sense, he experienced everything he'd ever been angry about, and its associated physical state. We saw that his frontal lobes were less active during this process. He was in effect hearing the 'noise' of his angry body and his angry memories, without the calming, judgement-forming faculties of the frontal lobes.

Talking formed a major part of his cure. Verbalizing angry feelings not only provides a window of time in which we do not act and do something we might later regret, it also rallies the sleeping frontal lobes into wakefulness. As we saw in the last chapter, the act of analysing one's own internal state requires activity in the medial frontal cortex. The act of putting that analysis into words calls upon further cognitive functions. The frontal lobes are kicked into life, the powerful feelings are interpreted and inhibited.

The important role of serotonin in the function of the ventromedial cortex of the frontal lobes has been documented for some time. Monkeys with high levels of this neurotransmitter are more cooperative and less aggressive. And it is of interest that people who are not depressed but who take the antidepressant drug Seroxat, which boosts levels of serotonin in the synapses, tend to show more cooperative behaviour and lower hostility to other people. Serotonin has also been found to play an important role in anxiety. A US National Institute of Mental Health study found key genetic differences in some individuals leading to decreased serotonin production. Around 70 per cent of the people who had this genetic feature also reported higher levels of anxiety on statistically evaluated tests.

Could such a genetic mutation have occurred by accident? Might it have had some positive benefit to our ancestors in the past? One perplexing thing about evolution is that a strength is also often a weakness.

Humans have the most complex brain of any species. But its complexity means that, while our brain is maturing, human offspring are dependent upon their parents for a substantial time – which strains the business of survival. A tendency to low serotonin and corresponding aggression might be a similar 'double-edged sword' – one which could have helped our ancestors to see off threats and secure food in a harsh, competitive environment, but which is less helpful in more stable, settled times.

To return for a moment to monkeys and their serotonin levels. One species of particular interest is the vervet. Vervets are small, weighing perhaps 5 kilos, feeding mainly on grasses. They are found in many parts of Africa and have been studied not least because of their very well developed social hierarchy. Females in the same tribe form a hierarchy, with the daughters inheriting their rank from their mothers – somewhat reminiscent of some of my colleagues in the House of Lords from the Scottish peerage, or the Daughters of the American Revolution. With the latter, however, any woman is eligible for membership who is no less than eighteen years of age and can prove lineal, bloodline descent from an ancestor who aided in achieving American independence. Vervets have the advantage they do not need to provide documentation for each statement of birth, marriage and death. High-ranking vervets take priority when it comes to the best food resources and the dominant males get the best sex. Colleagues in the Lords queue. One study of vervet monkeys who had reached the top of their hierarchy shows that they had higher levels of serotonin in their brain. I do not have any comparable biochemical data on the English Peerage and I know of no studies regarding Daughters of the American Revolution. However, one study does appear to show that American college fraternity officers at the top of their hierarchy have higher serotonin levels.[37]

Hoping

I think that personally I'm a bit of a casual optimist. My life has been spent in the belief that things will turn out satisfactorily in the end. A happy state of mind but not one, I think, that spurs a person on to the highest achievements. Many of my most successful friends are what I regard as pragmatic pessimists – prepared to take a dim view of things, unless surprised by the way they turn out. I think this is a personality trait that leads these friends to interpret information in a certain way, information that others might treat differently. They tend to say that the way they see things is the way things truly are. But of course, this belief is found in both optimists and pessimists – both groups believe the other to be looking at life in the 'wrong', rather than just a 'different', way. All this is hardly surprising, for as we have seen, the way our brain receives incoming information informs and modifies its very architecture. It takes a leap of understanding, even of faith, for anyone to consider that 'green' only means their version of green, not some universal state. Likewise for states of emotion – the physical structure of our brains and our very identities are bound together.

When assessing the importance of brain structure, one unresolved issue is the difference between the left and right sides of the brain. Measuring differences between the left and right brain is difficult and much of the work is controversial. But at the University of Wisconsin, Richard Davidson showed that abnormally low activity in the left prefrontal cortex seems to be associated with depression. Happiness to a large extent, it appears, is associated with more activity on the left side. Feelings of sadness also appear to be associated with more activation on the right. And people with a pessimistic viewpoint, or those undergoing a period of depression, show increased right-brain activity and, conversely, decreased left-brain activity. In

Wisconsin they have been following very young children and this fundamental 'left-brain optimist versus right-brain pessimist' distinction would appear to be established not long after birth.

Another important part of the stage in the brain's theatre of sadness and joy is the hypothalamus, and within it, dopamine, which is probably a leading actor. A 1950s experiment by James Olds and Peter Milner from Montreal observed rats after an electrode had been implanted into their hypothalamus. When the rats pressed against a bar, the electrode inside the hypothalamus received a small electrical stimulus. They seemed to enjoy the process so much that they would press the bar up to 4,000 times an hour, and preferred to do that than eat food.

This process can be reversed if rats are given a drug that blocks the action of dopamine. One typical compound is the anti-psychotic drug Haloperidol. This drug alleviates acute agitated states in psychiatric patients and is also used as a premedication before surgery – it helps people feel relaxed and comfortable. But the rat taking Haloperidol no longer receives pleasurable sensations, and it loses interest in pressing the bar. People given this and similar drugs to combat symptoms of psychosis – like hallucinations and delusional thinking – often report anhedonia, an inability to feel joy and, most commonly, an important symptom of depression. Dr Tonmoy Sharma at the Maudsley Hospital, London, gives a graphic description of anhedonia. 'Many people who go into mild depression can be cheered by "tea and sympathy". But in severe depression, anhedonia becomes a serious problem. It's worse than not being able to get any joy from life.'

A team at the University of Texas headed by Helen Mayberg has identified a specific area – the tip of the cingulate gyrus – which shows abnormally low activity in people with treatment-resistant depression.[38]

Antidepressants and therapy do not work for everyone so we need to know the underlying brain states. Given the complexity of the brain, and the protean nature of its extraordinary chemistry, it surprises me that the often seemingly blunderbuss therapy of giving antidepressive drugs in a relatively arbitrary way has been useful for so many people. But in years to come, I suspect people will look back at the therapies we currently offer for many disturbances of the human mind and compare these contemporary treatments with the medieval practice of seeking out witches and the barbarous treatments of cupping and blood-letting.

In our subjective world, differences of mood or attitude amount to little more than a view of life in general. I call myself a 'casual optimist' because, given the right information or circumstances, I can be happy, hopeful and joyful – but equally I can get depressed, feel that I am not achieving much and that my work is not highly regarded. But for some people, prolonged sadness takes root in the architecture of their brains, causing a predisposition that cannot even be treated by drugs. It was, to be fair, this aspect of severe depression and gross psychosis which spurred on doctors like Egas Moniz and Walter Freeman to go beyond the bounds of what we now would regard as acceptable medical practice.

Fear

Fear is arguably the most powerful emotion in our repertoire. Real fear results in massive changes in the body's systems which control blood flow, and indeed people can drop down dead from fear alone if the heart ceases its normal pulsation. The unique power of fear makes sense in evolutionary terms. Our ancestors developed all manner of impressive faculties like language and consciousness but they were by no means safe from predators or other perils. So fear assisted our survival. But if you are frightened, do you run like the wind, or freeze?

These are two competing motor responses, and animals may use either strategy, possibly because of a genetic predisposition. Also, their previous experience and the environment in which the fright occurs are important. Dr David Anderson's group in Caltech, Pasadena, have studied rodents.[39] Mice who displayed freezing behaviour after an initial fright had most neural activity in the lateral septum ventral and some parts of the hypothalamus, whereas mice who fled had more activity in cortex, amygdala and striatal motor areas, and different parts of the hypothalamus.

As we shall see, fear takes two routes in the brain. The first travels from the sensory input regions – our eyes and ears and noses – into the amygdala, hypothalamus and straight to the body, which immediately starts priming itself to escape. A second route has fear stimuli travelling more slowly to the frontal cortex, where they are assessed and subjected to reason. For instance, lying down in my tent in the jungle at night, I notice something apparently wriggling. I am immediately on full alert, thinking I have seen a snake. I am ready to bolt out of the tent. I look again. By now, information has travelled the second, slower route via my anterior temporal cortex and I can analyse the object as just a section of rope, which moves as my sleeping bag presses on it. My frontal cortex now 'points out' that there is no need to worry. I stop being afraid and my pulse and breathing return to normal, my blood sugar starts to fall, and normal blood flow returns to the organs that are not essential for a rapid escape.

Understanding the two routes of fear helps treatment of phobias. A key feature of such treatment is the gradual introduction of the source of their fear to phobic people. For instance, an arachnophobic is encouraged first to discuss spiders, then to look at pictures of them, see them in the flesh and finally even to handle them, starting with small ones. At each stage, they learn and reinforce that the object of their phobia is not dangerous. They are also asked to think about aspects of their phobia that are non-threatening – to think about the skill with which a spider builds its web, for instance, or the beauty of its construction. Inside the brain, related neurons are being teased and trained into a new direction. The quick fear route, which would have sent them into panic, is being superseded by the slower, higher-cortex route, which calls upon the new associations they have made. The message 'Not Dangerous' becomes ever more strongly imprinted in the new circuitry as it is tried out hundreds and hundreds of times, and the fear passes into the realm of reason.

While key areas like the amygdala and the cortex with its frontal lobes play a big part in all our emotions, a whole set of brain mechanisms and areas are called upon. A swift reaction to certain emotion-inducing stimuli is imprinted in the circuitry of our brains. However, a tiny baby, unless primed by a parent's fearful expression or voice, will not be afraid of a wriggling snake or of falling – until some of the connections are in place, it has not had a chance to learn such fearful responses. And so much else depends upon learning and memory.

To use the example of being startled in the night again, I may listen out for further sounds after being woken up by a loud crash, like the bark of a fox, and trawl through my memory to ask what that sound is. My memory will come up with a set of answers: 'Sounds like an animal – you've heard them before, barking dogs, mewling cats, etc.'; 'Sounds most like a fox – they're always hanging round your dustbins.' I draw upon my whole auto-biographical self – upon things I learnt last week and things I learnt as a child. If my wife then asks me what the noise was, I call further upon my memory to say, 'A fox.'

In that way, by comparing what I experience now with what I have experienced in the past, the contents of my memory, I can direct my response appropriately. In the next chapter, we will be looking at the way our brains develop, over time, to store memories about the world and ways of describing it.

It is interesting that a great percentage of our phobias are directed towards things that would, in our ancestral pasts, have posed a real threat: snakes, spiders, open spaces, people in crowds and heights, for example. Arguably, then, phobias are simply a primed survival response – but no longer appropriate to the conditions in which we now live. It seems that we are wired to avoid, or possibly wired to learn to avoid, the things that would have posed the greatest threat to us when we lived on the savannah. Psychologist Martin Seligman coined the term 'preparedness' to refer to this evolutionary aspect of human phobias.[40] He noted that there seem to be three types of stimulus at which we are primed to be afraid. First, anything that might be linked to disease – such as dirt in general, rotting flesh, maggots. Secondly, there are dangerous animals – for example, snakes and spiders. Thirdly, there is a category that Seligman called 'conspecifics' – that is, other humans who pose a threat. Under this heading we might class xenophobia – which is actually the fear of strangers, although the term is now more widely used to refer to a dislike of other nations or cultures.

Conditioning experiments by Arne Öhman of the Karolinska Institute have shown that we are indeed more predisposed to fear objects or animals in these broad categories. Tests have entailed teaching non-phobic participants to associate a given image with fear by administering an electric shock to them as they viewed it.[41] They can be conditioned to feel fear much more swiftly when exposed to images of snakes or angry human faces than when exposed to images of, say, a smiling face or the sun. Evolution seems very likely to have played its part – shaping our brains so that we are most likely to become afraid of the things that pose the greatest threat to our survival.

Changing the mood

I enjoy the pleasures of a decent white burgundy or a good red Chambertin, particularly after a stressful day. In doing so I entertain a form of behaviour that shares similarities with that of less fortunate people using needles and crack pipes on the street. There seems to be no human society that does not engage in some drug use, whether for ritual or purely recreational purposes. Many religions – including my own – sanction the use of certain drugs. In Judaism, wine has a celebratory value – its properties are well recognized by the psalmist: 'wine that gladdeneth the heart of man'. In Christianity it has important symbolic value and is a key part of the sacrament. And, of course, it induces feelings of sociability and relaxation, arguably providing better conditions for people to get together and reaffirm friendship and values. In many indigenous societies, substances obtained from cacti, coca leaves, mushrooms and many other plants are taken in order to induce hallucinations or trance-like experiences. Sometimes the users hope to be thus in touch with a higher power, or alternative reality. In the 1960s, Harvard University professor Timothy Leary purchased a massive amount of the synthetic hallucinogenic drug LSD to bring about radical social change through the mass altering of brain chemistry. His famous words – 'Turn on, tune in, drop out' – landed him with a thirty-year jail sentence. For centuries before this, writers and artists experimented with a variety of substances, summed up by Aldous

Huxley as 'a chemical vacation from intolerable selfhood'. Many artists, perhaps with some justification, have believed drugs enhance their creativity. Among writers alone, the list is endless: for example, Baudelaire and Yeats, who experimented with hashish, Rimbaud with absinthe, Coleridge, De Quincey and Keats with opium, and Balzac who required caffeine.

Using certain substances to alter our brain is so widespread and longstanding that it is arguable that it is a basic human drive, just as is living in groups, or establishing social hierarchies. Could drug use at some time have held any evolutionary advantage? Even today, the business of daily life is demanding and drugs that can temporarily relieve us, by making us forget or perceive the world differently, have some benefit in reducing stress. Some drugs help with the business of survival. The leaves of the coca shrub – the source of the drug cocaine – are chewed in South America to keep hunger at bay and banish tiredness at altitude. I have tried coca myself when climbing at 16,000 feet or more in the Andes in attempts to banish my altitude headache. I have to say it didn't work. Alcohol also can possibly produce temporary benefits – making us more courageous and daring at times of battle or attack, or keeping out the cold.

Drugs have a social purpose. Drinking alone is considered to be the hallmark of an unhappy, perhaps addicted personality. Drinking in company is more enjoyable, certainly – we laugh more, relax more. The same may be also true for many other drugs – including cannabis and cocaine. So drug use could act as a sort of 'social glue' – making group bonding easier and longer-lasting. Even substances used in a more formal setting, like the peyote cactus or the sacred 'herb' smoked by the Rastafarians, are used for a social purpose – worship. Any practice that unites humans and keeps them united has a survival benefit.

What do drugs do when we take them? Put extremely simply, they inhibit certain brain areas, such as the responsible, rational frontal lobes, and excite others, such as pathways which carry neurotransmitters like dopamine and serotonin. Drug usage generally has three different effects on brain chemistry and hence on behaviour and perception. Stimulant drugs, such as cocaine and amphetamine, promote feelings of euphoria, well-being and sociability. They also tend to produce a feeling of energy and accelerated thought-patterns. Depressant drugs, such as alcohol or tranquillizers, have the reverse effect – slowing down mental processes, enhancing relaxation. Psychedelic drugs, as well as promoting a range of mood changes, alter perception and sensation. Users might feel that time passes incredibly slowly or absurdly fast. Objects might seem out of proportion and hallucinations might take place. The experience of a drug is frequently affected by a person's basic mood or their tendency to depression or anxiety. The circumstances in which the drug is taken, the user's body weight and the number of times they have taken it before all modify the feelings produced. Genetic influences also play a part – many Asians, for example, cannot easily tolerate alcohol because they cannot metabolize it readily.

Over recent years, the use of cannabis has become so widespread that certain police authorities have experimented with 'decriminalization'. Press campaigns have argued that smoking cannabis is no more harmful than getting drunk, and might be less so – since it does not seem to result in a loss of inhibition, or increase in violence. In certain, much-publicized cases, judges have refused to prosecute people who have maintained that they use cannabis for the purpose of their religion or to relieve pain from arthritis or multiple sclerosis. Every weekend, young people from a range of social backgrounds take mind-altering drugs. There is not space within this book to cover the whole panoply of mind-altering substances known to man, or even the most commonly available ones on the London streets, so I shall restrict comment to five that are taken by mouth or smoked. They are frequently regarded as 'not serious': nicotine, alcohol, cannabis, Ecstasy and LSD.

Nicotine

One of the most enigmatic drugs of all is nicotine. It is one of the most abused of all drugs and the consequences of smoking tobacco are widely known and extensively ignored by millions of humans. Cigarette smoking is a major health hazard in the USA, where around one-quarter of the adult population smokes. Most smokers are of younger age, have lower income and often have a poorer education. What seems important is that regular smoking is nearly always started before people become adults. We should be more disturbed still that smoking among adolescents is on the rise and it is very common to see children – perhaps under twelve years old – smoking in the street in both Britain and the US. There is extensive research on the social aspects of smoking and it is clear that if it is started early in adolescence there is less chance of it being stopped later in life.

Nicotine has a very complex mixture of effects in the brain, both depressing and stimulating. It enhances some aspects of cognition – for instance, it appears to improve short-term memory. Much important research has been done in rodents where nicotine can be given (by infusion in solution) to particular parts of the brain involved in memory, such as the amygdala and the hippocampus. After such infusions, it has been widely confirmed that rats perform better at remembering the way through a maze. There is some evidence that people who smoke seem less likely to develop either Alzheimer's disease or Parkinson's disease. Tests in patients suffering from Alzheimer's, attention deficit disorder or schizophrenia confirm they had quicker reaction times and better memory after wearing a small nicotine patch on their skin. Studies like this, combined with animal studies that are helping to evaluate the underlying mechanism, may have therapeutic implications.

Another interesting effect, which is not fully understood, is that nicotine decreases sensitivity to pain. This is controversial and different studies have given conflicting results, possibly because researchers did not always take the sex of their subject into account. Animal studies are confusing because different species show different degrees of pain relief. Dr Jamner's group in the Department of Psychology and Social Behaviour in the University of California compared men and women – they gave electric shocks across the skin of thirty male and forty-four female smokers and non-smokers. During these experiments, the volunteers wore a skin patch that in some instances contained nicotine and in others a neutral substance. Needless to say, the volunteers did not know whether their particular skin patch contained nicotine or not. Men 'on nicotine' tolerated pain better – but there was no demonstrable effect in women. It is well known that nicotine also has a considerable effect on mood, but these researchers took this into account and concluded that the effects they reported were directly due to the pain-killing action of nicotine. Some of the same researchers have also done work on stress and nicotine. Smoking a cigarette increases the heart rate but, in tobacco users particularly, smoking has a calming effect. Is there a difference between men and women, which could be due to hormones? Women smokers, whether on the contraceptive pill or not, did not increase their heart rate or blood pressure under stress as much as non-smokers.

There was, however, no detectable difference between men who smoked and those who did not.

Nicotine, and smoking cigarettes, have a considerable effect on brain regions associated with how alert we are, some compulsive repetitive behaviour, and how we perceive various sensations. Tobacco addicts show activation of these regions during craving and when shown cigarette packets, a lighter or even an ashtray. This pattern of regional brain activity is very similar to that found with most other addictive substances. It may seem strange that the relatively mild effects of smoking – somewhat improved cognition, alertness, relaxation, for example – can result in such a powerful reward that so many users are so 'hooked' on the drug. But this, of course, is why studies on addiction generally are so relevant. Trying to understand the precise mechanisms of how nicotine acts in the brain, though extremely important, is horrendously difficult. It seems that there are numerous (at least twelve) different receptors to which this powerful tobacco product can bind in human and animal neurons. These receptors all have different effects on function, acting in different ways with various neurotransmitters. And in turn, of course, these receptors are produced by different genes. Genetic influences on the effects of cigarette smoking may be important in unravelling the complex issues raised by this addiction.

Alcohol

Among other actions, alcohol has potent effects on areas of the brainstem called the locus coeruleus and the raphe nuclei. These play an important role in both motor co-ordination and alertness, which unsurprisingly are two of the functions most impaired by having a drink or three. Substances like morphine have specific receptor sites within the brain – indeed our brain manufactures its own morphine in the form of endorphins, which are released in response to pain, stress and physical effort. Alcohol is different. It has no receptor sites of its own within the brain. Its molecules are electrically uncharged, which means that they are attracted to similarly uncharged regions within the brain – such as those that exist within the membranes of our neurons. As a result, alcohol has the potential to disrupt pretty much every brain function.

Alcohol also causes release of dopamine in the nucleus accumbens – the brain's reward system – which is why we perceive its initial effects as being pleasurable. This activity within our reward system also explains why alcohol use has what psychologists call a 'reinforcing effect'. In other words, we return to patterns of behaviour that give us pleasure. So however much you swear stoutly, as you gaze blankly into a fizzing glass of Alka-Seltzer, that you're never going to touch a drop again, science shows that you'll need will power.

Some of alcohol's effects may have less to do with its chemical actions on the brain and more to do with the very pervasive power of suggestion. Psychologists have performed tests on people who are receiving a dose of alcohol and others who believe they are but in fact are imbibing straight tonic water. They have found that believing you are drinking can in itself be sufficient to produce many of the symptoms of being drunk, such as sexual arousal and aggression.

Cannabis

In Britain, particularly among some younger people, cannabis is becoming almost as widely used as alcohol. Until the 1960s, when people like me began to travel to so-called 'hippie havens' like Morocco, India and the Far East, its use was largely confined to a small, rather sophisticated elite of musicians, painters and the like, and also to sailors in big ports. I remember my surprise at seeing the extent of its use in Afghanistan, when I arrived tattered, unkempt and malodorous in Kabul in 1963. But in 2003, a short walk through London's Camden Market attests to the different state of affairs now. The widespread availability of drug paraphernalia – papers, pipes, tins and fashion items bearing the distinctive saw-tooth design of the *Cannabis sativa* leaf – demonstrates the extent to which the use of this drug has become a 'normal' activity.

Cannabis is often presented as being a 'natural' and safe alternative to other drugs – but 'natural' doesn't equate to 'safe' or 'harmless'. Opium is 'natural' – it can be directly extracted from the heads of poppies. Cocaine comes from the leaves of a shrub. Many other substances are also directly obtainable from plants, some of them very potent poisons – hemlock, muscarine and atropine, for example, all of which, incidentally, have been used in one society or another (with great danger to the users) for their mind-bending properties. We should not be seduced into believing that 'natural' always means 'good'. Everything we put into our bodies has an effect. Even drinking too much water direct from the purest mountain spring carries a risk – in excess, it can cause death.

Cannabis use as hemp, marijuana and hashish has a long pedigree. The earliest references to the drug come from Chinese texts dating back to 2737BC, indicating that it was used medicinally to treat rheumatism, bowel disorders and, bizarrely, absent-mindedness. Excavations of graves in Western Europe and the Middle East have indicated that ancient peoples may have sprinkled cannabis seeds onto hot stones – possibly as a treatment for pain, possibly for ritual purposes. Much more recently, writers like

Baudelaire and his notorious Club des Hachichins (named after the hashishins, or assassins: trained killers in the pay of Arab sultans) took cannabis resin and reported transcendent experiences. (Alcohol has a medicinal history too. In the nineteenth century and early twentieth century, alcohol, usually in the form of brandy, was widely used on prescription in Britain for its medicinal value. We used to keep it on the ward at the London Hospital when I was a houseman there, and very useful it was to revive the physician after a difficult night-call.) Doctors only stopped giving tincture of cannabis for pain relief because it had such variable absorption when given by mouth and because there were no standardized preparations of the plant extract. Added to which, pain-killing tablets of drugs like aspirin, paracetamol and codeine were more effective.

Cannabis contains over four hundred separate chemical compounds, sixty of which are believed to have a direct effect on the brain. The most potent is THC – tetrahydracannabinol. Research in the 1980s and 1990s has shown that there are specific receptor sites for THC, distributed throughout the brain. The areas most endowed with cannabinoid receptors are in the hippocampus, cerebellum, basal ganglia and hypothalamus. In these regions, THC acts to reduce the firing of neurons. So we have a general idea of how the drug can affect various faculties. For instance, one of the roles of the hypothalamus is in regulating appetite – cannabis users often report feelings of great hunger. Motor functions – carried out by the cerebellum and basal ganglia – are impaired by cannabis. According to figures published by the American National Institute on Drug Abuse, 6–11 per cent of fatal-accident victims have THC in their blood. The memory – in which the hippocampus plays a big role – is significantly impaired by cannabis use. Rats exposed to THC for eight months showed the sort of cell loss in the hippocampus that is equivalent to what we would normally expect in rats twice their age. Cannabis compounds release dopamine and this action, and the pleasurable effects of the drug, can be blocked by drugs that prevent opiates having a similar effect. One such drug is naloxone, which is used by anaesthetists to bring people round after an anaesthetic.

But we do not fully understand how THC acts on the brain to produce many of the pleasant effects of cannabis, such as heightened perception and relaxation. There are cannabinoid receptor sites within the cerebral cortex, and any disruption to normal chemistry here would result in changes to thought processes. Cannabis may also have specific effects on an area of the basal ganglia called the substantia nigra – which is responsible for the production of voluntary action. Many users do report feelings of lethargy and a lack of will to do anything, which might be a result of interference to this area. But that said, not every user is affected in the same way. The American comedian Lenny Bruce said he avoided cannabis because it made him highly excitable and overactive. Some people experience real anxiety, are paranoid and feel persecuted. Other users have visual and auditory hallucinations that are not always pleasant. It is still not entirely understood why this should happen in some cases and not others – other than by saying that some individuals seem to have predisposition, particularly if they have suffered from some form of mental illness in the past. We have always used the term 'not a happy drunk' to describe people for whom alcohol's effects are chiefly negative, so it should not be surprising that there are some unhappy cannabis smokers out there, too.

In 1998, a House of Lords Select Committee examined the use of cannabis for medical purposes.[42] The inquiry was because a number of patients, mostly suffering from nerve disorders such as disseminated sclerosis, were taking cannabis to relieve their symptoms of pain, nausea, visual problems and trembling. What they were risking was prosecution and a prison sentence, particularly if, with legal supplies impossible to obtain, they grew their own plants. The committee was broadly very sympathetic to their plight and recommended to the government that clinical trials of cannabis compounds should be commenced urgently – for if the drug were used for medical purposes, of course, decriminalization would have been needed. Within a mere six hours of publication of the report, a government minister rejected the findings of the committee totally – demonstrating, I think, poor judgement. This predictable knee-jerk reaction was produced by concerns about any form of drug taking in our society – except alcohol and cigarettes, which are, of course, responsible for far more deaths and disease in the UK than any other environmental factor.

There is an amusing endnote to this inquiry. Every Select Committee calls a press conference when it launches a new report. On this occasion, just before the press conference, our parliamentary clerk, who ran the affairs of the committee brilliantly, pointed out that the press were bound to ask each of us on the committee whether we had taken cannabis. He suggested very tactfully that we should decide well in advance how we might respond to that question, clearly expecting that we would give an answer which might not be, how shall I say, too controversial. I admit that I was one of their lordships who had in the distant past some minor experience of the effects of this drug. As I was sitting there wondering what I might confess to, Lord Porter, the wonderful octogenarian – a brilliant academic from my university who won the Nobel Prize in 1967 for chemistry – leaned across to me and whispered, 'I've never ever taken cannabis.' And then, rather wistfully, 'What's it like?'

Ecstasy

Since the end of the 1980s, a synthetic drug called methyl-dioxymethamphetamine, or MDMA, has become popular among devotees of certain kinds of dance music. Known as Ecstasy, or simply 'E' and 'X' by its users, it rapidly gained a stronghold in youth culture. In January 2002, the *Observer* newspaper claimed that it had access to a police report in which it was stated that two million Ecstasy pills were taken in Britain every weekend. This is double the amount reported by the European drug monitoring centre in 1999.

The drug was first patented by the German pharmaceutical company Merck in 1913, though never marketed. It is believed that Merck intended to develop it as an appetite suppressant, though for reasons now obscure the drug remained forgotten. Between 1977 and 1985, a few experimental therapists obtained permission to test it in psychotherapy sessions. The recipe for the drug was obtained, it seems, by underground chemists in the USA during the 1980s. As the drug's popularity increased, organized criminals took over its production and supply on a mass scale. Some estimates suggest that the bulk of the pills that come to Britain, and which now retail at between £10 and £20 a time, originate in Dutch laboratories, where they are produced for between 10 and 20 pence.

Ecstasy's chief effect on the brain seems to be in binding to serotonin receptors – which prevents the neurotransmitter from being reabsorbed. This means there is an increased quantity of serotonin remaining in the synapses. In turn, this leads to the short-term effects of positive mood and amiability, as well as increased energy. The action of increased serotonin on the hypothalamus and the brainstem also accounts for the loss of appetite and muscle spasms seen in users.

Ecstasy also increases levels of dopamine, as well as blocking its reabsorption. This explains two more of the drug's effects. First, many users find it pleasurable and want to take it again. And secondly their experience of the drug sometimes results in an exaggerated feeling of meaningfulness and significance. To quote the young Ecstasy-user in the *Observer* article: 'It's the best, most positive experience of my life. It's life-enhancing.' High levels of dopamine have been observed in people suffering from mania and psychotic delusions – states of mind not dissimilar to those attained by mystics and users of certain drugs. People report sensations of almost religious intensity – seeing reality 'as it really is', perceiving themselves and the world around them as some beautifully ordered design of interlinked phenomena.

Compared to the estimated numbers of regular users, the number of Ecstasy-related deaths is relatively low. The police reported sixty fatalities up to 2001. Quite rightly, much is made of these deaths in the media – they are made more poignant by the fact that most of the victims are young people at the threshold of adult life. Whether all the deaths are due directly to the drug is open to question. There are very obvious risks associated with swallowing any substance whose origins are not known. Manufactured by individuals who may not even have a chemistry textbook, merely some dubious information from the internet, Ecstasy is by no means a safe drug. The high profits involved mean that there is a great risk of unscrupulous behaviour on the part of the sellers. Police drug squads regularly seize tablets that actually turn out to be for dog worming or fish-tank oxygenating, and others that contain high levels of other, cheaper drug compounds.

There have been calls to downgrade Ecstasy as a serious drug, thus reducing the penalties for possession and supply. Advocates of the change argue that there is no proven link between Ecstasy and violence – as there is with alcohol. They also argue that it does not lead users into violent crime to finance a habit – unlike crack cocaine or heroin.

For myself, the most compelling evidence against downgrading the drug comes from neuroscience. Andy Parrott of the University of East London has studied the short- and long-term effects of Ecstasy on brain chemistry. He has found that as many as one in five first-time users experience some memory loss. Regular use also interferes with the brain's continuing ability to produce serotonin – resulting in sleeplessness and depression. Dr George Hatzidimitriou and colleagues at the National Institute of Mental Health in Bethesda have done some important work in monkeys.[43] They found that alterations to serotonin neurons lasted up to seven years after Ecstasy usage, and that nerve cells that were damaged frequently regenerated abnormally. A survey conducted by Britain's National Addiction Centre found that one in four regular users of the drug were likely to develop a serious psychiatric disorder. Although the chance of any member of the population developing psychiatric problems is high, this figure is still above the national average of one in five. There should be no talk of downgrading the drug until more is known about its harmful side effects.

LSD

Woodcuts dating back to the Middle Ages depict a phenomenon known as St Anthony's fire in which whole communities became seized by a sort of collective brain fever, dancing in the streets until they dropped from exhaustion. Such episodes were probably triggered by the ingestion of a substance called ergot – a fungus that grows on rye. Among other things, ergot causes the contraction of blood vessels – which can lead to brain disturbances and circulatory problems.

One of the most wonderful men I ever met was Professor J. Chassar Moir from Oxford, my old boss – a brilliant surgeon and a rounded, modest academic. He was responsible for synthesizing the active principle in ergot, to make ergometrine in the 1930s. This drug contracts the uterus and its blood vessels so effectively and safely that millions of lives have been saved around the world. Severe haemorrhage after giving birth is now almost a rarity. By the 1940s, a Swiss pharmaceutical company was researching the possible therapeutic properties of ergot. Subsequently ergot was in some of the first effective treatments for migraine – some of which are still in use today. While attempting to isolate certain further constituents of ergot, a chemist called Albert Hoffmann inadvertently spilt some on his skin. His account of what next occurred has entered legend.

Forty minutes later, Hoffmann experienced dizziness, visual distortions and paralysis. He became so worried that he asked one of his lab staff to help him get home. The journey was undertaken by bicycle because of wartime fuel shortages, and by the time Hoffmann arrived at his house, he was undergoing a terrifying set of sensations. 'Familiar objects and pieces of furniture assumed grotesque, threatening forms. The lady next door [who brought Hoffmann some milk to drink] . . . was no longer Mrs R., but rather a malevolent, insidious witch with a coloured mask.' Hoffmann says that more terrifying than his distorted perception of the outer world was his sense of a radical change in himself. 'Every attempt to put an end to the disintegration of the outer world and the dissolution of my ego, seemed to be a wasted effort. A demon had invaded me. I was taken to another world, another place, another time.'[44]

It seems hard to believe, from Hoffmann's haunted

and haunting account, that this drug, lysergic acid diethylamide, could ever become at all popular. It sounds like the very worst sort of nightmare. But in the 1960s, drug use did not have the stigma attached to it that it has now. Drug-related crime was much less common, and Western society was coming out of a long post-war period of restraint and rationing. Young people had a bit more money and more time and there was a feeling of freedom in the air.

A research chemist called John Beresford, intrigued by Hoffmann's account, purchased an amount of the drug from the Swiss pharmaceutical company. Via Beresford, a Harvard University lecturer called Timothy Leary – who was then researching the effects of hallucinogenic mushrooms – tried some on a sugar cube. So impressed was Leary by his experiences that he ended up ordering another hundred grams. Throughout the 1960s, a time of wild optimism in many quarters, LSD was taken widely by devotees of the 'hippy' culture. Many saw their experiences while under the influence as being equivalent to the mystic states described by Eastern philosophers and religious figures. For a time it was even used experimentally in psychiatric hospitals for depression and schizophrenia.

Now I come up hard against a familiar problem, a recurrent theme throughout this book. St Augustine put it rather beautifully when he wrote, 'I cannot grasp all that I am.' I have only my brain to interpret my brain. And I can only convey my experiences using the terms of my own experiences. I have not taken LSD, and rely on the accounts of other individuals who have – whose own accounts, of course, can only convey what it was like for them. In the 1960s, there was a famous televised attempt to convey an acid-trip, using two grey-suited, well-spoken presenters, very much of the BBC school. The episode is hilarious for its upright, very correct approach to the material. As the drug takes hold of the presenter, he stiffly reports distortions to space and time, rather as if he were reading a serious news item. At one stage in the planning for the TV series *The Human Body*, it was suggested that I might like to do this experiment. In the end, we decided that to do so might seem irresponsible. Admittedly, the amounts of LSD – about 80–100 micrograms – now taken in the common street form of the drug are tiny compared to what Hoffmann and Leary took. Nonetheless, the effects of the drug can be very potent at even this low level.

LSD users report a range of mental phenomena. Visual and spatial distortions are most common – objects appear to have an aura, or to leave a trace when they move. Sometimes they are frozen in space, even when moving. Things may seem out of proportion – a friend spent an evening sitting in an armchair chuckling at the enormous size of his hands. Users also report swift and inexplicable mood changes – feelings of great happiness bubbling up for no apparent reason, sometimes to be replaced by foreboding, sadness, even terror. The drug is slow-acting, reaching its peak some four to five hours after ingestion – a factor that can be dangerous when people, thinking the drug isn't working, decide to take more. At the peak of its effectivity, people can experience a range of unusual sensations – such as of being weightless, or pinned to the ground.

There is a wealth of literature on the subject of the LSD experience – such as Aldous Huxley's *Doors of Perception*. Some acid users seem to have undergone what we might call a merging of self and other – such that they no longer experience themselves as being separate from the world outside. This in turn has many similarities to the states of mind that mystics have written about for centuries. On the other hand, plenty of people seem to take LSD and do little more than sit round feeling generally 'odd'.

Little is currently known about the effects of LSD on brain chemistry. Structurally, the drug is similar to serotonin itself – in the same way that the hallucinogenic drug mescaline, from the magic mushroom, resembles the neurotransmitter norepinephrine. If such drugs are administered to rats, there is a shutting down of neuronal activity in the raphe nuclei. This is interesting, because this is a brain area associated with waking and sleeping (see above). During REM sleep, when we have some of our most fantastical and vivid dream experiences, the raphe nuclei is similarly inactive. So, arguably, LSD acts to bring on a state of wakeful dreaming in our brains.

The locus coeruleus, the brainstem area next to the raphe nuclei, goes into overdrive in the presence of hallucinogenic drugs like LSD. When this area is stimulated in rats, they become excessively aware – easily startled by the tiniest of stimuli. So we could argue that LSD has a two-fold action – making us super-tuned to the sensory input of our environments, but also making our brains interpret this information in a dream-like fashion.

Degrees of addiction

LSD is not considered to be addictive – indeed, people seem to build a tolerance to it very quickly, such that they need to wait a few weeks between doses if they want to experience the same effects with a similar dose. Drugs like heroin and alcohol, on the other hand, are considered to have strong addictive properties. If long-term users are unable to find a supply of their chosen drug, they experience physical withdrawal symptoms. It is not thought that cocaine, Ecstasy or cannabis have these addictive properties, yet people do report becoming dependent upon them in just the same way. Crack cocaine is especially notorious for inducing powerful cravings.

So how can this be so, if our bodies do not create a need for the substance in question? The answer returns us to the brain's reward system – to the importance of dopamine in areas like the nucleus accumbens, and to the power of conditioning. Cocaine use results directly in an increase of dopamine in the nucleus accumbens – in other words, the drug taps deep into our personal reward structures. We also know that rats will self-administer any substance that has the same action upon their nucleus accumbens. This includes cannabis and Ecstasy. Whenever we indulge in any form of pleasurable behaviour, we are creating a neuronal pathway between our reward centres and any associated memories of the experience. The link between the two is so strong that heroin and cocaine addicts can report powerful cravings for the drug simply on being reintroduced to features of the environment in which they used to take it. For instance, the sight of mirrors and silver foil, teaspoons and needles, or even the streets where the drug was bought, can lead to real physiological changes which are associated with withdrawal, such as raised heartbeat and blood pressure. When ex-cocaine addicts were shown items of drug paraphernalia in one study, brain scans revealed activity in the prefrontal cortex and the anterior cingulate, areas associated with motivation. Their brains were gearing them up to go in search of the drug, thanks to the tiniest of clues from the environment.

So it is easy to see why so many people, having kicked a habit, start to use drugs again – especially when they find themselves back in the same environment. In 1971, epidemiologist Lee Robins studied 898 US soldiers in Vietnam. Half made heavy use of narcotics, such as opium, heroin and morphine. Around 225 reported symptoms compatible with addiction to these drugs. Robins did a follow-up study of these soldiers after they had returned to the United States, and found that 95 per cent of the addicts had stopped using drugs. This suggests that, for these soldiers, all the cues or triggers that caused them to want drugs were associated with Vietnam. Once they were away from these cues, their need for the drugs went away.

Broadly speaking, virtually all drugs acting on the brain that change perception not only have harmful side effects at the time, but can cause permanent damage. The irreparable harm may be partly because the constant reinforcement of connections changes the working of the mind permanently. Alcohol can be regarded as a paradigm. We may think that alcohol use is relatively trivial, and being tipsy a bit of a joke. Let us leave aside the lost working days caused by alcohol use. It is by far the most widely abused drug and one that has huge negative effects on the health services in nearly every country. In Britain over 15 per cent of people attending accident and emergency units have alcohol-related injuries. Around 25 per cent of men admitted as emergencies to hospital are there because of alcohol abuse; over 5,000 people a year die in England and Wales as a direct result of its effects. Drink costs the NHS around three billion pounds a year, resulting in more than 28,000 hospital admissions.

Apart from its effects on other major organs like the heart and liver, alcohol can bring about a most profound change in the brain. It may appear to be a stimulant, but it is mostly a central nervous system depressant. It interferes with the neurotransmitter GABA, and frontal cortical control is progressively lost. The details are not pleasant. Although initially there may be a feeling of happiness and well-being, alcohol produces loss of emotional control, violent behaviour, inability to co-ordinate movement, nausea, confusion, stupor (when the individual is barely rousable) and coma. Alcoholic coma is not a minor problem – it results in death in about 1 in 20 cases. Once people who use alcohol heavily are habituated to this drug after five or so years, blackouts, nightmares and hallucinations are common if drinking stops. Withdrawal also produces uncontrolled trembling of the limbs, delirium and epileptic attacks that do not always respond to anti-epileptic drugs. Prolonged use of alcohol results in brain degeneration, particularly of the cerebellum with loss of Purkinje cells, and nutritional disorders with visual disturbance, loss of memory and impaired intellect. While this is not by any means a complete list of the horrors, dementia caused by loss of cortical neurons, nerve damage and palsy are also well documented. Finally, it should be mentioned that in my own field of medicine, alcohol abuse has a profound effect on the baby in the uterus – including causing abnormalities of head and face development, and of brain function. Alcohol can have an effect every bit as severe as repeated sporting trauma from boxing, or, of course, a major head injury.

Perhaps, before finishing, we could do worse than return briefly to poor Phineas Gage for him to have the final word. Once Dr Harlow had had his skull dug up years after his death, it was kept carefully in Harvard. Recently, Hannah and Antonio Damasio published an interesting reappraisal of the injuries Gage sustained.[45] Having photographed the skull inside and out and taken rigorous measurements of it, then compared these with the measurements of the tamping iron, they started to assess the clues. Careful assessment of the bone lost from the cranium, a missing molar tooth and a small chip on the left facial bone gave useful information. With the help of three-dimensional modelling of the brain and appropriate computing, they and their colleagues built up seven possible trajectories for the rod as it passed through the brain. Because there had been no infection, this excluded any trajectory piercing the fluid-filled ventricles of the brain, thus eliminating two possible pathways.

Finally, by a process of elimination, careful deduction and knowing basic clinical information such as the fact that Gage had intact speech after the event (and, therefore, Broca's area must have remained intact) a complete picture was built up. Detailed knowledge of the configuration of a modern brain was obviously important. Using this remarkable forensic process, the Damasios were able to give a reliable best estimate of the precise path the rod must have taken. Their results clearly showed that critical damage had occurred to the prefrontal cortex in its ventromedial region. This is the area which at least in part controls emotional processing, social cognition and behaviour. But the areas to the side, which process other aspects of cognition such as language, arithmetical ability and our perception of personal space, would have been undamaged. Interesting, too, that the damaged region has a high concentration of serotonin receptors. Work in monkeys has shown that low serotonin levels are found consistently in monkeys that are aggressive and socially uncooperative. So perhaps, after all, like some of those patients undergoing prefrontal lobotomy, Phineas Gage had lost his soul through no fault of his own – beyond a brief moment of, until then, uncharacteristic carelessness.

chapter six

The learning mind

Holy Roman Emperor Frederick II (1194–1250) had a reputation as an amateur scholar. He filled his court with the famous astrologers and savants of the day. He wrote a book on the taming of falcons, and so insistent was he on the correct use of language that he once had the thumbs removed from one unfortunate scribe who spelt his name wrong – *Fredericus* instead of *Fridericus*. He was particularly fond of personally conducting experiments, which usually involved a strong element of cruelty. There are detailed accounts of Frederick written by Salimbene of Parma, an Italian Franciscan friar who died in 1287. Salimbene certainly recognized the Emperor's excesses but seems to have admired him. According to him, Frederick once selected two men, fed them handsomely upon a range of delicacies, and then ordered one of the men to go to sleep, and the other to go out hunting. The next morning, he had them both disembowelled in front of him, expecting to see that the man who slept had spent more time digesting. The court physician decreed that the man who slept had indeed digested his food better – but he may have made this up as everyone who worked for

Frederick II was afraid of saying anything to challenge him. In another bizarre experiment, the Emperor had two convicts sealed up in an airtight room. Frederick observed the room closely from outside, hoping to see the moment at which the men's souls left their bodies and escaped.

But Frederick II is better known for his interest in languages. He was inspired by an account in Herodotus of the Greek scholar Psammetichus, who wished to discover whether the Egyptians or the Phrygians – ancestors of the Greeks – had been the original race on earth. To this end Psammetichus seized two newborn children from their parents and had them brought up in complete isolation, with guardians who never spoke to them. According to Herodotus, the first word one of the children spoke was 'becos' – which meant 'bread' in the Phrygian tongue, and which thereby established to the delighted Greeks that they were the true inheritors of the earth. Frederick, unsatisfied by this account, decided to replicate the experiment. To make absolutely certain that his human guinea pigs would be exposed to no human language at all, he ensured that their care-givers were all deaf and dumb. The results displeased him considerably. Frederick had expected them to speak Hebrew – the first language, as he thought. But these children acquired neither Hebrew, nor for that matter Latin or Greek. Raised in these highly restricted conditions, they seemed to be lacking in intelligence, and unable to learn.

In Chapter 2, we dipped briefly into why Frederick's tinkerings were doomed to fail. The brain is a highly plastic organ – its neural connections group and regroup themselves according to the information that goes in. Functions like speech and language that are unexercised, particularly when we are very young, do not develop properly and their neurons become taken over by other faculties.

The brain in the womb

The brain goes through frenzied development in the womb – adding as many as 250,000 neurons a minute. And it grows after birth, reaching 80 per cent of the adult weight by the time the baby is two years old. At the same time, there is the sweeping, ruthless pruning process called apoptosis, which tailors the brain according to the needs imposed upon it by the environment. Apoptosis, a word coined by Hippocrates, is derived from the Greek meaning 'falling leaves', and is nowadays taken to describe the process of cell death. This process is present in all organs, particularly during early development; the most striking example of it that I can recall is the development of the hand. In the womb, our hands start off as paddles, but early on the webbing between the fingers dies away by apoptosis to leave separate fingers. All of us will be familiar with the growth of the tadpole, losing its tail and in the process developing into a frog.

And in the womb, the brain begins to develop after the first fortnight following conception. The first structure forms at around sixteen days – the neural plate. Over the next five days, this elliptical structure widens and flattens to resemble a circle with a groove cut into the top. The groove then closes over, so that what we have is a tube within a tube. The front end of this neural tube goes on to create the human brain, and its back end turns into the spinal cord. From around this point until birth, the brain will massively increase in weight – dividing and multiplying in just the fashion cells can be seen to do under a microscope. First the front end of the neural tube splits into three sections: the forebrain, midbrain and hindbrain. At seven weeks' gestation, each of these three structures divides again.

The early years

By birth, some faculties of the brain are present, but not yet fully connected. Much of the most active areas in newborns are related to survival at its most basic – breathing, feeding, sensation and movement. At the same time, other connections are present at birth that will not persist into childhood and adulthood. For instance, there are connections between the auditory and visual cortices which indicate that infants – if they could only tell us – are probably synaesthetes in their early years, hearing colours and seeing sounds. In most people, apoptosis prunes this faculty away before the hippocampus matures, ensuring that we cannot remember having it.

We tested the memory of some of our research team. One colleague could recall being terrified of a grandfather clock when lifted up to see it, at what must have been somewhere between two and three years old. Another recalled the shock of a goose entering the house and making a great racket – this can't have been after she was three years old, because she knows when her family moved from the countryside to the town. Significantly – although this is only a straw poll conducted over mid-morning coffees – these memories are clustered around a similar age. My own first memory – and I very much doubt that I am making this up – goes back to when I was around one year old. I remember being incredibly angry with my parents when I was left in a pram in a garden, in Crich in Derbyshire, on a summer holiday during the war. We only stayed in that vegetarian guesthouse twice – once when I was around twelve or thirteen months old and again when I was four – and I remember that garden very well, particularly the second time because my father gave me a balsawood aircraft to fly, which broke against the hedge on its first sortie. As I have never seen photographs of the garden, I think it likely that my memory is genuine and not reinforced by what I have seen subsequently, as so often happens with memories of early childhood.

As the hippocampus, site of the brain's long-term memory, does not reach maturity until we are three years old, it is very likely that most 'early memories' are false. People who say they can remember being in the womb, or being born, are probably mistaken. The human brain is like a complex building. It takes a long time for its various rooms and walkways to be constructed and furnished. It's a process so drawn out that, as we have already seen, some parts of our brain do not reach maturity until we are in our late teens, even in our twenties.

It is hard to say to what extent newborns experience the range of emotions that adults do. To the untrained ear, the crying of a baby sounds like an undifferentiated racket. But experienced mothers and nursing professionals feel they can often distinguish differences. A one-month-old baby can communicate varying messages, such as hunger, discomfort and fear, in specific ways. Many mothers report that the specific sound of their baby crying can induce a powerful bodily sensation in themselves. The American psychologist Jaak Panksepp suggests we all experience something similar when we listen to certain pieces of music and feel a shiver up the spine. Music most likely to trigger the 'tingle factor' (as one BBC Radio 4 programme called it) in listeners very often contains unexpected shifts of harmony or long build-ups to the completion of a certain musical phrase. Panksepp noted that the cries of human and animal infants, when separated from their mothers, follow the same pattern. I am not convinced by this description but it does seem that some mothers experience a drop in body temperature on hearing such cries, which is corrected when they are reunited with their infants. So even very small babies are able to express key messages in order to survive.

Other faculties seem to be present from day one of a newborn's life, or soon thereafter. Alan Slater at Exeter University examines babies when they are just one or two days old. His research indicates that, even at this stage, babies show a preference for looking at more attractive, symmetrical human faces. Like adults, they also show an instinctive interest in whatever is the most novel or unusual feature in their environment. In this respect, I remember with great fondness the late Dr David Baum, latterly Professor of Paediatrics at Oxford and more recently in Bristol, and first President of the Royal College of Paediatrics. When we were both in training at Hammersmith Hospital, David, who was then interested in the welfare of very small babies, invented the Baum nose. He would enter the neonatal unit wearing a large bow tie and a huge red clown's nose. Leaning over a newborn's cot he would very solemnly and very slowly move his head from side to side and then suddenly turn on a light inside the nose from a switch concealed in his pocket. If the baby moved its head from side to side to peer at his nose, David Baum would declare that this baby had an intact nervous system and was unlikely to be brain-damaged.

Throughout the first year of life, brain activity increases in the parietal, temporal and visual cortex, as well as in the basal ganglia and cerebellum. A full account of early brain development is beyond the scope of this book but we can see the effects of this growth spurt as the baby develops hand–eye coordination and spatial awareness. At this stage, one of the tireless favourites is 'peek-a-boo'. The baby's dawning understanding of the space it occupies becomes a source of fascination. The teddy bear has momentarily disappeared. Surely things can't just disappear? Or can they? As the bear remains hidden, a note of uncertainty and then anxiety creeps in. Perhaps, after all, the newly formed theory is wrong, and bears can disappear? Suddenly the stuffed toy is jumped back into view, and the baby laughs with a mixture of relief and joy. Through very simple games like this, babies learn to test out their theories and possibly to cope with acceptable levels of concern and anxiety.

Indeed, babies and young children test their theories on the world around them like tiny scientists. Peter Willatts at the University of Dundee performed a problem-solving experiment with six- to nine-month-old babies. A toy was placed out of reach on a tablecloth, behind a foam block, and then covered with a cloth. To get the toy, the babies had to devise and execute a three-step plan: pull the tablecloth to bring the whole assembly nearer, move the block out of the way and then lift the cloth. The six-month-old babies were incapable of doing any of this. The seven- and eight-month-olds could manage one or two steps of the sequence. It was not until the babies reached between nine and ten months that they could follow the whole sequence through. Dr Willatts, comparing his results to studies performed on the maturing infant brain, concluded that the planning and sequencing of action was only possible once the prefrontal cortex had reached a certain level of maturity.

Breastfeeding is best for early brain development and intelligence might be affected by whether or not a baby has been breastfed. Some mothers, for various reasons, either cannot or will not breastfeed. The standard formula milk bought over the counter is not an ideal substitute for it does not have the optimal ingredients for best nutrition. Dr Willatts considered that this might have implications for a child's subsequent learning ability when commencing school, so he compared children who as small babies had had bottle milk plus a special fat supplement he had researched, with those who had had standard formula milk. The particular fats he considers important are long-chain polyunsaturated fatty acids, present in breastmilk.

It turns out that babies who had had his supplement for the first four months were sharper and faster five or six years later. Six-year-olds who had been fed supplemented milk were not only more mentally alert and had a higher IQ, they were also better at solving problems – for example analysing and matching up photographs given to them under experimental conditions. So the moral is clear – breast is best.

The frontal lobes are the last area in the growing brain to become active – starting their journey at around seven months and continuing steadily until about two years. By the age of one, these fledgling frontal lobes are beginning to exercise control over the drives of the lower brain. Babies of this age can make a choice if offered one toy or another. Until this stage, most babies' attention can be diverted by any stimulus. But once the frontal lobes begin to grow, their little owner starts to develop his or her own priorities – fixating upon the fragile antique porcelain china, for example, rather than the slightly moth-eaten teddy bear being frantically waved in front of them.

Language begins to develop at eighteen months – although as we shall see later, even newborn brains are tailoring themselves to the communication-related elements of their environment. Wernicke's area – a banana-shaped sweep of neurons situated close to the ear in the left hemisphere – deals with language comprehension. It matures much faster than Broca's area, a smaller cone of material in the frontal lobe, which deals with language production. You can address quite complex instructions to two-year-olds, and they will understand you, without having the ability to respond. The shortfall between the two faculties might explain why children at this age – the so-called 'terrible twos' – can become so angry and frustrated. What I find interesting is that the ability to speak develops at quite a variable age. In the case of my own children, one refused to speak until fairly late (around two years old), one spoke at not much more than a year and one at around eighteen months. This in no way seems to be related to their ability or intelligence. It may be that some babies, being more or less contented with their environment, have a varying urge to communicate verbally.

The frontal lobes develop alongside the language areas, giving rise to the onset of self-consciousness. Children begin to understand themselves as separate entities, and to realize that other people have their own internal states. Language development both helps and is helped by this awareness: we use language to describe our own states, and to receive information about that of others. And it is only through being able to see yourself as a doing-I and a done-to-me that you can start to form basic subject-object-verb sentences. For instance: I (subject) hit (doing verb) the truck (object).

Although the two-year-old brain is very close in shape and weight to an adult brain, other faculties take years to fully develop – growing, shrinking back and growing anew in the meantime. For instance, an area called the reticular formation, a group of neurons that begins in the brainstem and projects up into the cerebral cortex and basal ganglia, plays an important part in maintaining attention. This only becomes fully developed at puberty – one reason why smaller children have a shorter attention span.

Puberty

Much of human communication is non-verbal. Our ability to read the facial expressions of others develops in yo-yo fashion – increasing at certain ages, shrinking back at others. Puberty seems to put the whole process on hold. This was shown in a study by Robert McGivern at San Diego State University.[46] His research team took a pool of volunteers, half of whom were pubescent and half of whom had not yet reached puberty. Black and white images of faces were flashed up on a computer monitor to each group for a very brief period – just one-eighth of a second. The volunteers were shown angry, happy, sad and neutral images, and were asked to say what kind of face they had seen while their reaction times were tested.

The results showed that the younger children, both boys and girls, were better at reading facial expressions than teenagers. And brain scans showed that a particular area of the prefrontal cortex was smaller in the teenagers than it was in the children. One explanation may be that the transition to adulthood, in brain and body, calls for a massive internal effort. As we might have to draft in an army of architects, builders, carpenters, painters and electricians to renovate a house, the developing teenage body and brain need to employ a raft of resources. Some of these come from outside – which is why teenagers have the ability to get through a week's worth of food shopping in a few hours. But other changes have to come from within, which means some economies have to be made. Certain brain areas shrink back to allow others the maximum resources – neuronal material, neurotransmitters, glucose – as they develop.

For some faculties to be acquired, others have to be lost, at least temporarily. In turn, we can begin to see the reasons behind some of the most perplexing and frustrating aspects of teenage behaviour. Of course there is a risk involved when the brain deliberately sheds some of its faculties. In fact, we see this kind of trade-off throughout nature: testosterone, for instance, gives us strength, but may also shorten lifespan. There is a quid pro quo in the growing brain. A reduced ability to read facial expressions may not seem particularly serious. But consider how important this is in sensing danger.

Teenagers – even those not driving – are the group most

at risk from road traffic accidents. Being permanently plugged into personal CDs and MP3 players to escape the chiding voices of their parents might provide a partial explanation. But MRI studies also indicate that the brain of a teenager may not have a fully developed capacity to assess threats. Deborah Yurgelun-Todd of McLean Hospital in the United States, in a study which admittedly leaves a number of important questions unanswered, flashed forty faces showing fearful expressions to adolescents aged between eleven and seventeen.[47] Younger teenagers reacted with fuller activity in the amygdala but less in the frontal lobes, while older females showed greater activation in the frontal lobes. Further studies indicated that adult brains showed stronger frontal lobe activity than that seen in older teenagers. Does this increase as we get older? I rather think it does. While I know of no data, it seems to me that we are much less likely to take risks as we age. When I think back to the rock faces I climbed in my teens and early twenties without really thinking twice, I am now horrified – I would simply be too scared to climb these again. And this probably applies to all kinds of decision-making in the way we conduct our professional lives.

We see a general change in behaviour with this shift in brain activity. When I was a boy, teachers and other figures in authority asked the perennial question: 'Why did you just do that?' to kids who had recently executed some inexplicably stupid or dangerous stunt, such as shinning up the school flagpole or setting light to someone's hair. There came the perennial reply: 'I dunno.' A feeble excuse on which brain science possibly sheds some light. The brain of a child or teenager predisposes them to be more impulsive, less able to think clearly about the causes and effects of their actions. As brain activity shifts, in later adolescence, towards the more rational regions of the upper cortex, we start to see that set of attributes we call 'maturity'. A calmer outlook follows with more stable moods. There is a decreased propensity to take risks, and a heightened ability to think things through before starting on a course of action. These are attributes handed to us by our frontal lobes.

Professor Andrew Newberg at the University of Pennsylvania performed brain scans on eight Tibetan Buddhist meditators to see what happens to the brain when it enters a trance-like state through meditation.[48] Because a meditative state would be difficult to achieve in a noisy environment (for example, using MRI), Professor Newberg employed SPECT scanning, a technique similar to PET scans, which uses radioisotopes to measure brain blood flow. The Buddhists were seated in a darkened quiet room and, so as not to disturb them, injection of the isotopes was through long tubing from an adjacent room. To complete the experience, to make it as 'natural' an environment as possible, the meditators burnt incense sticks in the scan room. Newberg's research shows first that meditation causes a shutdown in the posterior parietal lobe, the area responsible for giving us a sense of self in terms of our place in time and space. People report that, during meditation, they experience a merging of the boundaries between themselves and the world around them, as well as losing track of the passage of time. Secondly, there is heightened activity in the frontal and prefrontal lobes. The more this increases, the more parietal lobe action reduces. Thirdly, this pattern of activity continues in practised meditators, even when they are not meditating. This once again provides a sterling example of how regular performance physically alters the chemistry of the brain. Newberg has suggested that teaching teenagers to practise meditation might exercise and strengthen the frontal lobes, giving them enhanced control over their feelings and behaviour. I do not intend to get into the power of prayer here, but certainly it should follow that this may be one way in which prayer exerts such a profound effect on many religious people.

It is useful to note here that Deborah Yurgelun-Todd's experiments with fear-processing in adolescents showed the pattern differing from one culture to another. In Western society, the increasing trend is for a longer and longer period of dependency upon parents, and a corresponding abdication of adult responsibilities. In the Britain of today, the scarcity of jobs and the rising costs of housing mean that young people are choosing to remain in education and to stay at home with their parents until a later age. Arguably the introduction of tuition fees here, and the resultant crippling debts hanging round the necks of new graduates, will merely lengthen the period of dependency upon parental support. But in many countries of the world, puberty goes hand-in-hand with adult responsibilities – providing income, for instance, and rearing one's own family. Accordingly, we may see earlier maturation of the frontal lobes as the brain of such teenagers responds to the pressures placed upon them.

Sex differences

By the time we are teenagers, the orientation towards sex and gender has fully matured in the brain. Male foetuses produce a surge of testosterone, which in turn informs the development of their brains, producing male-typical characteristics and behaviour. This surge in hormones appears to inform many of the differences we see in later life – such as girls' superiority at speech, and boys' at spatial and systemizing tasks like building with bricks.

When puberty sets in, we see three principal differences between the male and female brain. First, a region within the hypothalamus known as the medial preoptic area is two and a half times larger in men than in women. This region contains more cells responsive to male hormones than any other part of the brain. Experiments in monkeys show that when this area is stimulated, male monkeys become interested in any female around them, providing she is on heat. Also, if this area is removed or damaged, they lose interest in sex.

Secondly, the bridge that connects the left and right hemispheres – the corpus callosum – is thicker in women than in men. As I have already noted, this might explain why women (and males inclining towards the more female-typical end of the continuum) are better at communicating and more able to express emotions and understand the feelings of other people. These appear to be tasks that call for a shunting of information between the 'feeling' right brain and the 'analytical, vocal' left.

The results of this suggest that women's brains might be wired more efficiently for the task of social interaction. Studies by Professors Ruth Campbell and David Skuse at University College, London, have indicated that the root cause of this might be the female X chromosome.[49] Girls born with Turner's syndrome lack one of the paired X chromosomes and sometimes experience problems with social behaviour in later life. They may, for example, find it hard to read body language, and they can have difficulties in following the direction of another person's gaze. Work at the University of New York, Buffalo State School of Medicine and Biological Sciences has also indicated that the male and female brain may process information from human faces in different ways. Pre-pubescent boys use more of the right brain and girls more of the left, suggesting that girls are innately better at detecting more subtle nuances of facial expression – making them into better communicators.

The third difference between the male and female brain does not become apparent until we are much older. As the ageing process takes its toll – most visibly upon our reproductive systems, and upon the eyes, skin and teeth – there is a slight decrease in the size of our brain. Males tend to have a slightly bigger brain than females but they tend to lose brain tissue sooner. The loss is not great in either sex but a greater percentage of tissue loss in men occurs in the frontal and temporal lobes – areas concerned with thinking and processing emotions. Hence our archetypal image of the grumpy, fussy old man. Women's tissue loss occurs more in the hippocampus and parietal areas. Accordingly, they tend to be more forgetful in old age, and worse at locating objects.

The brain's growth to adulthood and beyond looks less like a flower emerging from a bud, and more like the action of a glacier, carving out the contours and idiosyncrasies of the landscape along its inexorable march – a bulge here, a cavity there, a strength in one area and a loss in others. Disorders, in a general sense, seem to emerge when the process is either too pronounced or incomplete. For example, it has been suggested that synthaesthesia might emerge as a result of inadequate cell death, though I am unaware of any real evidence for this. Perhaps left-handedness might have the same origins. And there is evidence that autism might emerge out of the brain 'trying' too hard to establish gender identity.

Simon Baron-Cohen's research in Cambridge has already been mentioned. Autism is more prevalent in males – and he feels that this disorder is an extreme form of male-typical brain chemistry. Male autists tend to have high levels of testosterone and develop male sexual characteristics earlier than their counterparts. Cooperation and sharing in play situations, understanding other people's emotional state and an early predisposition to be interested in human stimuli (to be interested in a person, say, rather than a clock) are traits more evident in girls than in boys. This difference is most clearly pronounced in autistic boys. The same goes for communication skills:

girls score higher at detecting emotions from people's eyes, and autistic boys score worst of all.

Boys as we have already seen – whether autistic or not – show a general tendency towards what we call 'systemizing'. In play, they show a preference for toy vehicles, construction systems and weapons – objects with a definite purpose, which can be controlled according to a set of unchanging rules. Later on, boy-typical hobbies like stamp collecting or train spotting express a degree of preference for the classification of items over interaction with people. In adult life, this tendency shows through in the predominance of males in professions like maths and science, engineering and construction – disciplines that are more concerned with 'how things work'.

Anyone who has ever shared the business of driving and map reading with a spouse knows that there are also crucial differences in the way men and women understand spatial location. When boys and girls are asked to draw a map of a journey they have done once, the boys' maps tend to be more accurate. They centre upon the route taken from A to B, while girls' maps are more concerned with significant landmarks along the way. I became particularly aware of this once when attending a conference in which a female colleague and I were both staying in the nether regions of a large, meandering hotel complex. We became quite lost – and I remember trying to mentally retrace my steps in terms of various turnings left and right, without any success. My colleague, in contrast, got us back safely, because she remembered that our corridor was graced with a particularly bad imitation of a Matisse.

There are many people with degrees of autism who are entirely integrated into society – they hold down jobs, they have relationships and families of their own. But for some, the effort of doing this is exhausting, because they have learnt how to cope with other people by systemizing

– in effect, they have developed impressively vast lists of the things people do. These lists work 90 per cent of the time but, of course, humans are not entirely predictable, so a constant updating is needed. For that reason, some autistic people shy away from environments or professions that call upon them to interact with other people. At the same time, they can excel in the manipulation of predictable systems. They can demonstrate what Baron-Cohen calls 'islets of ability' – brilliance at chess, or drawing, or at memorizing and classification – which seem essentially male, as distinct from female, abilities. Maths and engineering are professions in which people with varying degrees of autism can flourish, though, of course, this is not the same as saying that everyone who does well in these professions has some degree of autism.

But using Baron-Cohen's theories as a guide, we could begin to make some interesting, brain-based conclusions about other groups of people. The shy, anorak-wearing hobbyist, uncomfortable in social relationships and happiest at home with his neatly ordered collections and routines, is certainly an extreme archetype of the male personality. But it is present to a greater or lesser degree in most of us – the hero of Nick Hornby's *High Fidelity* and his companions at the record shop were really just passionate systemizers of music and albums. They had the advantage that the object of their obsession is widely appreciated by the general population – whereas people who apply the same enthusiasm to steam engines, stamps or vintage wines are likely to be labelled, and sometimes cruelly dismissed, as 'nerds' or 'geeks'. The evidence suggests that such traits, and the extent to which they are present, boil down just to a surge in chemicals at a crucial time in our brain's development. As Baron-Cohen himself points out, by understanding the states underlying such differences in personality, hopefully we might in time learn to be a bit more tolerant.

There's an analogy to be drawn between the way each person's personal brain geography shapes their identity, and the way physical geography and climate affect the cultures and outlooks of whole nations. Countries with harsh desert environments and a corresponding scarcity of resources like water and food often have a tradition of showing hospitality to strangers. For a Bedouin, to turn away a visitor without feeding and watering him and offering him a bed for the night would be unthinkable. This acts as a form of insurance – the Bedouin gives, knowing that were he in the same situation, he would be sure to receive. The nature of the terrain shapes the behaviour of the people within it. In cooler, more abundant countries, there is less risk to human life, and therefore less need for such a tradition. In the same way, each of us is dependent for so many of the little quirks and idiosyncrasies of our identity upon the geography of our neurons.

That note brings me to a more contentious issue of brain and behaviour. For a long time, scientists and others have debated the issue of homosexuality, and whether it stems from bodily differences or from one's early experiences. An autopsy study performed on the brains of male homosexuals by neuroscientist Simon LeVay in 1991 showed that an area of the anterior hypothalamus was twice as large in heterosexual men as gay men – twice as large, in fact, as it was in heterosexual women.[50] LeVay came in for some heavy criticism, some of it from gay rights groups, who felt that the association of homosexuality with a variant of brain architecture could result in heightened discrimination. Not discouraged, LeVay, who is himself gay, went on to report that the corpus callosum – thicker, as you will recall, in women than in men – is often thicker in homosexual than in heterosexual men. And a further study done under the auspices of the National Institute of Mental Health has suggested that there may be a specific gene, passed down the maternal line, which influences sexual orientation in men.

Some people might at this point say: 'Is that it then? We know the answer, just like that?' But it is not so simple. A genetic or brain-based predisposition to homosexuality might not necessarily result in an adult who is orientated towards their own sex. The psychologist Daryl Bem pointed out in 1996 that biological differences merely produce a *tendency* towards certain social experiences. He coined the phrase 'the exotic becomes the erotic' to mean that sometimes we deal with the awkward feelings associated with being different from others by turning them into pleasurable feelings. And as we now understand, experiences can in turn feed back into the plastic brain, to reinforce and heighten original differences. So possibly a child with certain traits of homosexual-type brain chemistry might or might not go on to become homosexual, depending upon how his family and other children behaved towards him, and depending upon the extent to which his brain adapted to his perceptions of difference.

Now you're talking: the development of language

One fascinating and intricate way in which our plastic brain works is in the acquisition of language. What do we mean by language, and why is it there? We have already looked at the nature and purpose of emotions. Emotions may be defined as a means by which we communicate our own internal state and those of others to our higher brain. Frowning, for instance, is a way of telling both yourself and the people looking at you that you are angry. So is language just a more abstract tool for the same thing?

The short answer to that is yes. Chimps use language in exactly this way – in the wild they have a set of some thirty-six sounds. Other animals have greater or lesser vocabularies. But for all animals, except ourselves, one sound means one thing. They do not combine sounds into strings, or sentences, to create different messages. Human communication is crucially different. We use combinations of sounds called phonemes (such as 'da' or 'ba') to create an endless variety of potential meanings. In other words, for animals a sound is intimately linked to its meaning – a certain type of bark means 'danger', another type is used by mothers looking for their young. For humans, there is no innate connection between a sound and its meaning. The phoneme 'da', for instance, could be Tyneside dialect for 'dad', or Russian for 'yes', or it could be combined in German with the phoneme 'zu' to mean 'to it', or it could be combined with yet more phonemes, 'sta', and 'rd' and 'ly', to make 'dastardly' in English. A phoneme has no inherent meaning in human communication – it is just a building block of a word. You could not expect to tell the design of a vast Roman mosaic from looking at one tile. In the same way, a phoneme is the smallest unit of a wider pattern, but unlike a mosaic, in which tiles eventually conjoin to form a set image, human language can use phonemes to convey an infinite variety of meanings.

There are some unique elements of human physiology that make this use of sounds possible. Our vocal tract has evolved to have a very sharp right-angled bend and a larynx lower than that seen in other primates. This refinement creates a larger space in which the tongue can move along two axes and create two resonant cavities at the same time. By being able to filter and shape the sounds emerging from the throat, these cavities allow for a variety of contrasting sounds – like vowels and consonants – to be made in rapid time.

This refinement is at the expense of other faculties. Most primates have their larynx much higher up the vocal tract, allowing them to breathe and drink at the same time. By contrast, every bit of food and drink we swallow must pass over the trachea, or windpipe. It takes time for this feature to develop in humans – human infants are born with the ability to breathe and drink simultaneously and this remains for their first three months or so. But as the capacity to use language develops, this ability is lost. The better we become at talking, the more we are at risk of choking.

If language involved a change and a risk, it must have had some clear evolutionary benefit for us. So why do we need language, and how it did it come about? There are two main arguments. The first suggests that language, like consciousness itself, was merely the crowning glory of millions of years of selective breeding. The Darwinian view is that we continually select for adaptations that make us better at surviving and reproduction; this argument suggests that language was the ultimate survival device. With words, we could communicate sophisticated messages necessary for survival within ever bigger and more complex social groups. It would enable us to hunt more effectively and protect members of the group.

The second view suggests that language was merely a by-product of a wider expansion in other skills. Specifically, those skills would have been the production and use of tools. The argument is that to see a need for a tool, identify how to make it, make it and then use it implies the capacity for sequenced thought. There is evidence that the regions of the brain that control sequenced hand movement and speech rely upon the same mechanisms. This school of thought argues that we became predisposed to develop language as our brains developed through tool use. Language, like making and using a tool, is a sequence.

Furthermore, the basic unit of language, the sentence, conveys meaning via a three-step sequence: subject (I), verb (make), object (it). The words in brackets might give some clue as to my next point. The very process by which one would manufacture a tool implies some awareness of a key concept – a self, separate from others and capable of doing. So I, Ug, a caveman, am not the same as this piece of stone, or this bison which I wish to kill, and by doing something to this piece of stone, I can kill this bison. Such strings of reasoning, it is argued, lay at the heart of our development of language.

Many theorists have seen language as the driving force behind a range of further innovations. By enabling accurate communication, for instance, we could argue that language enabled us to live in ever more complex groups – culminating, for example, in Nineveh, according to the account in the Book of Jonah 'a great city . . . three days' journey across'. Through its capacity to combine sounds to convey an infinite variety of meanings, we could say that language 'freed us from the moment', enabled us to think in the abstract, to create art, to invent, to dream.

Language also allows for the establishment of codes of action and morals, and for the 'how we got here' accounts, such as the story of the creation in Genesis or the moral codex of Hammurabi. These can then be passed accurately from one generation down to the next. In this way, the human group establishes and confirms its own identity – a process which leads us from being simple hunting and child-rearing packs into being tribes and eventually nation states. By enabling an analysis of our own internal states and those of others, language also allowed for social change. The anthropologist Jack Goody suggests that a further mutation of language, writing, enabled people to critically review their own history and the history of their groups or tribes, thus facilitating dissent and revolution and progress. It is difficult to understand why writing seems to be so recent in human history. If we mean by writing a device for expressing linguistic elements by a series of man-made visible marks, writing is not much older than 5,000 or 6,000 years. But we know that humans from 30,000 or 40,000 years ago drew pictures, and there is recent archaeological evidence that primitive stylized engraving on bone might go back 75,000 years or more. So why did it take so long to develop written words? Is it because, until man started to build towns and lay down roots in one place, there was no pressure to use engraving for distant communication? Whatever the answer to this question, some people believe that the emergence of language shaped the architecture of our brain, and even suggest that the fundamental differences of left and right brain emerged as a consequence.

When a modern human hears someone say a sentence, a complex flurry of actions occurs inside the brain. First, it has to recognize that what it is hearing is language, as opposed to, say, the noise of a babbling brook, or a pneumatic drill. The evidence suggests that we have a hard-wired predisposition to do this, but our brain depends on the right sort of environmental stimuli to get this process up and running.

Many of us will admit to feelings of jealousy towards small children who, after only a few weeks in a new country, are able to chatter away like the natives. This is because of the plasticity of the brain's language circuits – a plasticity that exists only during a certain time frame. Patricia Kuhl, of the University of Washington at Seattle, has done work with newborns which indicates that we are all born with an innate ability to detect the difference between spoken language and other sounds, whether they are generated by humans or not.[51] We also seem, as I mentioned earlier on, predisposed to recognize the unique pattern of phonemes that make up our mother tongue.

But by the time we reach the age of a year, we can no longer recognize the phonemes and speech patterns of languages to which we have not been exposed. The circuitry inside our brain has undergone a growth and pruning process, such that it can only recognize the phonemes of the languages we hear on a daily basis – and usually that is whatever language our mother speaks. We dismiss all other sounds as non-language. Interestingly, this 'window of opportunity' for language may also apply to certain other faculties. A study led by Olivier Pascalis of the University of Sheffield has indicated that, during the same six-month period, we are able to recognize different human and monkey faces. But after that point we can only differentiate between human faces.

Hard-wired grammar

In the early stages, acquiring language is as much about what goes in as about what's happening inside the brain. George Hollich at Johns Hopkins University has suggested that the way mothers in particular and adults in general speak to babies enforces the process by which they learn the specific phonemes of their own language.[52] Phrases like 'Oozat ickle pwitty babba den?' are examples of 'Motherese', a simplified but exaggeratedly rhythmic form of speech common to all human groups, which helps the baby to learn.

Dr Hollich believes that infants understand much more than we think. 'If you consider the almost infinite number of words, as well as the potential mappings between words and meanings, learning a language ought to be impossible,' says Hollich. It has often been thought that children learn language step-by-step from one word to the next. The research at Purdue indicates that infants may be learning words and grammar simultaneously. For example, in one study, babies are shown a series of short animated sequences on a screen, such as an apple repeatedly colliding with a flower. Then another screen will show the apple in one corner and the flower in the other. A voice asks, 'What hit the flower?'

Of course, these babies can't talk yet, so the length of time they stare at a given image indicates their comprehension. If the baby looks longer at the apple it suggests the question is understood. If the baby looks at the flower, it suggests they can identify the word 'flower' but don't fully understand the question. By fifteen months, children look directly at the apple. This is surprising because most experts predict that infants should look at the object specifically mentioned in the question. Infants will do this, but only if the question is, 'Where is the flower?'

The position in which mothers instinctively hold their babies may also help language acquisition. About 80 per cent of mothers hold their babies on their left side, regardless of whether they themselves are left- or right-handed. So sounds tend to enter the baby's left ear – and thus will tend to be processed in the right hemisphere. The right hemisphere is not, in most humans, where our language faculties sit, but it does develop earlier than the left side. It also is better at interpreting the rhythmic and melodic sounds that parents use when talking to infants. The theory is that this establishes a loop of non-verbal communication between mother and child, as sounds from the newborn will enter the mother's left ear first, and travel to her own right hemisphere.

When babies start to make babbling sounds, they use the left, language side of their brains, and correspondingly make more movements with the right side of their mouths. The babbling of a six-monther born to French-speaking parents will sound different to that of one born to Japanese parents – because by this stage, babies only babble with the set of phonemes specific to their native tongue. Even deaf babies, born to parents who communicate with them via sign language, will make hand movements identical to the babbling of normal babies – suggesting that the seemingly nonsensical speech of infants is in fact a crucial rehearsal for the full-blown use of language.

The first two years are the most vital for language acquisition. The University of Chicago psychologist Janellen Huttenlocher has shown that the frequency with which parents speak to children up to the age of two has major consequences for their language use throughout the rest of their lives.[53] The more language they hear, the greater will be their vocabulary – regardless of whether they understand the words they are hearing. It is another classic example of the 'use it or lose it' principle – brains that get exercise grow ever stronger. So it is not surprising the experiments of Frederick II didn't work.

However, the plasticity of the brain is very important. As we have seen, in 95 per cent of people the left brain has assumed the responsibility for language by the time we are five years old – but evidence suggests that the right side can take over if problems occur at a sufficiently plastic stage. Children who have had the left side of their brain removed can still produce and comprehend speech – although they may have problems in using complex grammar or the future tense. As is the case in so many areas, the extent of the recovery depends upon the age at which the damage occurs. A young child can 'catch up' after the damage caused by a brain injury. His 65-year-old grandfather, on the other hand, might never regain the ability to speak after a left-brain stroke.

The evidence from stroke patients reveals just how compartmentalized a process our use of language is – many areas are involved, each with specific purposes. For instance, if stroke damage affects Broca's area of the left brain, then people are unable to speak, but they don't lose their ability to understand what is said to them.

So what specific structures of the brain are involved

when we produce and comprehend language? The business of distinguishing speech from non-speech noise occurs as an interaction between the thalamus and the auditory cortex. It proceeds from there into the main language processing areas. These are chiefly on the left side in most people, and sit surrounding the auditory cortex. As we saw earlier, two of the most crucial centres – Broca's and Wernicke's areas, dealing with language production and comprehension respectively – were first identified around 150 years ago. Thanks to increasingly sophisticated developments in brain-imaging technology, we now know that other areas play an important role – notably the Silvian fissure, the grooved part of the brain dividing the temporal and frontal lobes.

Further to the ability to detect the sound units of our own language is the ability to distinguish between one phoneme and another. There is a tiny part of the left brain that lights up only when we hear consonants – if this is not functioning normally, people show difficulties in distinguishing words that depend upon consonants for their meaning. For instance, we need to distinguish between 'p' and 'b' sounds to understand the difference between the words 'pat' and 'bat'.

Once we can distinguish one phoneme from another, we then begin to learn how to combine them into larger units, called morphemes, to convey meanings. A morpheme can be a whole word, like 'cup', or just a part of a word, like the suffixes '-s' and '-ly' in English. 'S' can make a noun plural – for instance, book/books. Or added to a noun, it can denote possession – as in 'the cat's basket'. Added to an adjective, like 'quick', the suffix '-ly' turns it into an adverb, to qualify an action, as in 'he walks quickly'. So morphemes have two functions – they can stand as words, and therefore convey ideas in their own right. And they can modify the meanings of other words.

The system of rules by which we combine words into

phrases, and phrases into whole sentences, is called syntax. The English language has an 'analytic' syntax, meaning it relies on the order of words to signify changes in meaning. For instance, 'John kisses Mary' conveys a different meaning to 'Mary kisses John'. To show who's doing the action, and who has it done to them, we English speakers put the subject – the kissing person – first, followed by the verb, and then finally the object. In other languages, such as Hebrew, Latin or German, 'synthetic' syntax is used – meaning that changes tend to be made to the words themselves, such as adding extra morphemes to the front or the end of a word to denote whether it is the subject or the object of a sentence.

In the growing child, this ability to combine phonemes into morphemes and morphemes into sentences is one that develops in tandem with vocabulary. The first form of language we use is mimicry – babbling as a rehearsal of the sounds and patterns of our mother tongue. This gives way to the acquisition of real words – generally occurring at about the age of one. As vocabulary increases, babies will naturally start to combine words into rudimentary sentences at around the eighteen-month mark. And although these sentences may be grammatically incorrect, there is a slowly emerging logic – which suggests that a process far deeper than mere mimicry is going on inside their brain.

Linguists are particularly interested in the mistakes children tend to make at the age of three or four – errors which increase in frequency as their ability to convey complex meanings grows, and which also, interestingly enough, they wouldn't make at an earlier age. For instance, children, having learnt that the suffix 'ed' conveys the past tense, will apply the rule to all verbs they come across – so they might say 'I runned' or 'I holded', whereas at the age of two they might have correctly used 'ran' and 'held', because at this stage they were purely imitating their elders. Or, having learnt that the suffix 's' makes nouns plural, they might say 'mouses' or 'mans', instead of 'mice' and 'men'. So children move quickly beyond a simple imitation of their parents' speech. An in-built ability to learn the rules of speech is switched on when the vocabulary becomes big enough to need such rules. Children may apply them rather enthusiastically in the early stages – in the same way as it is hard to prise a child off their bicycle when they have just learnt to ride unaided. This is a crucial defining feature of human language – that we apply a set of rules to a set of sounds in order to create an infinite variety of meanings. Chimps taught to communicate via sign language never develop this capacity.

Ever since the groundbreaking 1950s work of linguist Noam Chomsky, the scientific community has accepted that this ability – a universal language acquisition device – is built into human brains from birth. So in a sense, Frederick II was not entirely wrong. His mistake was in believing that a specific tongue – like Hebrew or Latin – pre-exists within the brain, like the operating software that comes with a new computer. But the capacity to acquire and use a language – any language – does exist. What the meddling emperor failed to understand was that this 'program' has to be activated by exposure to sounds.

To date, we have not isolated any specific site in the brain responsible for this piece of software. Studies of certain injured patients suggest that damage to the more frontal regions of the brain's language centres – such as Broca's area – is associated with difficulties in grammar and syntax. We also know that patients with damage to Wernicke's area can preserve an ability to construct grammatical sentences, while talking incomprehensible gibberish. The effect would be not entirely unlike the stories created by the wonderful comic 'Professor' Stanley Unwin, who died last year. Here he is telling a very famous children's legend which you might recognize:

Once in a long far awow, in the Germanic land, there was a great city called Grubbelsberg or something like that, with an Obermeister-Bergelmasty who was in charge. Now there they had a surfeit or rat-suffery, where all they used to creep and out and gnaw sniff and gribble into the early mord (and the late evage) there, biting the bits of the table, also the tea-clothers; and when people were asleep in their beds, so these rats would gnaw into the sheebs and also the whiskers of those who was dangly hoaver.

There was a great suffery. Not only this: the larder, foodage, all the fine things of the world was not enjoying by the peopload them-selves on account of these rats doing a sniffy most (and the chew-chew and stuffle down their ratty grebes); because these were the fattest raps and also over-producey in this great lovely city which was otherwise . . . tsk tsk.

From *The Pidey Pipeload of Hamling*

Stanley Unwin claimed that his fascination with nonsense language was learnt from his mother who, when she tripped up one day, told her small son that she had 'falloloped over and grazed her knee clapper'.

Broca's area might be the brain's grammar centre, but scientists are by no means agreed on this. It may even be the case that there is no specific unit at all – merely a pattern of neural firing, taking place in a range of different regions of the brain. The problem is made more difficult by the fact that, where language use is concerned, no two brains are quite alike. Researchers have been able to pinpoint sites within the brain that con-trol features of language as narrow as the names of fabrics, vegetables and gemstones. But the size and location of these areas differ in everyone – unsurprisingly when you remember that our brains are shaped by what we do and what happens to us in our daily lives. So we would expect to see some clear differences in the naming centres of, say, a market gardener and a Hatton Garden diamond dealer. And if this is true of the brain's naming sites, the same may also be true for grammar.

Alfonso Caramazza, psychologist from Harvard, suggests there are independent brain systems for grammar, meaning and word form, each with their own locations and components within the brain. There is evidence for this from studies of brain-injured patients – for example, victims of strokes that have affected one part of the brain. Some can name objects, but not supply any of the associations that go with them. For instance, they might be able to look at a picture of a church and name it 'church', but not be able to say what they thought it was there for, who used it, what went on inside, etc. Similarly, other people could look at the same picture and be able to tell you that it was used for worshipping God, people went inside and thought and prayed and sang, that the insides smelt of wood, wax and incense. But they would be unable to come up with the word 'church'.

I mentioned my puzzlement at the late development of writing during Palaeolithic times. It may be relevant that there are separate sites for reading and writing in our brain. One argument to explain this is that the brain's plasticity leads it to coopt neurons that might have been used for other purposes at earlier points in our evolution. They still could be coopted elsewhere if we do not activate them by learning to read and write. A rather cheap example of this process would be that my handwriting used to be fairly neat (even though I am medically qualified) but has become appreciably worse since computers replaced the pen and paper.

As is so often the case, the evidence that these skills sit in a specific corner of the brain comes from examining the brain in its disordered, rather than its healthy state. For instance, there are people who can write, but not read, and others who can read, but not write. Brain scans performed upon people with high verbal IQs show that they have activity in the superior temporal gyrus when reading, and in the middle temporal gyrus when naming objects. Conversely, people with low verbal IQs have the opposite arrangement. As we saw in our brief foray into the issue of dyslexia, people with this disability seem to have particular problems with the fast processing of both visual and auditory information.

So, to recap, the ability to acquire and use language is hard-wired into the human brain. When an infant makes its first sounds, it is making use of a distinct facility, one very different from the laborious way we, as adults or teenagers, must pore over text books and vocabulary lists as we struggle with a new language. This is why, even after years of practice, we never quite acquire the same level of fluency in a second or third language. Translators and interpreters, experts in their field, always translate from the second or third language back into their mother tongue, because this is the language in which they have the greatest degree of fluency. We develop our first language and we learn others. The learning of a second or third language involves the same parts of the brain we used when we uttered our first words. But because it is a learning process, it also calls on other brain areas, the same ones we use when we first begin typing on a computer keyboard, for instance, or learning the route from home to our place of work. These facilities of the human mind would be classed as memory, which is our next subject.

A rooted sorrow

Canst thou not minister to a mind diseased,
Pluck from the memory a rooted sorrow,
Raze out the written troubles of the brain
Macbeth, Act V Scene 3

Shakespeare was here unwittingly glancing upon some of the hot topics of brain research. When we think about memory, some key questions immediately emerge. The guilt-ridden Macbeth is talking here about the memory of the terrible deeds he has committed. We all shudder, sometimes, as we remember acts in which we would rather not have played a part or words we would rather not have uttered. Other memories, of course, can evoke quite a different reaction from our emotional palette – joy, sadness, anger and fear. But there is another class of memory that seems to have no clear emotional element – the memory we dip into to recall a name, for instance, or how to change a tyre. So do these two types of memory occupy different areas of the brain? What processes are involved when an event enters our memory, and when a piece of learning becomes 'second nature'? And if they are so rooted in the architecture and the chemistry of our brain, how is it that we can forget things over time? Why is it that my ninety-year-old mother can recall playing tennis and losing in Brook Green, Hammersmith, as a schoolgirl but not what she ate for lunch yesterday? And what is a 'memory' anyway – does it occupy a bit of shelf space inside the brain until it is recalled, or is it the act of retrieval itself that makes a memory?

It is a modern cliché that all of us over a certain age remember what we were doing when President Kennedy was assassinated. Over time, other world events have surpassed that tragic event in magnitude. But clichés often contain an element of truth. A brief trawl through your more personal memory banks will prove the point further – the memories associated with your most intense emotions are likely to be those you recall best.

All I think I am saying here is that certain memories stand out because they are unusual in comparison to the unremarkable routines of daily life. Studies have shown that the human mind selects novel stimuli – if you note how often and for how long an infant gazes at an object, you will see that it looks more frequently and for longer times at unfamiliar objects. This selectivity also works within our memories. You have probably driven to work more times than you have sat an exam, for instance, which is why you can recall the latter better than the former. If you were to remember every journey into work, your brain would quickly become clogged with detail, like Shereshevski, the hapless Russian 'Memory Man'.

But what is interesting about our memory of key traumatic events – whether we saw them unfolding on the news, or whether they occurred to us personally – is how much detail seems to enter the record. When we talk about September 11, it is significant that we remember *what we were doing* when we heard the news – a detail not relevant to the unfolding events in New York. We can often recall a host of other elements – how we learnt the news, what we were wearing that day, the weather outside. We call this the 'flashbulb memory': the ability of the brain, during intense experiences, to capture a moment in vivid and exceptional detail.

Memory and emotion

The anthropologist Victor Turner argued that the rituals and ceremonies conducted in tribal societies make great use of the interaction between emotion and memory. Initiation rites, for example, are performed to ensure that young boys entering manhood learn the particular mythology and customs of the social group, and their own rights and responsibilities as adult males. Like all adolescent males everywhere, they might be more interested in other concerns, probably of the female variety.

This, in Turner's view, is why rituals came into play. In ritual, the familiar becomes unfamiliar. People behave in a

Which of these figures is most evoked by the word 'kiki' and
which by 'bouba'? See what Professor Ramachandran suggests
the great majority of people decide on page 97.

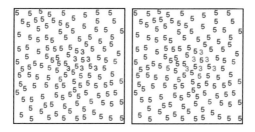

The numbers in the two boxes above are laid out identically.
To most of us the mass of 5s and 3s is a jumble and it takes a
few moments to identify the 3s. A person with synaesthesia,
seeing 3 as red and 5 as blue, will immediately pick out the
triangle of 3s on the right (see pages 99–100).

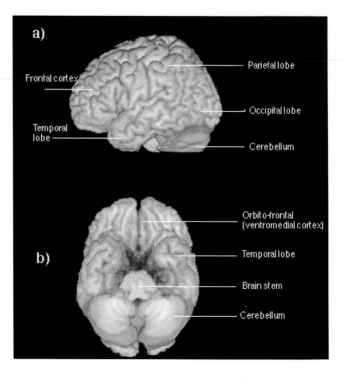

Images from MRI scans of the outside cortex of the brain:
a) the surface of the left hemisphere;
b) the brain from underneath showing both hemispheres.

An artificially coloured scan of the right hemisphere seen
from inside (medial view).

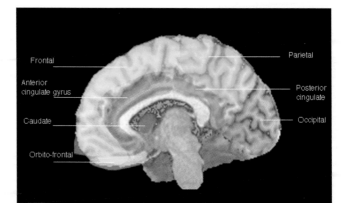

We thought that Imperial College physicists would be more introverted than Butlin's Redcoats, who relish being photographed. Introverts salivate more than extroverts (page 310), and our physicists had no difficulty licking sufficient gummed tape to seal these fruit boxes.

Building a bridge above a precipice 2000 feet up was a television metaphor for the initiation of a synapse: a) The first contact with the neuron on the other side is tentative – a grappling iron is thrown. b) I prepare to cross the synapse, hoping we've got a secure connection. c) This isn't quite as much fun as it looks. d) As more lines are brought across, our synapse becomes secure. e) Planks are laid between the ropes. f) The synapse is now so well used that our cameraman, the redoubtable Chris Hartley, can cross repeatedly and film an image.

OPPOSITE

Nick Murphy and I debate who goes into the shark cage first, as a two-ton great white hovers expectantly. Scan (a) shows a cross-section of the brain in the region of the alarm centre, the amygdala. Compare this with scan (b), which shows activity that is likely to be happening in the amygdala, and in the temporal cortex, during my frightening experience.

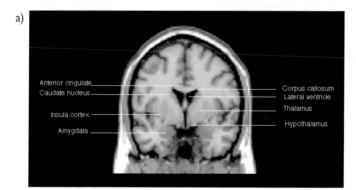

a)

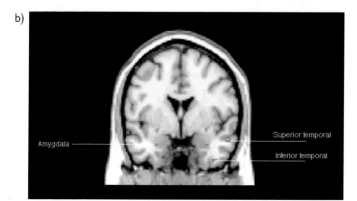

b)

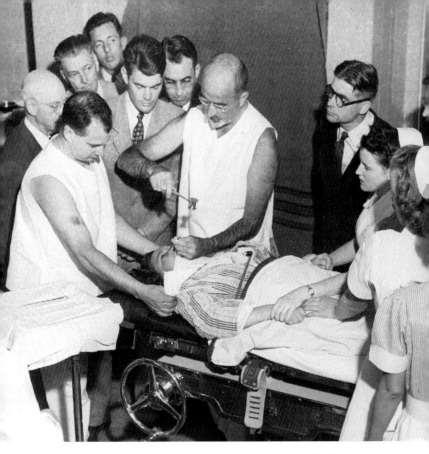

Dr Walter Freeman performing a lobotomy in 1949, in which an ice pick is inserted under the upper eyelid of the patient to cut the connections in the front part of the brain. © Bettmann/CORBIS

OPPOSITE

De Kelder cave in South Africa, inhabited by man for millions of years. Primitive man used flints as knives for well over a million years, and certainly used sticks as tools or weapons, but our hominid ancestors only relatively recently – perhaps 100,000 years ago – thought to put the two together to make something much more formidable: an axe, or a spear.

Richard Wiseman, ace psychologist and member of the Magic Circle, is able to gauge by reading vocal clues, body language and my facial changes whether I am lying about what I did last night – an increasingly useful skill in criminal investigations (see page 403).

A huge proportion of our brain is used when socializing. In fact, social interaction is one of the most taxing parts of human existence. This piece of filming, a party scene that took three days to make, was on screen for around 40 seconds. Television is a voracious beast.

strange manner, they move in odd, stylized ways, they might wear face paint, masks, peculiar clothes. Songs are sung and words uttered that are never used in daily life. Different foods are eaten. Those undergoing the ritual are assaulted by strange sights, frightening and bizarre symbols. Their senses are sometimes rendered exceptionally keen through ordeals of starvation, pain and sleeplessness. We can see this to a lesser degree in the rituals still prevalent in the Western world. A wedding involves certain restrictions – such as the belief that the groom must not see the wedding dress, or glimpse his bride-to-be on the night before the nuptials. During the ceremony, we behave in certain, restricted ways. We put on our best clothes, mothers are expected to cry, the bride and groom undergo the 'ordeal' of the marriage vows. The ceremony is made meaningful because it is different from daily life. Indeed in the Hebrew language, one aspect of the word holy or sacred – *qadosh* – has an association with the meaning 'separate', or 'set apart'.

In our pasts, and in some tribal societies today, the 'set-apartness' of the rituals is more pronounced, more extreme. And Turner's theory was that such a battery of remarkable, fear- and wonder-inducing experiences served to ensure that the learning process was thorough and unforgettable. In other words, the deliberate oddness of the ritual tagged it with strong emotions in the minds of those undergoing it. And in turn, that meant that knowledge being communicated through the ritual became deeply etched in the memory.

Turner's interest in ritual may be related to the fact that his mother was an actress, and he grew up in a family where the theatre was a feature of daily life. Some of his theories on the importance of the unusual in learning are replicated in the techniques of some dramatists. The Marxist playwright Bertolt Brecht used the same concept in his *Verfremdungstechnik*, or alienation technique.

Brecht, in accordance with his communist ideals, believed that the theatre's role was to teach. And that people learnt better, took the messages home and did not forget them if they witnessed something unusual. Brecht sought to do this by subverting the normal conventions of theatre – he kept the audience very close to the actors and it was often possible to see the stagehands moving furniture. Instead of enjoying some cosy bit of entertainment, the audience became 'alienated' from the play on the stage, not wrapped up in it. A clear example of this is seen in his musical play written with Kurt Weill, *The Rise and Fall of the City of Mahagonny*. At first it seems just like another boring night at the theatre, then suddenly a ten-ton truck roars onto the stage; the actors jump off the back of the lorry and start to build a set. I have never forgotten the production I saw aged seventeen; the novelty-detection system of the viewer's brain tags such an unusual experience with a higher than ordinary level of emotion. Thus a distinct memory is formed.

How is a memory formed inside the brain? The formation of all memories – tinged with high emotion or otherwise – takes place via a process called long-term potentiation (LTP). It seems memories are formed at the instant they happen. The first stage lasts less than a second. When a neuron receives an intense signal due to incoming information, its interior floods with calcium. This alters the balance inside the neuron, priming it to fire much more vigorously and for longer if the same pattern of input comes its way again. The calcium also encourages the neuron to make growth changes, developing new connections to other neurons.

The dendrites – where information is received by a neuron – swell in response to the calcium flood, maximizing the number of synapses with which the neuron can communicate. The swelling goes down after six hours or so, but as this occurs, the growth of new dendrites will be cementing the pattern for as much as two days after the event. Meanwhile, regular pulses of chemical activity will pass down the circuit, keeping the original experience alive and preserving its memory in the circuitry of the brain.

The LTP process can be seen all over the brain – frustrating news for those who might hope that there is a specific place where memories are formed and stored. But we can nevertheless talk about certain regions as being particularly significant for the human memory. The hippocampus is particularly endowed with LTP-responsive cells. Its location, at the back end of the sensory processing trail inside the brain, suggests that its purpose is to capture and preserve incoming information. Furthermore, the hippocampus seems to be particularly sensitive to acetylcholine, a neurotransmitter that plays a role in suppressing prior activity in order to divert and focus attention. It has been argued by the neurobiologist James Bower, formerly at the California Institute of Technology, that acetylcholine 'cleans up' each memory, suppressing the irrelevant information and preserving what is most salient.[54] It may be that when the LTP activity is particularly intense – for instance, following on from an intense emotional experience – the acetylcholine clean-up job can only be partial. That is why we can remember certain events with 'flashbulb' detail – preserving the most seemingly irrelevant information.

So we can now understand why certain events enter the memory as vivid snapshots, and others more dully or seemingly not at all. A novel and intense experience, like witnessing or playing a part in an accident, triggers a correspondingly fervent sequence of action in the brain. But an experience we undergo every day is treated with greater selectivity. When the clean-up procedure goes to work within the hippocampus, only what is most novel enters the record. If nothing about my Tuesday mid-morning coffee break is remotely noteworthy, I may not, over time, be able to recall anything about it that distinguishes it from any other mid-morning coffee break.

Hippocampus and amygdala: their role in memory

We know much about the role of the hippocampus in preserving memory from studies of people with damage to that area. In one famous case nearly fifty years ago, a man with severe epilepsy underwent brain surgery. The surgical procedure resulted in the removal of roughly two-thirds of his hippocampal tissue. Ever since then, this unfortunate man has been unable to lay down any new memories. It is hard to conceive what this must be like. One analogy is said to be that of the goldfish that never becomes bored of swimming around in a small bowl because its memory span, we are told, is no longer than seven seconds.* In normally functioning humans, our consciousness is a stream – each moment is connected to the one preceding it and to those that follow. With a damaged hippocampus, each and every moment might appear anew. Our surroundings would appear permanently strange, the people around us as eternal strangers. We would be unable to formulate and execute plans of action at even the simplest level because we would forget what our objectives were. As I write this, a team at the University of Southern California, led by Theodore Berger, are attempting to create the world's first artificial hippocampus – using information derived from millions of cross-sections of rat brains, programmed onto a silicon chip. One day, future generations may just possibly possess the

---

* There is some evidence that fish have a much longer memory than just a few seconds.

technology to replace the damaged tissue of human brains with similar devices – ensuring that the loss of faculties like memory can be overcome.

The amygdala also plays a role in memory. Here the brain 'tags' incoming information with an emotional value. The frontal cortex uses this tag to devise and direct the appropriate plans of action. If the amygdala decides to tag the job interview we are facing as 'serious', for instance, we behave in a serious manner and keep jokes to a bare minimum – if we want to be employed. But information from the amygdala also loops back into the hippocampus. So, when it comes to recalling the interview, we also recall the emotions associated with it. The more intense the emotions, or the more novel they are in the context of our everyday lives, the deeper they are etched into the memory.

The emotional content of a memory helps us to learn important messages. In our evolutionary past, this would have had an added survival benefit – directing us away from harm and towards safety, without the loftier realms of the brain's thinking, higher cortex becoming involved and slowing things down. Neuroscientist Joseph LeDoux calls this the 'quick and dirty route'. When, for instance, we see a snake – or a long, wriggling thing like a snake – the amygdala's direct connection to the processing centre in the thalamus results in an almost instantaneous perception of fear. A conscious awareness of what we have seen comes later on. LeDoux argues that human brains have evolved with this early, emotional hair-trigger because it would have been vital to us in the past. Predators like snakes strike extremely quickly. There's no benefit whatsoever in us standing round, hand on hip, fingers on lips, with our brain pondering what we might have seen. We need some mechanism that gets us out of harm's way, and fast. Thanks to the mechanism, we can go on to live another day – to survive and reproduce.

So the amygdala adds emotional colour to our memories.

But does this mean that our memories are sitting, like little cans of film, in the archive of the hippocampus? Some people believe that this seahorse-shaped nubbin of brain tissue is less like a storage unit and more of a super-efficient assembly line, which constructs memories as and when they are required. Antonio Damasio has argued that when we retrieve a memory, parcels of information come together from separate regions all over the brain. These so-called 'convergence zones' are situated next to or near the neurons that were involved in registering the initial event. Damasio has identified various regions that act as convergence zones for the uniting of sensory information about people, perceptions and emotions. The hippocampus, under this theory, would simply be the structure that brings the process together and filters out what is not required.

Damasio's view suggests that our brain does not store whole sequences of events, like reels of film, but breaks them up into their smallest constituent elements. So my memory of my first snowball fight is actually composed of little experience units, like 'wellington boots', 'cold hands', 'wet bottom', 'broken window' – which are all stored separately from each other, next to the neurons that first felt the cold, or saw the broken window. They can be used, alongside other units, to make up an endless combination of other memories – in the same way as we use combinations of sounds to make words. So the 'cold hands' unit might also go into my memories of skiing trips, 'wet bottom' might go into memories of my children when they were small. The hippocampus straddles these separate units of experience, splicing them together like a tape-editing machine to form different memories.

Episodic and procedural memory

If I suffered damage to my hippocampus, then I might expect to lose my memory of that first, ill-fated snowball fight. But what else would I forget? Interestingly, the man who had much of his hippocampus removed did not forget 'everything'. He could remember his life as a young man before the operation, and skills he learnt during that time. He could also master new skills, such as new tunes on the piano, or writing while watching his efforts in a mirror.

This is because, in the exercise of learnt skills, he was using 'procedural memory'. In patients with Alzheimer's, we can often observe the same phenomenon – they may have suffered severe damage to their hippocampal tissue and be in a state of permanent bewilderment, but at the same time, certain 'islets of ability' endure, enabling them to perform quite complex actions which they learnt in the past. Clearly, memory does not just capture events and experiences, but also skills and sequences of action.

This distinction has led some neuroscientists to discriminate between 'episodic' and 'semantic' types of memory. Episodic memory equates to what we have described above, as our memories of events and episodes. Semantic memory is the sum total of what we have learnt, distinct from how or where we learnt it, or any of the emotions associated. It seems that the hippocampus is chiefly important for episodic memory. In a case described by Endel Tulving and Gene Schachter, a patient with severe hippocampal damage was unable to recall any significant event of his own past, but his semantic memory remained intact. He could describe with accuracy and detail the route he used to walk to school, and the details of changing a tyre, even though he could not recall anything that had ever happened at school, or any time he might have changed a tyre.

This is one of the reasons why, in the murder-mystery

tale, the culprit feigning amnesia is so wont to be sniffed out by the investigating Poirot, Morse or Lord Peter Wimsey. People faking amnesia tend not to understand – having been denied access to this book – the difference between episodic and procedural memory, and thus will fake a general blotting-out of personal, autobiographical memories and learnt skills. So, for instance, they might pretend to have forgotten their names and also how to make a cup of tea – when in a genuine case, only the former would be beyond recall. They also fail to understand that in genuine cases of amnesia, the memory loss is fragmentary, and there is a continuous 'bubbling up' of little remembered details over time.

Amnesia and the case of the crime writer

Someone who became well acquainted with amnesia – to her great personal disadvantage – was the British detective writer Agatha Christie. On 3 December 1926, she disappeared from her home in Berkshire, leaving her car abandoned off-road in rural Surrey. For eleven days she remained missing, causing intense speculation in the press and a search involving police officers from three forces. Finally, on the evening of 14 December, her husband Archie – who had been having an affair, and was under suspicion of having murdered Agatha – identified a woman staying at a hotel in Harrogate as his wife.

Agatha Christie refused to speak to the press about her disappearance. Archie himself told them it had been due to an attack of amnesia. Throughout the rest of her life, the crime novelist maintained a strict silence on the subject. The press decided that the whole saga had been an elaborate stunt, designed to gain publicity for her next book. Agatha Christie, who had formerly not been shy of publicity, lived thereafter as a recluse. So what was the truth behind her disappearance? Was it, as her husband told the newspapers shortly after she went missing, that she was researching the phenomenon for a book? Was she simply trying to drum up media interest, or trying to seek revenge on her unfaithful spouse? Or did she genuinely experience a loss of memory?

One interesting indicator of the truth is that Christie signed her name in the Harrogate hotel register as a Miss Teresa Neele. The woman with whom her husband had been conducting an affair was called Nancy Neele. Some conspiracy theorists have suggested that Christie chose the name Teresa as an anagram of the word 'teaser'. Others have quite reasonably suggested that her motives might have been two-fold: a bit of publicity for the book, and one in the eye for her spouse. But Christie's subsequent withdrawal from public life might suggest that her plight was more genuine. It could be argued that she withdrew into obscurity because she was ashamed that her stunt had gone so awry. But it could equally be said that she was hurt, quite honestly offended that, instead of treating her amnesiac episode with sympathy, public opinion dismissed her as an attention-seeker.

Incidents of genuine amnesia can often involve a patchy retention of significant details. For instance, in one case, a preacher by the name of A. Bourne went missing, and was later found to be prophesying hell fire and damnation under the name of A. Brown. In another case, a doctor prompted an amnesiac woman to use a highly practised motor skill – to dial a random telephone number – and when she did so, she found herself speaking to her mother. So Agatha Christie's choice of pseudonym – a name similar to that of her husband's lover – may not have been a conscious trick, but a fragment of the truth bubbling up through her forgetfulness. Possibly it was the extreme stress surrounding the disintegration of her marriage – her husband married the real Ms Neele three weeks after divorcing Christie – which led to the bout of amnesia in the first place, and the choice of moniker was an unconscious way of signalling this.

As the case of Preacher Bourne/Brown indicates, amnesia does not result in total loss of memory. People may forget elements of their identity but not the skills they have learnt. So our memory for things we have learnt is separate and distinct from our autobiographical memories. Accordingly, there's a different type of activity within the brain as we master new skills – like riding a bike, or playing a musical instrument.

Second nature: the chemistry of procedural memory

Martha Curtis was a musical child prodigy who began playing the violin at the age of nine. She had suffered from epilepsy all her life, and as she grew, experienced seizures more and more frequently. In spite of extensive drug treatment, Martha had as many as four attacks of unconsciousness a month, sometimes collapsing on stage in front of an audience. In 1991 at the Cleveland Clinic, she had the first of three major operations. Eventually she had nearly 50 per cent of her right temporal lobe removed, a huge gamble for a musician because it is this area that is believed to be active when playing a musical instrument. To the astonishment and gratification of her surgeons, Martha's violin-playing was unaffected by the three operations. In fact she seemed to be able to master and learn pieces that she had hitherto considered too complex. Her doctors at the Cleveland Clinic concluded that, since she had been suffering from epilepsy from a very young age, Martha's brain had compensated for her faulty right temporal lobe. Being plastic at that age, it had coopted other brain areas into use, so that when Martha learnt and practised her violin-playing, the right temporal lobe was effectively bypassed.

Although Martha is clearly exceptional, every brain has a built-in element of plasticity. It adapts itself to the input. And we can see this process at work most clearly when we look at how we learn a skill. Many of us will be able to remember, with effort, the painful process of trial and error by which we first learnt to ride a bicycle, or how to drive a car. But very few of us remember this when we perform it as a matter of routine in later life. At a certain point, after repeated practice, we seem to be able to ride or drive without thinking particularly hard about what we are doing. We do it 'automatically'; it has become 'second nature'. Occasionally lorry drivers say they are unable to recall the route by which they arrived at their final destination. Driving from one city to another has become so familiar that they seem scarcely conscious of doing it.

Inside the brain, we can see a corresponding shift from the higher cortical areas of processing to the lower rungs of motor and balance as we learn and master a skill. When it is new to us, we are consciously thinking about it – remembering what to do at stage A and stage B, monitoring our progress. But over time, we start to think about it less and less, to the point where we are barely conscious. Once the procedural memory is stored in this lower level, it becomes permanent. This is why we don't forget how to swim or ride a bike, even though it has been a number of years between attempts. I find this with the physical skill of skiing. It was horrendously difficult to learn, but now comes totally naturally once a year, after a few seconds of hesitant starting. I do notice, however, that with age I have become more cautious about which slopes I tackle.

The reason for this shift from the higher conscious levels of the brain to the lower, unconscious regions is one of economy. The ever-plastic, ever-adapting brain sends procedural memories down to the basement as soon as it can, in order that it can be freed up for new tasks. If it did not, there would be a limit to what we could learn in a lifetime. And there isn't. There are limits to other elements of the equation that determine what we learn and what we don't – for instance, the degree to which a certain skill is important to us, the degree of motivation. But the human mind has the potential to continue to learn.

In the mid-1980s, some interesting experiments were done at Washington University in St Louis. Researchers made PET scans of volunteers as they read aloud a list of nouns. In a second experiment, they were asked not only to read the nouns, but to come up with a corresponding verb – for instance, to pair 'chair' with 'sit'. By comparing the PET scans, the research team, led by Marcus Raichle of the Mallinckrodt Institute of Radiology and Michael Posner of the University of Oregon, was able to isolate the area of the brain that was involved in thinking of the right verb to say.

But then the researchers noticed something new. When volunteers had been through the list a few times, the map of brain activity looked radically different. The overall pattern was one of markedly lower intensity. But paradoxically, when a word-association task was new, an area, the insula cortex, was dormant. It became active only when the task had been practised a few times. What was even more odd about it was that for years neurologists thought the insula cortex was purely involved in somatic and visceral sensation from, for example, the bowel, and taste perception – playing no part in the production or comprehension of language.

What the researchers had seen was a new task becoming a habit. The first time one of their subjects had to find an associated verb for a hitherto unseen noun, there was a raft of activity across the brain – from the cingulate cortex at the front to the cerebellum at the back. But as the skill became practised, there was a marked shrinkage of neuronal activity. The brain effectively turned down its own dimmer switch, so that it was using only the minimum areas and energy.

These results confirmed something that Richard Haier at the University of Southern California, Irvine, had also found. His work with PET scanners had shown that there was no link between the level of activity inside the brain and a person's IQ. A 'busy' brain was not, in his findings, a clever brain at all – in fact, it was rather the opposite. It seemed that a clever brain was one that could shrink back to the barest levels of energy consumption in the swiftest time – thereby freeing up its circuitry for other purposes. It's not what you know, but how quickly you can master it.

This brings me to another important aspect of memory. Because, if we can see a gradual shifting of information from the upper to the lower cortex as we commit things to memory, this implies that we are endowed with two different sorts of memory. We call this the distinction between the short and the long term.

We are all aware of this distinction in our daily lives. I might say that my knowledge of French, or of Freud, such as it is, resides in my long-term memory. My knowledge of the sentence I have just devised in my head and am now typing, on the other hand, resides in my short-term memory. The very construction of those labels expresses the chief difference between the two faculties – one holds information for a long time; the other empties itself more rapidly. Back in the 1950s, psychologist George A. Miller published a paper entitled 'The magical number seven, plus or minus two'. Within it, Miller put forward his thesis that the short-term memory of humans seems to be able to hold onto roughly seven units of information – such as words, names or numbers. And this idea is still largely accepted. The short-term memory of the average human is too small to encompass a London telephone number – one reason why there is always outrage when BT announces the arrival of yet another digit.

But the example I use above raises a question, because there are some strings of numbers I use most days – phone numbers of friends and colleagues, my PIN number for my credit card account – just as there are skills I call upon every day, such as typing or driving, which I can execute with the bare minimum of conscious effort. I don't have to keep them in the limited space of my short-term memory – regular use has ensured that they are stored in my long-term site.

So how, and why, does information transfer from one type of memory to the other? As we saw above, factors like the novelty of the incoming information and the intensity of the emotions attached to it help determine how indelibly a memory is stored. So does the frequency with which the information is encountered. I can reel off sums like 'seven times seven equals forty-nine' in an instant, not because I am actually doing the maths inside my brain but because I am revisiting a little pattern of words stored in my long-term memory, etched there after my having spent hours reciting multiplication tables, dirge-like, at school. The process of long-term potentiation has ensured that, whatever else I might fail to remember from my classes, some elementary maths will remain stuck inside my brain.

The shift from short to long term

Recent research conducted with mice and flies suggests that short-term memory uses proteins that are already present in the synapses. But in order to shift this information into the long-term memory, new proteins have to be manufactured. The creation of these proteins is controlled by a further protein called CREB. CREB would seem to be involved in many of the situations where the brain has to 'get used to' new conditions – for instance, in adjusting the body clock, or in developing a tolerance to drugs. Memory disorders are extremely common: we lose some memory with age, and the ravages of Alzheimer's disease are increasingly significant in our society. So a better understanding of the molecules involved in memory formation is of enormous importance. Many scientists, and some pharmaceutical companies, have an increasing interest in enhancing learning and memory with drugs, and so studies on CREB are a key area of research.

CREB is present in the nervous system or brain of many animals, from shellfish to humans. The fact that it has been present throughout evolution underlines its potential importance. In Los Angeles, Dr Alcino Silva has been working with mice.[55] They were placed in a chamber where they were given a small shock to the foot. Their memory was assessed later by putting the mice back in the same chamber and measuring the time they remained motionless, presumably in fear of receiving more shocks. Dr Silva then compared their response with that of mice born with a modified gene that blocked the action of CREB when they were given an oestrogen, tamoxifen.

When tamoxifen was administered to these mice before they entered the chamber, the length of time they remained motionless was unaffected two hours after shock training but was considerably shortened when measured twenty-four hours after training. This suggests that CREB-associated genes do not play a role in the creation of short-term memories but that they help in forming long-term memory, created after about a day and which can stay in the brain perhaps for ever.

Dr Leslie Ungerleider, at the American National Institute of Mental Health, used MRI to see what happened inside the brain of people as they attempted to 'hold on' to a remembered image of a face over increasingly long time periods.[56] When the delay was short, there was marked activity in the prefrontal cortex and temporal lobe. But as the delay grew longer, the temporal lobe firing shrank, leaving only the prefrontal cortex at work. In other words, Ungerleider has pinpointed what may be the site of the short-term memory – somewhere within the frontal lobes. Ungerleider's studies built upon earlier work performed with monkeys. Using electrodes or radioactive pellets to 'knock out' the hippocampus or the prefrontal lobes, the researchers were able to note the distinct roles of these areas in long- and short-term memory. A monkey with a damaged hippocampus can hold a memory for about ten seconds, but no longer. It also loses its memory if there is any distraction during this period. But a monkey with a damaged prefrontal cortex had no working memory at all. In the English language, we talk of certain ideas or information as being at the 'front of our minds', and indeed the neurological evidence does suggest that this is where the brain activity is happening.

So we recruit different areas of our brains to help us remember information in the short and long term. And a variety of factors can determine whether the information gets shunted downwards for permanent storage, or whether it remains with us for only a few hours. Michael Merzenich of the University of California at San Francisco showed that motivation plays an important part in determining how deeply a memory is implanted. He studied the brains of monkeys as they learnt to associate a certain pattern of spinning wheels with a reward of food. He discovered that when a reward was entailed, many more neurons were coopted into the learning task. The same is surely the case for human learning. One of the reasons why my classmates still remember their multiplication tables is because we were all motivated to win the weekly prize.

Studies have also shown that our recall can be quite strongly influenced by the setting in which we learn things. Alan Baddeley, of the University of Bristol, asked volunteers to learn lists of unfamiliar words while weighted down with scuba equipment at the bottom of a swimming pool. He discovered that their recall of these words was more accurate when they were tested in the same circumstances. It's possible that we could all exploit this most curious aspect of our memories – and we don't necessarily need a load of diving equipment and a swimming pool. If you sniff a powerful aroma, like lavender oil, as you learn some fact, you may be able to recall it more efficiently if you sniff the same substance again when you are tested.

Sleeping on it

The ability of our brain to store information in the long term may also be affected by what we do after the event. A joint Belgian and Canadian study headed by Pierre Maquet demonstrated that when we are trained to do a certain activity, our brain continues to repeat the related pattern of neuronal activity during rapid eye movement (REM) sleep. REM sleep is the time when people dream. The work of Avi Karni and Dov Sagi at the Weizmann Institute in Israel has further shown that REM sleep improves the learning of a task.[57] They measured the length of time it took four subjects to learn how to recognize an embedded patch of stripes within a pattern, and tested them both before and after REM sleep. If they were woken before or during REM sleep, they failed to master the task. In other experiments conducted with rats, they found that interrupting REM sleep completely thwarted the learning of a new task, yet interrupting non-REM sleep just as frequently did not. So a good night's sleep, like emotion, acts as a kind of cement for our memories.

Dr Allan Hobson's group at Harvard have recently done some interesting work that shows the potential of the 'power nap'. Apparently, even a nap improves learning. Sara Mednick and Robert Stickgold, with colleagues at Harvard, have demonstrated that midday snoozing isn't such a bad thing.[58] It reverses the morning irritation, boredom and frustration with mental tasks that we all often feel, while a 20 per cent overnight improvement in learning a motor skill is largely traceable to a late stage of sleep that early risers miss. Overall, their studies suggest that the brain uses a night's sleep to consolidate the memories of actions and skills learnt during the day. In one experiment, subjects were asked to report the horizontal or vertical orientation of three diagonal bars against a background of horizontal bars in the lower left corner of a computer screen. Their scores on the task worsened over four daily sessions. A thirty-minute nap after the second session prevented any further deterioration, and a nap lasting one hour boosted performance in the third and fourth sessions to morning levels. The researchers wondered whether this effect was due to 'burnout' of brain visual circuits involved in the task. So they engaged a fresh set of neural circuitry by switching the location of the task to the lower right corner of the computer screen for just the fourth practice session. Subjects experienced no burnout and performed about as well as they did in the first session – or after a short nap.

How might a nap help? Recordings of brain and ocular electrical activity monitored while napping revealed that longer one-hour naps contained more than four times as much deep or slow-wave sleep and rapid eye movement (REM) sleep as the half-hour naps. Since a nap hardly allows enough time for the early morning REM sleep effect to develop, it would appear that a slow-wave sleep effect is the best antidote to burnout. The researchers suggest that neural networks involved in this task are refreshed by 'mechanisms of cortical plasticity' operating during slow-wave sleep. So when the boss catches you with your feet up – with the commodity markets going haywire – you tell him or her that you are merely restoring the plasticity of your brain.

Macbeth is obsessed by sleep but the power nap might not have been ideal. Having killed Duncan and the servants while they, and everybody else, were sleeping, he longed for prolonged sleep to provide relief from the state of mental torture into which he'd flung himself. His decline, from a war hero into a paranoid coward, indicates how powerful the human mind can be, and how dangerously it can distort our view of reality. As I shall explain below, the memory is far from exempt from this. We can remember, without knowing that we are remembering. And conversely, our memories can make us think we remember things that never happened.

The buried past: unconscious and false memories

In an experiment conducted by American psychotherapists in the early 1970s, a group of strangers were placed in a room and asked to pair off. Without speaking and without thinking too deeply, their task was to find a partner who reminded them of some close member of their family, or alternatively some missing element in their family. The newly formed couples were then told to speak, to compare notes as it were, about their backgrounds. They were then told to double up once again, teaming up with other couples who they felt shared the common attributes they had just identified.

These people formed units that replicated their experiences of being within their family group. They might discover, for example, that all four of them came from families where there had been difficulties in expressing anger, or love. Or they all had come from families shaken by bereavement, divorce or abuse. Even those who hung back and were reluctant to form either initial pairs or eventual foursomes had more than just this trait of reticence in common. When they did form foursomes eventually, they discovered that each had been through early and significant experiences of rejection – been abandoned, adopted, or brought up in orphanages or foster homes. What the therapists may have been seeing was the action of unconscious memory.

The work of social psychologist Robert Zajonc has indicated that our reactions to people differ depending upon whether we have seen them before, even if we do not consciously know that we have seen them. In one fascinating experiment – in chicks, not humans – he measured pecking rates (a form of mutual recognition) in chicks who had been housed together for just sixteen hours when they were one day old.[59] The chicks had been treated with a dye colour so those chicks who were strangers to each other and those who had been companions could detect each other. The chicks could discriminate between a stranger and a previous companion within one minute of meeting. It seemed that the opportunity to peck during the initial housing together was a prerequisite in producing social discrimination. In a famous human experiment, volunteers were shown a quick-moving display of faces. Secondly, they were shown another display – containing some of the faces they had seen before and some new ones – and asked to rate their attractiveness. Although the first assembly of faces was shown so fast it would have been impossible to register them consciously, these people consistently rated the twice-viewed faces as being more attractive. Later they were exposed very briefly to a view of a particular person. They were then asked to join forces with a couple, unaware that one of the couple was the person they had glimpsed momentarily. This couple were engaged in a staged dispute and the subjects were asked with which member of the couple they agreed. The subjects consistently sided with the people they had seen before.

For a long time neurologists have thought that we can be guided by unconscious memories. People with Korsakoff's syndrome, the result of brain damage caused by long-term alcoholism, suffer from amnesia; they are unable to form or retrieve any new conscious memories, so that patients cannot remember recent past events. There is evidence that their recent experiences are not totally forgotten – they are unconsciously stored and this storage may influence their behaviour without them understanding why. In 1911, the Swiss psychologist Edouard Claparède rather unkindly hid a small pin in his hand so that when he shook hands with a woman patient with Korsakoff's disorder, he pricked her palm. When reintroduced to Claparède a few days later, the patient had no conscious recollection of having met him before but refused, nevertheless, to shake his hand, feeling that something bad might happen, although she could not explain what or why.

Years later, Antonio Damasio was treating a patient with significant damage to his hippocampus. To test whether the man could still learn new information, Damasio developed a test in which the patient was exposed to social meetings with three different people. It was known as the 'good guy/bad guy' experiment because the situations were of three types. The first was pleasant and rewarding. The second, 'control' experience was neutral. And the third experience was rather nasty – in that the poor individual found himself stuck in a room with an abrupt, unfriendly experimenter, who subjected him to an array of mindlessly enervating psychological tests.

After the five days over which these experiences took place, Damasio then showed his patient a set of photographs of different faces – which included the three different experimenters. The patient was asked questions like, 'Who, out of these photographs, would you approach if you needed help?' and 'Who do you think is your friend?' Despite claiming that he did not recognize any of the faces he was shown, the patient consistently picked the 'good guy' as a friend, as someone he would approach for help. Damasio concluded that, in spite of the obvious damage to the learning and memory structures of his brain, the patient was still able to form and access quite a different type of 'memory' – unconscious 'memories' with an emotional content. In other words, as we have seen in other contexts, we can learn with our emotions.

Brain scientists in one corner, psychoanalysts in the other

Psychoanalysis – sometimes seen as the arch-enemy of brain science – rests upon exactly this view of the workings of our memory. It argues that we consciously remember only a small portion of our past, just as we are consciously aware of only a small portion of what is going on inside our brain. Psychoanalysts believe that our actions and thought-processes are governed strongly by unconscious processes. By delving into them, by being encouraged to verbalize them, we bring them into the realms of our conscious control.

Memory has been one of the traditional battlegrounds between the psychoanalysts and the brain scientists – the extremists of the former believing that experience and environment are everything and that the only cure is through talking, and the extremists of the latter maintaining that the root of every disorder lies in brain chemistry, which could be altered through surgery and drugs. The growing understanding of the interplay between the plastic brain and the environment to which it adapts itself has not yet resulted in any long-lasting truce between the two disciplines. But in some instances, recent findings in neurological science concur with things Freud proposed over a hundred years ago.

Both Freud and Jung believed that traumatic experiences create a splitting effect within the mind – the result being that people are plunged into destructive internal conflicts between different aspects of themselves. An example might be the fate of Wilhelm Reich, a psychoanalyst who some considered a natural successor to Freud. He died in prison in the USA in the 1950s, having been hounded by the American government for supposed communist sympathies. Reich had a number of extremely strange ideas, among which was his conviction that aliens were trying to contact the earth; he may have been nearer the mark with his view that the government had mounted a conspiracy against him. Freudians would point out that Reich experienced an early trauma – having caught his mother having sex with his tutor. The young Reich informed his father, and his mother committed suicide. So, in later life, it is proposed there was a 'splitting' of Reich's psyche into good and bad aspects. What he envisaged, in the form of invading spaceships and conspiring government agents, was actually his own guilt at his actions as a boy. All this seems fanciful. Perhaps Reich had good conscious reason to feel the way he did, particularly as various letters about Reich's 'activities', signed by J. Edgar Hoover to his secret agents, are still in existence.

Curiously, research conducted at McLean Hospital in Massachusetts has indicated that the corpus callosum –

the bridge between the two halves of the brain – might be smaller in children who have been subjected to abuse or neglect at an early age. In later life this may possibly lead to reduced 'traffic' between the two hemispheres. Such people might find themselves supposedly less able to verbalize their feelings or memories, or achieve a balance between the left-brain 'thinking' and right-brain 'feeling' parts of their personalities. In other words, what we would see is an individual of more disparate parts, which might even seem to be in conflict with one another – rather as Freud suggested.

Memory – and in particular the memory of trauma – is one arena in which scientists and psychotherapists seem destined to slug it out for a long time to come. The debate has become especially contentious since a number of people undergoing therapy 'recalled' years of buried memories, including recollections of traumatic abuse. In some high-profile cases, families torn apart by abuse allegations have sought compensation from the therapists concerned. There is even a group called the False Memory Syndrome Foundation, which offers support to families who have suffered in this way. But some evidence from brain science suggests that the two phenomena might be equally possible: first, that it is possible for people to repress or 'bury' traumatic memories, and secondly, that people can have memories that are quite false.

It is clear that human memory is highly selective and subjective. Last year, Tessa Livingstone, the producer of the BBC's *Child of Our Time*, and I filmed our own experiments into false memory. We tried to deliberately create false memories from childhood. We asked a number of adults if they could remember when they were in a hot-air balloon when a child. Not surprisingly, none could – after all, balloon travel is a bit of a luxury that few of us have ever experienced, including our subjects. Using family snapshots, we then made a collage photograph of people in the basket of a hot-air balloon, pasting in child-hood snapshots of our subjects when they were children. We used Photoshop, the computer program, to do this. Many of the 'photos' we created were very crude – and to our eyes, obvious forgeries. When we then showed these photos to our adults, most looked rather doubtful and still could not recollect their 'aerial journey'. Nevertheless, within a few days a surprising number of these adults telephoned or returned to tell us to say that, on reflection, they did remember the trip, sometimes quite well. One woman even recalled that shortly after the ascent she felt so frightened that she asked to be landed on earth so that she could be let out of the basket.

We have seen already that the degree of emotion associated with an event can determine how well it is remembered. But in traumatic memories, the reverse seems to happen. People who have been through deeply traumatic experiences seem to remember only fragments, or to have blanked the whole thing out. One explanation could be that the overload of the amygdala during the original event is repeated whenever a recall of the event is triggered. This produces emotions so intense that the person involved has difficulty in consciously processing the memory, and even more difficulty in verbalizing what they are remembering. A further argument is that the stress hormone cortisol may interfere with the process of long-term potentiation.

Whatever the mechanism, the younger a person is when they experience a traumatic event, the greater the chance that they will bury some or all of the memory. But what about people who have undergone repeated trauma – such as abuse victims, or the survivors of concentration camps and 'ethnic cleansing'? All the evidence would suggest that the human mind certainly buries isolated, stand-alone events of high emotional stress, but if such experiences were repeated, then they ought to remain within the grasp of our memories. Yet the incidence of memory loss is just as high in these cases.

One answer to this may be that someone who is repeatedly traumatized becomes adept at repressing the painful memory. This presumably would offer some psychological protection. The work of Chris Brewin at University College, London offers an alternative spin on this notion.[60] His work indicates that some people who suffer persistent memories of traumatic incidents may often have a less efficient interplay between the short- and the long-term memory. Possibly, people with a more efficient mechanism are better able to suppress their traumatic long-term memories in order to get on with their daily lives.

There's another false memory experiment you can try on yourself. Read the following list of words out loud: sour, candy, sugar, bitter, good, taste, tooth, nice, pea, chocolate, cake, heart, tart. Now put this book down for a moment and try to write down as many of the words as possible. A version of this experiment was first used by memory researchers Henry Roediger and Kathleen McDermott in 1995. They wanted to see how easy it would be to induce a false memory in normal healthy subjects. So let's look now at the list you made. If it includes the word 'sweet', then you have experienced a false memory. But it is easy to see how the conditions were favourable to you coming up with this word – because all the words in the list were related to it.

Stranger still, our brain seems to know a false memory from a true one, even if we don't realize it. Daniel Schachter at Harvard performed PET scans on volunteers as they looked at lists of words, some of which they had seen before, others of which were new to them. As they looked at each list, the subjects were asked to say whether they had seen certain of them before. Lists that the subjects had seen before activated the hippocampus and the language centres of the brain. However, lists that the subjects thought they had seen, but in fact had not, also activated the orbitofrontal cortex. This area would seem to be a sort of anomaly detection system – pointing out when something is 'not quite right'. Studies at the University of California, Berkeley, have shown that people with damage to the orbitofrontal cortex may lack an ability to regulate their own behaviour – telling inappropriate jokes, acting overly familiar with strangers and using crude language.

One factor that influences how accurately we remember things is consistency. Studies have shown that we revise our memories in order to make them fit in with what we later know. Michael Conway of the University of Durham asked a group of student volunteers to take a 'study-skills' course that claimed to boost learning and recall. Compared to students who had not undertaken the course, they showed no improvement – indeed, their final exam performance was slightly worse. But all the students reported that the study-skills course had been helpful to them. There is an explanation for this. At the outset of the course, the students were asked to rate their studying skills. They were also asked to do so at the end of the course, and to recall how they had originally rated themselves. In describing their earlier performance, the students recalled themselves as significantly worse than they had actually been. In other words, they had revised their memories to be consistent with the view that, since taking the course, they had become better at studying.

Losing it: ageing and memory

Parents and teachers will know something of the scorn young people sometimes show whenever their elders have a lapse of memory. As we advance through life, we may

299

become annoyed at ourselves as it becomes increasingly hard to remember to post a letter or where we put our car keys. Many people begin to worry quite seriously at this stage, seeing this forgetfulness not just as a sign of ageing, but as a possible indication of Alzheimer's disease. In actual fact, the patterns of memory loss are different.

For most of us, the problem is not that the memory has been wiped clean, but that we experience difficulty in retrieving it. Given the right cues, or even equipped with our own method of remembering, we can come up with the right answer. Research has indicated that older people can perform just as well as younger counterparts in memory and cognitive tests, but they respond better to comfortable conditions and a relaxed schedule. Given the pressures of time and stresses in the environment, their ability becomes impaired.

Learning 'how to remember' can also be particularly effective for older people. Making strange or unfamiliar associations can help to nudge memories into the light. For instance, I have difficulty remembering my car numberplate, LD52ZIR. I recall it by associating 'LD' with a Lethal Dose, which is just over 50 per cent of a rare chemical compound, ZIR. Incidentally, my rare chemical 'ZIR' does not exist.

We filmed Andi Bell, who is the reigning World Memory Champion, for the television series accompanying this book. Andi uses a combination of the association technique described above and another trick called the 'Method of Loci' to perform his memory feats. On the day we filmed him, he memorized the entire order of ten decks of cards in twenty minutes. On the face of it, this seems an impossible feat. He then explained to us exactly what he was doing.

First, he was marrying each card to a particular pictorial image in his mind's eye. For instance, the three of clubs was associated with Mickey Mouse, because of its resemblance to the cartoon character's prominent black ears. The use of imagery helps with recall because it adds an additional trigger-facility – a little 'hook', if you like, that enables us to fish the memory out with greater ease. Studies have shown two further things: that a new memory linked to an older, familiar one has greater ease of recall. Secondly, the more bizarre the association between the old memory and the new one, the better the likelihood of recall. As we know, the human brain is attracted to the unusual and the novel.

The second stage of Andi Bell's memorizing process is to attach each of the pictorial images to points along a memorized route. This is what we mean by the Method of Loci – which is thought to have been devised by Simonides in 500BC. Invited to give a speech at the dinner party of a Greek grandee, Simonides temporarily left the building – perhaps, like most public speakers from time to time, he forgot his notes – or possibly he just wanted to get away from the banal conversation. While he was outside, the building fell down, killing all the guests and crushing them so badly that identification was impossible. Simonides proved invaluable to the concerned relatives because he was able to remember where each guest had been sitting around the dining table. Impressed by his own powers, he devised a technique whereby he visualized a room he knew well in great detail, and then placed items he wished to remember within certain locations around that room.

Andi Bell uses the streets of London as his particular 'memory room', and he spends a lot of time pounding the highways and byways of the capital, carefully memorizing features and landmarks along the route. His method not only gives him another trigger-device, but also ensures that he gets things in the right order.

The World Memory Championship, held each year in Britain, involves the competitors learning lists of thousands of words and random numbers, and quoting at length from novels and poems. Dr Eleanor Maguire of University College, London, decided to test the lead competitors to see how their brains differed from those of ordinary folk. She and her colleagues took the eight finalists in the Memory Championship, and two other volunteers who had been studied before and found to have an unusual ability to remember things. Dr Maguire's team then compared this group of memory champions with ten other people whose education and job were comparable. This control group did not profess to having an especially good memory but were regarded as average. All had at least two to three hours of intelligence tests and other mental assessments, as well as MRI and fMRI scans. Maguire found that there were no physical differences in the brains of these memory champs, but they all used the techniques similar to those I have described. 'Although more research is needed, it may be that we all have the potential and neural capacity to improve our memories,' says Eleanor Maguire. So a better memory may be a question of practice using the special techniques I have described.

A research study in 1968 by Ross and Lawrence showed that undergraduates taught to use the Method of Loci were able to recall 38 out of 40 items on a list. Twenty-four hours later, they could still recall 34 items. Other studies have suggested that the Method of Loci can increase recall by between two and seven times.

I thought it might be interesting to put Andi Bell's methods to the test. With the camera crew filming, I tried to learn a list of thirty unrelated words in four minutes, by a combination of what I call marriage and burial. That is, I first 'married' each word to an image, and then I 'buried' it in a specific place within an imagined familiar location – my house in London. As I moved in my mind along a specific route around my house, I married and buried more of the words. For instance (words to be learnt in capitals):

WASHing an APPLE in the garden

A JET flying round MADLY in the hall

In the dining room an ACTRESS is smeared in MARMALADE

They tested me a few hours later, and I showed 70 per cent recall. This did not seem a satisfactory result to me so I tried to develop my own strategies. I was then able to recall another list of thirty random words in the right order after just three minutes of attempting to memorize them. Using the rhythm they made in my mind and associating them musically, I had 100 per cent recall. So there may be different methods we can use to enhance our minds and different techniques may help different subjects.

As I stated earlier, memory problems are not a sure-fire indicator of senile dementia. In fact, the statistics indicate that no more than 10–15 per cent of the population will develop Alzheimer's disease. And its effects on the memory are different and unmistakable. People with normal age-related memory loss tend to experience an increase in those 'tip of the tongue' memory moments, where they know they know something, but cannot quite recall it – usually because it is not an item they need to recall every day. People with Alzheimer's disease, in contrast, forget the names and functions of objects they encounter in their daily lives.

Exercise the brain

It was once believed that we are born with a finite number of neurons, and that from birth this declines. Nowadays it is understood that the actual volume of tissue lost due to ageing is very small. What does suffer significant loss is the basal forebrain, the area that supplies acetylcholine to the hippocampus. It is argued that this affects the plasticity of the brain, making it increasingly hard for the brain of an older person to adapt to changing circumstances. A key protein, called PP1, in mice – which seems to have a role in deleting memories – has recently been identified.[61] By blocking its action, researchers found that they could boost the learning and memory capacities of mice. The American National Institute on Aging (NIA) has also indicated that hormones play a role in age-related memory loss. Older men with higher levels of testosterone perform better at visual and verbal recall. A similar process seems to be at work for their wives – post-menopausal women who received oestrogen replacement therapy were able to maintain their verbal memory and proved better at learning new facts. This is particularly interesting because oestrogens, as we saw with the CREB-deficient mice, may play a part in the action of this protein.

Randy Buckner and his team at Washington University, St Louis, have shown that there are two basic types of age-related memory deficiency – so-called 'under-recruitment' and 'non-selective recruitment'.[62] When there is 'under-recruitment', older people are less able to use the specific areas of the brain that usually aid memory. 'Non-selective recruitment' is when older subjects tend to recruit parts of their brain that are not useful in tasks that require memory processing.

In an important scanning study in 2002, Dr Buckner compared activity in the brain of people in their twenties with healthy adults aged seventy to eighty. Each was asked to memorize a series of words while being scanned by fMRI. Their brain blood flow was measured in three specific areas of the frontal cortex. In the first study, the younger and older adults were shown words and asked to try to remember them for later. Dr Buckner says, 'As previous studies had shown, we confirmed that the older adults did not recruit the critical frontal regions as much as the younger adults.' The older people also showed non-selective recruitment of parts of the cortex that would not aid memory processing.

Dr Buckner then tried to see if the problem could be resolved by helping the older people with a trick that would recruit the appropriate frontal cortex. This time he gave them a list of words and asked them to put them in categories – such as whether the word represented something 'concrete or 'abstract'. When they did that, the older adults showed increased activity in the appropriate frontal regions, and their memory performance improved. This approach helps under-recruitment, but not, it seems, non-selective recruitment.

The increased blood flow they identified reflects greater activity in regions of the brain that are used during the tasks. Of the three areas of the frontal cortex Dr Buckner and colleagues imaged, two were in the left hemisphere and one on the right. The left hemisphere is generally dominant in language processing, so contains regions that you would expect to show increased activity while memorizing words; conversely, the right hemisphere should remain inactive. The third area the researchers studied was the prefrontal cortex, highly concerned with memorizing verbal material.

Aids to memory (such as putting words in categories) had no effect on non-selective recruitment in the older people. The young men and women used the left frontal area when they memorized words; that, it seems to Dr Buckner, streamlines their information processing. But older people used both left and right sides, even when they used strategies to improve their memories. What is most encouraging about these studies is that these healthy old people still had parts of their brain they could recruit to solve tasks – the frontal cortex on the 'wrong side' had not atrophied or degenerated, it was just 'under-recruited'. So the reduction in processing capacity of the brain with ageing is not completely irreversible. It is likely that rehabilitation therapy to improve cognition might be helpful to keep healthy older people living an active, independent life.

The next important area for this research is to see what happens in unhealthy older people – those with conditions like Alzheimer's disease. Currently, it is unclear whether there are cognitive tricks that could possibly be used by older people which might result in better retention of memory and slowing down the processes of brain deterioration.

A older study, undertaken in 1993 by Richard Mohs and team at Yale School of Medicine, showed that people with more education showed less age-related memory loss.[63] Years of schooling equip people with a sort of 'memory toolkit'. It gives them better methods to learn and recall data. Efficient learners are on the lookout for patterns and make increased use of tricks like categorization.

But there is no reason to suspect that you are doomed to lose your memory because you left school at the first opportunity. Even a daily walk to the shops can do the brain some good – research has shown that when adults aged sixty to seventy-five perform a gentle walking programme, their memory, judgement and planning abilities improve compared to adults who do other forms of exercise. And what does your brain think of your morning jog? At Nihon Fukushi University in Aichi, Japan, researchers studied seven healthy students who jogged a prescribed course for thirty minutes a day, three times a week, for twelve weeks. These students were tested for specific aspects of mental ability on three occasions during the three-month period. For comparison, seven sedentary non-jogging students did similar tests. After the twelfth week of the study, the joggers scored nearly 30 per cent better than the couch potatoes. There are a number of other studies that suggest that regular physical exercise enhances memory. Studies using MRI scanning might help to evaluate any brain changes due to exercise, though it seems that in general, physical exercise increases brain blood flow quite substantially. The evidence would suggest that this is valuable in older people, too. It is never too late to start exercising your brain.

chapter**seven**

A question of character

When Philip Larkin returns to his old Oxford college, he learns that his room-mate from the past now has a son who is old enough to be an undergraduate himself. This is quite an unwelcome surprise and it leads to poignant consideration of his own existence – what is it that has made the lives of two close friends, contemporaries at the most formative stage of their life, end up travelling down such divergent paths? Larkin, a solitary, never married and never had children. The theme of needing human company, but also feeling suffocated by it, runs through his poetry.

So what made Larkin tread his path? To some extent, our lives' journeys are dependent upon chance. The right job offer here, the sudden illness or little windfall there. In Larkin's case, he always maintained that he fell into librarianship because the War Ministry wrote him a stern letter, shortly after he had left university, asking him what he intended to do with himself. Exempted from military service because of his poor eyesight and fearing a life of dull oblivion in the civil service, he responded to the first advert he saw of a job that seemed bearable. It just happened to be a sub-librarian's post in Wellington, Shropshire. But is that really all there is to the matter?

> . . . Where do these
> Innate assumptions come from? Not from what
> We think truest, or most want to do:
> Those warp tight-shut, like doors. They're more a style
> Our lives bring with them: habit for a while,
> Suddenly they harden into all we've got.

'Dockery and Son', 1963, Philip Larkin, *Collected Poems*
(Faber and Faber 1988)

Our personality is the driving force behind the experiences we undergo and the evidence suggests that chance may play only a small role. Some brain and psychological research strongly points to the fact that many of the key elements of our personality are present from birth, identifiable in the way our brain responds to the world. And they are certainly firmly, and largely irrevocably, established by the time we are of university age. I am convinced that this is the case with my own younger siblings, and certainly with my own children. The artistic child drew exquisite drawings and showed a visual sense of the colour of the world from the age of three. My most patient offspring stood in his drenched cot aged one when a burst pipe flooded his room at night; when we opened the door in the morning there was a small, cold, water-logged figure, in a totally sodden Babygro with hair plastered around his head, standing in a puddle of cold water – not crying, but simply waiting to be rescued. And the performer, well, he danced in the streets when he heard buskers, strutted even harder when he noticed people watching and was positively flamboyant to attract their attention if he thought they'd stopped watching. And each from his or her first year of life showed a different aspect of reasoning and sociability – aspects which to my mind are clearly identifiable today. So, in other words, Philip Larkin didn't have quite as much choice about becoming a librarian and a poet as he thought – his brain probably made him do it.

The licking test

For the BBC television series of *The Human Mind*, we undertook a little test in London's main fruit market, Spitalfields. After getting our victims to fill in a simple personality questionnaire, we asked two groups of people – a platoon of student physicists at Imperial College, and a troupe of Butlin's hospitality staff – to compete in licking yards of two-inch-wide gummed tape. This was the competition between the White Coats and the Redcoats. You will not be surprised to learn that Imperial College physicists tend to be serious introverts. They spent most of the time in the market looking as if they would rather be in a darkened room. The Butlin's Redcoats, on the other hand, looked as if they had momentarily come in off the beach, and were clearly extremely extroverted. As you can see from the photograph in the plate section, the physicists found it well-nigh impossible even to smile at their predicament, while the Redcoats were grinning from ear to ear and competing with each other to get on camera. Various psychological studies under controlled conditions have shown that introverts produce much more saliva than extroverts. So, after a quick splash of lemon juice on each contestant's tongue, we expected our physics students to be able to lick far more packing tape to encircle boxes full of oranges than the lively Butlin's group – even though, being extrovert, the Redcoats were certainly trying and competing much harder.

The British psychologist Hans Eysenck was among the first to categorize personality traits. He defined what he felt were a number of specific, different human personality types, and believed that these types could be linked to specific biological traits. The degree to which a person showed extroversion or introversion, he maintained, was linked to levels of arousal in the brain. In others words – guess what – our brain chemistry influences who we are.

It is worthwhile making a pit-stop here to briefly discuss what I mean by terms like 'personality' and 'extrovert'. 'Personality' in my definition is typified by the characteristics, emotional responses, thoughts and types of behaviour that are constant over time and circumstances in a particular individual. Eysenck devised a scale of personality upon which there were three extremities: Extroversion (E), Neuroticism (N) and Psychoticism (P). Character traits associated with type E might be, for instance, sociability and risk-taking. Traits associated with N might be guilt and excessive concerns about health. P types might be more likely to be aggressive and impulsive. Psychologists, over time, have improved on Eysenck's model – most notably Drs McCrae and Costa with their 'Big Five' scale of personality type.[64] These are:

OPENNESS TO EXPERIENCE
imaginative vs down to earth
variety vs routine
independent vs conforming

CONSCIENTIOUSNESS
organized vs disorganized
careful vs careless
self-disciplined vs weak-willed

EXTROVERSION
social vs retiring
fun-loving vs sober
affectionate vs reserved

AGREEABLENESS	soft-hearted *vs* ruthless trusting *vs* suspicious helpful *vs* uncooperative
NEUROTICISM	worried *vs* calm insecure *vs* secure self-pitying *vs* self-satisfied[65]

Eysenck asserted that people with an extrovert personality are continually in search of cortical arousal. They are more ready to go to parties to meet new people, they seek out new experiences such as adventure sports or foreign travel and, like our Redcoats, are likely to seek occupations that engage or entertain other people. They enjoy and value close bonds with others, being warm and affectionate. They also tend to be dominant socially, enjoying leadership, and are frequently assertive. Extroverts are much more ready to be exhibitionists, and have a sense of potency in accomplishing goals – they are more likely to become inmates of the Big Brother house. Introvert people, by contrast, prefer to stay alone, or in small groups of familiar people and in familiar surroundings. They are much less likely to act on impulse and generally take a cool, sober look at life. Like our Imperial College physicists, they may be most happy in a deserted basement laboratory on a sunny Bank Holiday weekend. Eysenck argued that introverts already have a predisposition to be highly aroused. Introverts don't seek out additional arousal and if, for example, human company is thrust upon them, they experience what is to them an unpleasant level of arousal overload.

The theory is that their extrovert counterparts are chronically underaroused. They have to go rooting out experiences that will provide the 'reward' of arousal; they are quite frequently in pursuit of a 'fix'. Philip Larkin, then, was on this reckoning an introvert with an inbuilt 'desire' to avoid cortical arousal. Poems like his 'Vers de Société' (1971) beginning with the lines:

> My wife and I have asked a crowd of craps
> To come and waste their time and ours

suggest that he viewed social occasions as a form of torture.

Hence our Spitalfields lemon test. The introverts salivated more because they were more sensitive to arousal. An extrovert, with his chronically low levels of cortical arousal, would need to seek out more than a drop of lemon juice to get his brain's reward centres going. Accordingly, when we filmed the test, we found that the introverts were able to lick at least 10 per cent more packing tape than the extroverts, and sometimes considerably more.

The motivation factor

Other scientists since Eysenck have attempted to add a degree of refinement to this scheme. For Professor Jeffrey Gray, who occupied the professorial chair in London previously held by Eysenck, the degree of cortical arousal is just one factor. Gray suggested there are two opposing systems within the human personality. The *behavioural approach system* (BAS) consists of those brain structures – such as the hypothalamus and amygdala – that lead us to approach certain stimuli in search of a reward. For instance, we might approach attractive people at a party in pursuit of flirtation, flattery, acceptance and sex.

Opposite it is the *behavioural inhibition system* (BIS), linked to the frontal lobes, which is sensitive to punishment and therefore controls and suppresses behaviour that might lead to danger or pain. So the BIS makes us stop at home to check our bank balance and tax liability while the BAS inclines us to the spending spree or an evening in the pub with our friends when we should be working for an exam. In this model, the extrovert is simply someone whose BAS is stronger than their BIS. Introverts, in contrast, have a stronger BIS than a BAS, and hence they flinch from experiences, for fear of a possible negative outcome.

This idea is supported by some work with PET scans which has shown that frontal lobe activity seems slightly greater in many people with introvert characteristics – in other words, their brain predisposes them to be more in control of their behaviour.[66] Dr Debra L. Johnson's study at the University of Iowa showed that introverts appeared to have heightened activity in both the frontal lobes and the anterior thalamus – an interplay possibly crucial for problem-solving and making plans. Extroverts, on the other hand, seem to show more activity in the anterior cingulate cortex, temporal lobes and posterior thalamus – structures involved in sensory processing. Perhaps this could help explain these differences in personality.

The 'social' element of this equation is interesting, because it suggests that we all treat other humans at some level as 'raw data', to be avoided or sought out depending on our degree of extroversion or introversion. A study by Turhan Canli and John Gabrieli at New York University looked at the response of the amygdala of introverts and extroverts when they were shown happy and sad faces.[67] Extroverts showed increased activity in the amygdala when they were shown happy faces, but both types responded with equal intensity to sad faces. Extroverts, in their search for reward and/or arousal, possibly understand that they will get more of the desired result from a happy person. Perhaps we also need to ask why, if – as Gray's idea suggests – extroverts seek reward rather than thinking of possible consequences, they are not equally excited by the sight of any human face, happy or sad, regardless of the possible consequences of contact with its owner? Perhaps the extrovert brain is more discriminating because of its increased need for arousal.

Dopamine's influence

Sensitivity to dopamine – the neurotransmitter involved in the brain's internal reward system – seems to play a part in determining whether we are extroverts or introverts. Extroverts seem particularly sensitive to the action of dopamine, especially in the nucleus accumbens area of the brain. As mentioned in Chapter 2, people who exhibit thrill-seeking behaviour (something that often comes with the extrovert personality) can have a mutation of a gene which results in dopamine being prevented from bonding to neurons. However, it remains highly controversial whether this dopamine gene is seriously altered in people who are outgoing and risk-taking extroverts.

To use a metaphor, dopamine is the currency that is fed into the slots of the nucleus accumbens – the fruit machine – in order to obtain the jackpot: the feelings of pleasure. Only, the theory goes, with extroverts, there is less money to play with and accordingly less stimulation. Consistently under-rewarded by their environment, they are compelled to go in constant search of ever-increasing sources of stimulation. So they engage in more social activities, more novel and unusual situations and sometimes more cocaine or gambling.

It has been suggested that people who exhibit criminal behaviour are more likely to have the mutation that prevents dopamine bonding. This is controversial but the notion ties in with our general understanding of reward centres in the brain. Extroverts certainly show a tendency to think about the rewards rather than the possible negative consequences of their actions. Nonetheless, it is a long way from finding a genetic link. Whatever the truth about the genetic background, professionals working in the courts and prison systems note that many of the offenders they deal with on a daily basis are not so much 'bad' as unable to see a link between what they do and what happens to them. The explanation 'I didn't know I'd get caught' is trotted out more often than we might think. It may be worth pausing to consider that activities such as scuba diving, bungee jumping and sky diving are excellent ways for the executive thrill-seeker to get his dopamine fix at the weekends. But they are options unavailable to people on lower incomes. Perhaps we should not be entirely surprised then that joy riding, petty theft and drug abuse are crimes more typically associated with people from poorer backgrounds. They might be less the strategies for personal gain or revenge on society and more the negative consequences of the dopamine-deficient brain's hungry quest for arousal.

Testosterone's influence

Testosterone is another substance that has powerful effects upon our character. For instance, girls born to mothers who have high levels of testosterone exhibit more male-typical 'tomboyish' behavioural traits like tree climbing, and a liking for competitive games. This is because testosterone is linked to increased levels of curiosity and aggression. A key difference might be whether you view the prospect of a white-water rafting trip as exciting or terrifying. If it's the former, then you may have high levels of testosterone.

A famous French experiment involved injecting one group of cows with testosterone and comparing their behaviour to that of uninjected beasts.[68] Cows in general are relatively timid animals and most breeds are not particularly adventurous. They often show various signs of fear if made to enter unfamiliar areas. They hover around the entrance, staying immobile and hugging the walls. All this is quite familiar – I show these traits myself at most parties. When cows are faced with novel objects, frightened animals delay rather longer before they approach them. They also spend less time studying unfamiliar objects. Fearful cows tend to feed less, become unsettled more easily and run further distances from a possible threat. So we can measure the anxiety levels of individual animals by analysing these reactions.

The French scientists confirmed that, with a good dose of testosterone, heifers on steroids entered unfamiliar fields or pens faster, spent less time in the doorway, kept their heads high indicating they did not feel threatened, and hugged the walls of a barn less. They fed for longer, were more ready to explore a maze and were less frightened by humans. In another experiment, the researchers looked at two different breeds of cow.[69] Heren cows, which are bred for their fighting ability and have naturally high levels of the hormone, were significantly less fearful and more ready to explore their environment than were Brune des Alpes cows, a dairy breed. Brune des Alpes cows have average blood levels of testosterone – that is, lower than the Heren cows – and in this research were more easily stressed.

It is inaccurate to think of testosterone as 'the male hormone'. It is produced by the testis in large quantities, but it is a hormone important for the function of the ovary, too. And androgens, the 'parent' hormones that give rise to testosterone, are produced by two small glands next to the kidneys, the adrenals, in both men and women. But excess testosterone produces characteristics in the body, brain and behaviour that we might call 'male-typical'. At the purely physical level, these would include a low degree of facial fat, resulting in a pronounced jaw and prominent cheekbones, a deep voice, large hands and greater upper-body muscle strength. At the psychological level, we also see heightened curiosity and sex drive, aggression and competitive behaviour.

Rather than seeing the profound differences between the two sexes as two opposing sides of a coin, we are better off considering the gender 'divide' as being more a continuous line of varying states, a spectrum. A tendency towards one or other end of this continuum has little to do with sexual orientation. No one has yet found any definite link between heterosexuality or homosexuality and increased or decreased testosterone levels. Yet differences in personality may still be linked to the levels of certain chemicals sloshing around inside our brains and bodies at some point during our past development or present activity.

It seems that differences in hormone levels can clearly hold sway over whole aspects of our character and the lives we lead – influencing the jobs we do, our choice of partner and our hobbies. But not all movie stuntmen have high testosterone, and not all librarians have it in short supply. The functions of testosterone are many and the effects various. I have already pointed out that older men with high levels of the hormone tend to show better visual and verbal recall. The various studies that have been done give inconsistent and variable results, but Daryl O'Connor at the University of Manchester was able to improve the verbal fluency of a group of men by giving them weekly injections of testosterone.[70] After just four weeks, they showed a 20 per cent improvement, but they also performed worse with a spatial task, which involved putting a peg into a specific place on a board. So while testosterone turned them into chatterboxes it isn't terribly clear whether they were better at washing up and ironing – and certainly not better with a clothesline. Personally, I find this experiment slightly surprising – not least because in my experience it is extremely hard to find men who are prepared to have injections at all, let alone with a substance that might shrink their testicles. But apparently the research was originally geared to trying to find a male contraceptive hormone – it may surprise you that giving men testosterone also tends to reduce testicular activity and to lower sperm count. This is because if testosterone is injected into a man it interferes with the natural stimulus to the hypothalamus – a feedback mechanism. The hypothalamus detects the testosterone and is 'fooled' into believing that this hormone is coming from the testes. Therefore it reduces its activity and stops stimulating the pituitary. In turn, the pituitary gland reduces its output, which normally stimulates the testes or ovaries. Without adequate pituitary stimulation, the testes tend to reduce activity and to shrink.

The Winnie-the-Pooh test

We can find brain-based evidence for other traits of personality besides extroversion and introversion. Research indicates that people with a tendency to be happy show increased left-brain activity, and their sad counterparts have heightened activity in the right. This link can also be seen in the cases of people who have suffered damage to one or other hemisphere. People with damaged right brains can be prone to bursts of laughter and euphoria, whereas people with damaged left brains may be given to crying or feelings of despair. And if a sequence of happy and sad video clips is shown to normal, well-adjusted people, you may observe the varying action of the left and right brain. Generally, unpleasant sights are a bit more likely to increase right-brain activity, and pleasant ones left-brain activity. It has even been suggested that people with more active left brains are better able to cope with the physical effects of stress.

This aspect of the brain's wiring may influence the way people think. Therapists maintain that anxious people tend to link their problems together into an overarching schema of doom. When the red reminder bill plops onto the doormat, they do not just register it as a problem to be solved. They also feel heightened regret that they have not been promoted, concern about their recent row with a close friend, worried that they've gained a little weight recently. They tend to link each separate islet of concern into a wider context which reinforces their sense of guilt and feelings of inadequacy. 'I am all wrong', such a person might conclude, 'because of this red reminder bill and this and this and this.' As we have seen, the right brain may possibly be better at grasping the bigger picture and may form wide conclusions from a disparate array of inform-ation. A therapist's rationale, based on evidence or not, might be to try to train the individual concerned to use a 'more left brain' approach. To learn to treat each problem as a separate entity, rather than thinking about the sum total of all the problems.

Philip Larkin showed a tendency that might be thought a 'right-brained' view of life. He once said that sadness was to him 'what daffodils were to Wordsworth', and that he believed 'most people are unhappy'.[71] Most of us would agree that such an outlook leads to a rather painful passage through life, but many creative people – including Larkin himself – would also argue that it is precisely this disposition that enables them to cut through the trivia of daily life and shed light on the reality beyond, however uncomfortable that may be. As Montherlant once wrote, 'happiness writes white' – meaning that only bland statements proceed from a state of joy.

To go from the subdued to the sublime, it should be obvious that if all the categories outlined by Eysenck and modified by McCrae and Costa are accurate, they should pass the Winnie-the-Pooh test. Either that or, heaven forbid, A.A. Milne, generous benefactor of the Garrick Club in London, got it all wrong. So – on these definitions – we have Tigger, sociable, fun-loving, wanting to try new things, affectionate and clearly an *extrovert*, with Owl, retiring, sober, distant and reserved, an *introvert*.

On the *openness to experience* scale, Rabbit is down to earth, routine, conforming, while Roo is independent, loves variety and shows the start of imagination. On the *conscientiousness* scale, Kanga is the most organized, careful and self-disciplined; Tigger on the other hand (who, of course, is also an extrovert par excellence) is the least conscientious – disorganized, careless and self-willed. It does not take much intelligence to see who I would elect to the *agreeableness* slot. Pooh is soft-hearted, trusting and helpful; probably there is no animal in the Hundred Acre Wood who is ruthless, suspicious and uncooperative – after all, Owl, who is perhaps closest, does give Eeyore's tail back without a fight.* And the *neurotic* – well, as it happens, the most intelligent animal, the introverted Eeyore, of course, is worried, insecure and self-pitying. Presumably Christopher Robin himself would be the calm, secure and self-satisfied one – but he does not really count as he is away so often. That really leaves only Piglet, but Piglet is still very young indeed – hardly yet self-aware. One may hope that with Pooh's influence he may, in time, emulate

* There is a point about Owl here. You will recall that, outside his house, he has a notice that reads 'PLEZ CNOKE IF AN ANSR IS NOT REQUID'. Believe it or not, there is a very similar notice on an important house in London – the Speaker's House in the House of Commons – which says: 'PLEASE DO NOT RING IF ANSWER IS REQUIRED'. I leave you to work that out.

him – a triumph of environment over genetics.

But of course, while we think of people as types – and hence A.A. Milne's delightful analysis – we have all these attributes in varying degrees. All of us will score somewhere along each scale. Moreover, the degree to which we score will change to some extent, depending on the environment around and within us and our mood at the time.

Left-brainers and right-brainers

A simple search on the internet can furnish you with a list of hundreds of famous personalities, fictional and otherwise, long-dead and living, who suffer from a seemingly gloomy view of life. The same does not seem to be true of the pathological optimist. Arguably, fictional characters who always look on the bright side start to grate on us after a while, seeming not only bland but even stupid and irritating.

A good example would be the main character in Eleanor H. Porter's children's book *Pollyanna*. Nowadays we use the term 'a Pollyanna' somewhat pejoratively to imply that someone with irrepressible cheerfulness might also be a bit naive and simple. From a psychological point of view, perhaps we are just jealous of people with that peculiar state of mind. What is remarkable about men like Philip Larkin, Winston Churchill and Tennyson is that they achieved greatness while apparently struggling against a state of mind (and brain) that told them to give up, that often affirmed everything was pointless, that frequently filled them with despair and dread.

Pollyanna, as you might recall, is a story about a little girl. I do not know whether I find the story more irritating than the character of the little girl, but Pollyanna always saw the good side of people, and her positive personality meant that eventually things always turned out for the best. *Pollyanna*, a moral fable of a certain American type, was published in 1913 and, for reasons I find totally mystifying, received immediate praise and success. The story is that the orphaned Pollyanna arrives in Beldingsville to live with her remote but dutiful spinster aunt. In spite of the lukewarm reception she experiences, everyone with whom she comes into contact is enlivened by her cheerful and infectious optimism. The miserable, the isolated, the sick and the awful – all are entranced by this little girl's enthusiasm and zest for life. But Pollyanna is then hit by a motorcar in the street (pretty advanced for 1913) and she loses consciousness as a result of her head injury. As she gets better she finds that she has lost the use of her legs, and learns that she will never walk again by overhearing a conversation between her aunt and a visiting medical specialist. A bit of a specialist myself, I can say that this is not the only fallible medical opinion ever given. She eventually does find a way to regain her past happiness – but I really won't expect you to read the book. Basically, Pollyanna appears to devise a sort of homegrown version of Cognitive Behavioural Therapy, called the 'glad game'. She employs this mental strategy in order to try to find something, in every situation, no matter how sad, that could make her happy.

I do not know what the cynical, child-hating Philip Larkin might have written about *Pollyanna* – I would be bemused if he ever read it. But a study in 1988 suggested that Pollyanna-type thinking, as well as being delusional, is a good strategy for mental health. Shelley E. Taylor of the University of California, Los Angeles, noted that people with the most robust mental health also appeared to demonstrate three types of illusory thinking.[72] They rated themselves in an unrealistically positive light relative to other people. They believed themselves to have more control over their lives than they actually do. And they had an unrealistically positive expectation of the outcome of events. By contrast, people with poor mental health showed a correspondingly more accurate and realistic attitude.[73]

This suggests that mentally healthy people maintain their illusions by actively filtering out information that would challenge their unrealistic views. There is an obvious difference between a mental illness like depression and a tendency to view things in a negative light. But presumably a tendency to be negative could lead to clinical depression. This stuff warns against any oversimple notion of a world divided into happy left-brainers and gloomy right-brainers. In light of Dr Taylor's findings, a happy brain seems to be one whose owner acts rather like an unprofessional scientist who is determined to prove his hypothesis regardless of the data to hand. He skips gaily over the fine details, ignoring an inconsistency here, placing undue emphasis there.

There is, of course, a limit to this talk of character. Namely, that we categorize and ascribe personalities and attributes to people on a superficial understanding of them and fail to understand the complexity of the human brain. It is summed up best for me by the approach of a man who I think had an extraordinary grasp of what it is to be human. I suppose the long tussle with the insanity of his wife, and the sadness and guilt which that caused him to feel, resulted in his constantly writing about the human mind – and the close relationship he saw between sanity and insanity. There is no doubt that Luigi Pirandello was one of the most significant playwrights of the twentieth century. One of his most important contributions was to take a long hard look at character in the context of how it is portrayed on stage – he turns the drama on itself, often a feature of good art in many different media.

His trilogy of plays about the theatre is extraordinary. The first of these, *Six Characters in Search of an Author*, is a masterpiece. It is Pirandello's definitive statement about the inconsistency of the theatre. In it, he points to the fact that when we watch an actor on stage, he is acting and not real. So his central theme (to which he returned repeatedly in his writing) is the conflict between reality and illusion. We see the actor, we may empathize with him, but, of course, he is pretending to be someone he is not, in a situation that he has not experienced in precisely the same way as the character he is playing. And, of course, this is exactly the paradox we experience when we judge another person in real life – we are effectively using our brain to judge theirs.

In *Six Characters*, we, the audience, start by entering an auditorium; the curtain is already open. A rehearsal is in progress on stage – incidentally, and ironically, another play of Pirandello's, *The Rules of the Game*. There are tedious arguments between the Actors and the Director about how to interpret what the author intended. The Leading Actor is protesting about having to wear a cook's hat. The Director replies, 'Yes, my dear fellow, you have to wear a hat and you have to beat eggs on stage *and* you have to represent the shell of the eggs you're beating – now, get on with it.' In the midst of this trivia about the nature of being an actor, Pirandello delivers a stunning *coup de théâtre*. Out of the gloom from the back of the theatre, six Characters, all dressed in black, slowly but dramatically make their way through the auditorium, and sombrely mount the stage uninvited. The Director protests he cannot be interrupted – he and the Actors are trying to portray real life and cannot be disturbed in rehearsal. The lead character, the Father, pleads, 'But let us be your drama.' The story that the Father is going to tell about what happened to him, his wife, his stepdaughter and three children, he promises, is much more real than any play could ever be. Unfortunately, he says, their author stopped writing about them and they are therefore held in suspended

325

animation throughout time with their tragedy 'unresolved'.

In spite of himself the Director is fascinated. He and the Actors watch the Characters and see them act out their story. It soon transpires that the Father, an introverted, remote figure who loves his children but is estranged from his family, has visited a brothel. The Mother is left short of money and her daughter, the Stepdaughter, is working in this same brothel as a 'novice' (possibly a virgin, we do not find out) prostitute. The Father, during his visit, does not recognize his stepdaughter and he attempts to take her clothes off to seduce her. At this point in the play within a play, the Director stops the action. He protests they cannot portray such lurid action in a theatre even in the enlightened times in which they live – 1916. So, using the identical dialogue, the Actors then act out their interpretation of the short scene they have just witnessed. But instead of being poignant, it is now close to prurient. The Director, dissatisfied, takes over the role of the Father and acts out yet another version of this clandestine meeting between the Father and the Stepdaughter – and turns it close to farce. The Characters protest vehemently that this is not at all as it happened. As the drama proceeds, we, the audience, are thrown further into recognizing that what is being portrayed on stage cannot be portrayed accurately. It isn't real life. It is drama transformed into melodrama by actors, and a parody even more remote when done by the Director. And we, the audience, each perceive it differently again. And this is the essence of perceptions of personality – we are using our human mind to understand another person's mind. We cannot be objective. There is a key speech by the Father in which Pirandello sums up an indelible insight into personality and human consciousness.

My drama lies entirely in this one thing. In my being conscious that each of us believes himself to be a single person. But each of us is many people, many people – according to all the possibilities of being that are within us. For with some people we are one person, and with others we are somebody entirely different. But we always have the illusion of being one and the same person with everyone we meet. But it's not true. It's not true. And we find this out very clearly when we are, by some terrible chance, caught in the middle of doing something we shouldn't be doing. And we are left dangling, suspended in mid-air. And then we see that every part of us was not involved in what we were doing and it would be a dreadful injustice for people to judge us by this one act, as we are held there, as if we were suspended throughout eternity with one's personality summed up in a single, uninterrupted action. Now do you see the treachery of this young girl? She caught me in an unrecognizable situation, doing something I shouldn't have been doing, with her . . .

Victor Meldrew, the archetypal grouchy old man of the BBC1 comedy series *One Foot in the Grave*, was a great hit with audiences – not least because he tapped into a phenomenon of which we all have some experience. A psychoanalyst might say that we enjoyed watching Meldrew, not because he reminded us all of the moody men we knew, but because he spoke to the 'grumpy man' inside all of us. To see him 'out there', on the TV screen, the object of laughter and ridicule as his grouchy tendencies reached increasingly bizarre heights of excess, was in some way fundamentally reassuring to us. The laughter of the audience was the laughter of relief, a great unconscious cry of 'Thank heaven, it's not me.'

In the mood . . . for shopping

It's worth briefly considering what we mean by the word 'mood'. To do so, it is helpful to compare it to other classifications of behaviour, such as disposition, or emotions. We could say that disposition, like personality, is something that is constant, and observable over time. So Eeyore, for example, would be said to have a gloomy disposition – whatever happens to him, this aspect of him remains the same.

Moods, on the other hand, are transient. They are related to specific times, situations and triggers. So a piece of music might put me in a good mood. This might spill over, so that I continued to be in a good mood for a few hours afterwards. But then the feelings would fade and my more constant disposition would take over again. Emotions also fade and change over time, but they are a more conscious, communicative form of behaviour. Fear and joy exist to show both myself and others that there's something out there to be feared or glad about. I can also, providing that my frontal lobes are sufficiently well developed, show a degree of control over my emotions. I can stop myself feeling angry. I can suppress my joy if someone close to me has had some bad news. Moods are more shadowy. It is hard to consciously alter one's mood, and at the same time it is possible to have one's mood changed without being aware how it is happening, or even that anything is going on.

For decades now, supermarkets have used very subtle techniques to influence the mood of shoppers, and the science of 'retail atmospherics' even has its own 'learned' journals and publications. But why should it really matter what sort of a mood we're in? If I want some herrings and a packet of hooks, and the shop has herrings and hooks, I will buy them, won't I?

Retail atmospherics takes a rather different view. Its proponents base their theories on one indisputable fact of human behaviour. When we are in a positive mood, we are open to new experiences. When we are in a negative mood, we avoid them. If you translate this wisdom into the setting of your local supermarket, then a good mood will make you want to stay in the store longer, maybe have a snack at the cafeteria, explore different areas of the store, take a risk by buying new products. A good mood also makes us spend money because, in the Freudian interpretation, we seek a 'pay-off' to any pleasurable experience. Just as satisfactory sex ends with an orgasm, we seek some ultimate climax to a pleasant shopping experience. So we spend money. Other psychologists think the process is more about reciprocity – that is, having been given a pleasant experience by the supermarket, we want to say thank you, and so we buy things.

It would take nothing short of a tranquillizer dart to make me want to thank Mr Kwik-Save as I pick my way down crowded aisle 33 towards the peanut butter – especially when it turns out to be the wrong brand or, in my case, not kosher. But research has shown that there are a number of ways in which the retail environment can be manipulated to put customers in a better mood. Perhaps it even works on me.

Some of the more obvious tactics are music – personally, a xylophone rendition of 'Mr Sandman' never quite does the trick for me, however – and smell. But studies have also shown that the layout and general condition of the store have a powerful effect. A team at the University of Göttingen tested this in 1997 using two IKEA stores. The first store had been newly decorated. Its layout had been specially designed for ease of access and manoeuvrability. New goods were artistically displayed in unusual and interesting ways – for instance, a new armchair was suspended from the ceiling with a spotlight trained upon it. Products were arranged in a manner that was intended to be inspiring to the customers. For example, chairs were displayed, not stacked up boringly in the 'chair department', but included within whole suggested décor schemes for different rooms.

The second store was the complete opposite. It was shabby and in need of a good coat of paint. Its aisles were cramped and narrow, the layout confusing. Goods were just stacked up and arranged purely functionally, with no concern for aesthetic appeal. The Göttingen team found that customers in the first store spent longer over their shopping. They spent more money, they showed greater willingness to have a snack in the cafeteria and they made more spontaneous purchases – that is, they walked out with more than they intended to buy. The latter is a phenomenon that certainly affects me – despite my general lack of enthusiasm for the shopping experience.

I have to say that I have behaved extremely badly in a large IKEA store in London. (And to be fair to IKEA, things appear to have greatly improved there in recent years.) Perhaps the worst, but most popular, time to visit is a Monday Bank Holiday. Many years ago, on a tight budget, I went with my wife and two children to get a set of bookshelves. The atmosphere in this part of the store was lovely, the assistants helpful, and soon we were going back to get a bigger trolley, and then a bigger one still. Eventually we had a massive trolley, piled high with all sorts of 'assemble yourself and don't worry about abrasions' furniture. The trolley had a mind of its own and kept careering into displays and into small defenceless children. Anxious mothers stared wide-eyed in terror as they saw this juggernaut trundling towards them. If I dragged it behind me so I could see innocent bystanders over my shoulder, packages fell off the back and my diminutive offspring were left whimpering in the aisle, holding their feet. Increasingly slowly, we made our way to the check-out area, crowded and dark, with mysterious boxes piled in a chaotic fashion some twelve feet high on either side of a long gloomy corridor. Eventually we came to a complete halt, reminiscent of a pile-up after a crash following an Aston Villa match just beyond the nearest exit on the M1. There we encountered gaunt, haggard people, some with apparently starving children (judging from their pitiful cries), who had given up all hope of ever seeing their nearest and dearest again. No music played here in the silent, shadowy dark. I am ashamed to say that, with hardly a glance at each other, my wife and I wordlessly abandoned the trolley and slunk away.

For the *Human Mind* television programme we tried the effects of mood-manipulation for ourselves. Two identical twins were our unwitting victims. In the morning, one twin was submitted to various situations that should induce a positive, happy mood, and the other twin to experiences to induce a negative mood. The negative-mood twin went to a quiet room in a gym near a shopping centre. First, she was made to read and sign a serious-looking consent form. Then she read a series of statements from a laptop screen. She was told to imagine saying to a friend things designed to make someone feel bad, such as, 'I sometimes feel so guilty about the hurt I have caused my parents.' While reading it, she listened to some sad music – Barber's Adagio for Strings. After being left alone for a while, she was taken to the shopping centre and sent shopping for thirty minutes. Later, the other twin went through the same basic procedure, except that this time the statements were designed to induce a positive mood, such as, 'I feel I can do just about anything.' She also had the joy of listening to lively music. Thereafter, she was similarly given half an hour to go shopping.

The mood induction worked impressively. The sad twin bought less, looked at fewer items and went into fewer shops. She also felt less satisfaction with what she bought and did not really like the trainers she'd purchased. After just twenty minutes of shopping, she became distressed and didn't want to continue. Eventually she was taken back down and given the positive mood induction procedure to put her back in good spirits. Meantime our happy twin appeared more enthusiastic and more energetic. She bought lots of items, and went into several different shops, looking at lots of different products. She even bought a couple of gifts for her sister – it turns out, of course, that we are all likely to be more generous when we are in a positive mood. The twins were reunited and both were sent home happy after we had explained how we had manipulated them.

The eyes have it

This little demonstration shows that 'mood', unlike something hard-wired like extroversion or introversion, is a changeable commodity. It can be influenced by a few subtle events, yet it has quite concrete effects on the choices we make in our lives. We do not even have to be aware of the events, in order for our mood – and resultant behaviour – to be altered. I regularly come out of the supermarket with more goods than I intended to buy. Some police stations have experimented by putting pink in the clink – coloured lightbulbs in the cells – which is thought to have a calming effect on distressed and violent prisoners. What is presumably happening here is an interplay between the sensory cortex and the limbic system – one that does not have to enter the conscious realm of the frontal lobes in order for it to have an effect on our behaviour.

But while the mood of anybody can be preyed on unsuspectingly by the subliminal trickery of the supermarket or the constabulary, some people show more marked tendencies with their moods. That is what is meant by the observation that personality is something that remains constant over time and regardless of circumstances. Some people are irritable, whatever happens to them. David H. Zald of Vanderbilt University undertook PET scans of eighty-nine subjects.[74] The team specified that they wanted only right-handed volunteers for the test

– remember that a right-hander has a dominant left brain.

The subjects were given a questionnaire, which asked about the extent to which they had experienced unpleasant moods during the previous month. In the PET scans, Dr Zald and his team found that people who reported a tendency towards bad moods showed increased activity in the ventromedial prefrontal cortex. This little area – just behind the right eye in right-handed people – is understood to play a part in mood regulation. Studies in animals have indicated that the ventromedial prefrontal cortex has some role in controlling heart rate, breathing, stomach acidity levels and sweating – bodily functions that are also implicated in mood. It is not yet known whether heightened activity in this region is the consequence of bad moods, or the cause of them. But the next time you find yourself in a rotten temper, maybe you should try rubbing your right eye.

No, this is not a serious medical recommendation. But it is peculiar that some healthcare professionals seem to have helped people cope with high levels of anxiety by teaching them a simple eye-movement technique. EMDR (eye movement desensitization and processing) is a method first tried by Dr Francine Shapiro from Palo Alto, California. She noticed that, following her distressing cancer diagnosis, her feelings of anxiety were alleviated when she rapidly moved her eyes from side to side. Dr John Spector at Watford General Hospital now teaches the technique to people who have recently experienced extremely traumatic situations. The underlying theory is that negative, anxious, distressing feelings are situated within the right brain, and these feelings, if too powerful, cannot be processed or verbalized because those skills lie instead within the left brain. If EMDR works, its protagonists claim that it does so by stimulating right-to-left brain communication, ensuring that the person doesn't become overwhelmed by their state of anxiety.

Apparently, brain monitoring has shown that there seems to be a normalization of brain-wave patterns after only three sessions, with right-hemisphere activity dampening down and left-hemisphere activity gearing up.

Brain chemistry and depression

So many features of personality depend upon underlying brain states. Techniques such as Cognitive Behavioural Therapy and even just regular practice at trying to put oneself in happy situations can effect a partial rewiring job within the brain, replacing old, undesirable patterns of thought and behaviour with newer, more helpful ones. Even something as simple as a regular laugh-in with friends seems to be able to set up a loop of positive feedback – so that by pretending to be happy, we can actually become more happy.

But though we have the ability to bring certain aspects of our mood and behaviour under conscious control, we do not undergo dramatic character change as a result. Even if I became noticeably more cheerful, people would not suspect that I had undergone a personality transplant. But there can be circumstances in which people do undergo rapid changes in personality. Depression, schizophrenia, bipolar disorder and strokes can all have very powerful effects on personality, leaving the sufferers, and their families and friends, feeling very much as if they have become 'someone else' – rather like Phineas Gage after his accident.

A relative of a depressed patient once described her to me, very aptly, as 'looking like herself, but as if she'd had all the air let out of her'. Depressed people can often seem to be lacking in buoyancy in this way – listless, weary, unconfident. What can be even more distressing is the speed at which someone can enter this state, from seeming to be perfectly well.

It is worthwhile to make another pause here to explain some terms in greater depth. Earlier on I discussed certain personality types, and considered the brain chemistry that might have shaped them. But we enter more murky territory when we begin to talk about depression as an illness. If Philip Larkin had taken a course of anti-depressants, would he have turned into the life and soul of the party? The answer, obviously, is no.

Psychiatrists talk of two principal types of depression. Reactive depression occurs in response to distressing circumstances, such as job loss, infertility or bereavement. Endogenous depression (meaning 'arising from within'), on the other hand, can occur without any triggers. But in both cases, the condition induces a recognizable change in the mood, thoughts and habits of the sufferer. These changes include disturbed sleeping patterns, loss of appetite and sex drive, poor concentration, persistent or recurrent negative thoughts, feelings of loss, hopelessness and guilt. Depression also has a beginning and an end. Except in rare instances of a condition called persistent anhedonia, an untreated case of depression will alleviate itself in time. So depression is quite evidently a different beast from the more generalized negative or gloomy disposition. But it, too, is created by alterations in brain chemistry.

Wayne Drevets of the US National Institutes of Health is a key figure in establishing the neurobiology of depression and has emphasized how valuable brain scans have become in understanding what may be going wrong. Michael Posner at the University of Oregon has suggested that there are four areas of overactivity in the brains of depressed people. The prefrontal, outside edge of the frontal lobe, which enables long-term memories to remain within conscious access; the upper middle part of the thalamus – known to stimulate the amygdala; the amygdala itself – centre of emotions; and the anterior cingulate cortex – the area of the brain that becomes active when we concentrate on things. Posner suggests a scheme of action something like this. My overactive amygdala sends 'sad' feelings up towards my prefrontal lobe. This then consciously looks for a reason to attach to the emotion. It asks, 'Why do I feel sad?' and then rummages in the long-term memory to find some possible cause – for example, a bereavement. This business becomes hard after a while – other things arrive to claim my attention – but the feelings of sadness do not go away, and nor does the troublesome urge to look for a reason. So then my anterior cingulate cortex kicks in to shut out any other claims on my consciousness and ensure that my attention is focused on the sad feelings and my suggested reason for them. Small wonder I can't just 'snap out of it'.

Possibly our poet, Philip Larkin, in addition to having a negative-inclined outlook on life, might have suffered from a recurrent form of depression. His poem 'Aubade' (1977) describes his abject state of mind on waking every morning in the early dawn light to contemplate the certainty of his eventual death. This phenomenon, of waking early and having difficulty getting to sleep, accompanied by persistent negative thoughts, is a well-documented symptom of endogenous depression.

Larkin also stated in an interview with the *Observer* that his parents were 'rather awkward people . . . not very good at being happy. And these things rub off.'[75] His words possibly indicate an awareness that he might have inherited this tendency, though, of course, his rather depressive trait might have equally been produced by his childhood environment. But genetic variants can lead to low levels of serotonin in the brain and this, it has been shown, can be the cause of behavioural traits such as aggression and anxiety. And these are aspects of behaviour seen in some depressed patients.

Genes again

What about the genes which play a part in influencing our moods and attitudes to life? Some recent research, conducted by Dr Mary-Anne Enoch at the American National Institute of Alcohol Abuse and Alcoholism, has suggested that women may have a brain more likely to be hardwired to experience greater levels of anxiety.[76] It is known that women in general have lower levels of an enzyme called COMT, which some scientists believe is responsible for regulating mood. The NIAAA study has linked lower levels of COMT to a mutation of a certain gene (called Val 158 Met) – and women with this mutation score particularly high on questionnaires designed to assess their levels of anxiety. Interestingly, the same results are not seen in men. However, the findings of this study are somewhat controversial, in that scientists are by no means in agreement as to the functions of COMT. The current evidence suggests that it affects dopamine regulation in the prefrontal cortex, and changes in the COMT gene may increase susceptibility to schizophrenia and other severe disturbances.

There is some powerful evidence that depression may pass from one generation to the next. Serious depression possibly affects as many as two million Britons each year and perhaps ten million Americans. Some people are repeatedly thrown into the darkness of despair and for a proportion of families this is a recurring nightmare. An important study by neurologist Wayne Drevets and colleagues at Washington University School of Medicine has found that the brains of people with a family history of depression often show a characteristic abnormality.[77] An area called the subgenual medial prefrontal cortex, part of the forebrain the size and shape of your finger located behind the eyes and between the brain's two hemispheres, is consistently smaller. The post-mortem studies he has done – studies that would probably be condemned in Britain since the government's unsupportive attitude to the storage of brain tissues after post-mortem – have shown a most important finding. Although the brain of depressed people is smaller in this region, it has roughly the same number of neurons. The loss of bulk is because the supporting cells, the glial scaffold, is reduced. It would seem that the glial cells are not just acting as a glue but have a more important role. It is known that these cells store sugar as an energy source and they may also be important for the maintenance of adequate serotonin. Depression that occurs in association with this loss of cells may be associated with reduced levels of this neurotransmitter. Whatever the precise mechanism, these findings of cell loss are quite different from those in individuals with a single attack of depression. This is one reason for considering a possible genetic cause, running in families. But depression may run in families for another reason. As we know, the plastic brain responds to its surroundings – so a child surrounded by depressed adults may develop the brain of a depressed person itself.

A contender for a 'depressive' gene may have recently been identified. In February 2003, an American firm called Myriad Genetics, Inc. announced that it had found a single gene, known as DEP-1, which seems to be linked to depression. This firm has a strong commercial interest in offering people screening for possible cancer genes, but it is broadening its net. It is believed that the DEP-1 gene has no effect on serotonin or noradrenaline, usual suspects in disorders of brain chemistry (along with dopamine and acetylcholine). Rather, it seems to operate within an entirely novel pathway, which is still under investigation. Myriad Genetics have done their research on four hundred Mormon families in the Salt Lake City, Utah, area. Mormons are wonderful subjects for genetic research, because they are fascinated by genealogy and most keep excellent detailed family records that stretch back many generations. They also tend to have large families – many do not use contraception – and marriage is mostly to a small population – other Mormons. So any genetic tendency for a particular trait or disease carried by these people is likely to be more obvious and more common. It will almost certainly be a number of years before this research leads to any new antidepressant products becoming available on the market. But it is interesting that the founder of the Mormon Church, Brigham Young, believed that he held the answer to mankind's problems. In the case of depression, he might just have been right.

But how could a tendency towards depression have managed to enter our genes? It is hard to understand how such a condition could have any evolutionary advantage. Randolph Nesse, at the University of Michigan, has written extensively on this subject and argues that presumably at a former point in our history reactions associated with depression might have helped us to survive.[78] When dogs wish to avoid danger, such as conflict with other dogs, one strategy they adopt is to render themselves slower, smaller and seemingly lifeless. They flatten themselves to the ground, mute their responses and stay submissively still. Possibly depression in humans is a remnant of this inhibitory survival response. Depression often occurs as a by-product of high stress – enforcing a slowing-down of body and brain. Could it perhaps have also been the means by which the body and brain could repair and recover?

Depression can possibly also be considered as a faulty sub-set of emotion – a method by which we unconsciously signal to others that we are very much 'not okay', effectively asking for help. A study at Ohio State University has shown that men who are abusive and aggressive are actually less likely to display symptoms of depression or anxiety, even though they will freely report high levels of stress in their lives. Perhaps violent acts become the only means by which such men can signal their distress and seek help.

There are many other reasons for a sudden change of personality. It can happen as a result of a brain tumour, or after a left- or right-brain stroke. In *Mapping the Mind*, Rita Carter quotes the example of a High Court judge who, after a right-brain stroke, became a great crowd-pleaser. His courtroom rang with the sound of laughter, and defendants and lawyers grew increasingly aghast as they saw stiff penalties being handed out willy-nilly for minor offences, and serious criminals sometimes allowed to walk free. Eventually the judge was required to retire, a state of affairs about which, as Rita Carter notes, he was also blithely content.

We saw in the previous chapter how brain injury can result in a personality change. Phineas Gage became a different person. Another interesting example is the case of Unity Mitford, who it is said was a changed person – prone to fits of anger and depression – following a gun-shot wound to the head. Unity Mitford was a notorious aristocratic socialite, sister of the brilliant novelist Nancy. As the daughter of Lord Redesdale, the second Baron, she had a privileged existence, being educated at home by her mother. Her parents certainly held right-wing political views and supported the British fascists. In 1936 another daughter, Diana, married the leader of the British Union of Fascists, Oswald Mosley. Both Diana and Unity travelled to Nazi Germany and Unity was flattered to meet Adolf Hitler, as well as many of his immediate governing circle, Himmler, Goering and Goebbels. Hitler apparently told the German press that Unity was 'a perfect specimen of Aryan womanhood', and some people consider that Unity was, in some way, in love with Hitler. When Britain declared war in 1939, Unity was still in Germany, with her world collapsing; she tried to kill herself with a shot to the head. Six months after the war was declared she was repatriated to Britain in an injured state on a special train, hired by her father. But some papers, which came to light just recently in the Public Record Office, imply that when she arrived back in Britain in January 1940, she was carried off this train on a stretcher but with no outward signs of any injury.

Guy Liddell, one of Britain's wartime spymasters and in charge of B Division of the Security Service (MI5), was involved with investigating espionage and conducting counterespionage operations. He was insistent that Unity be searched on arrival on her private train. He later minuted: 'We had no evidence to support the press allegations that she was in a serious state of health and it might well be that she was brought in on a stretcher in order to avoid publicity and unpleasantness to her family.' The Home Secretary, it is said, stopped the search. The reasons for this intervention have never been satisfactorily explained. The present Lord Redesdale, Unity's cousin, and a friend and colleague of mine in the House of Lords, has been quoted as saying: 'I love conspiracy theories but it goes a little far to suggest Unity was faking it. But people did wonder how she was up on her feet so soon after shooting herself in the head.'

Sabotage of the self

In all our explorations of the human mind, one observation seems to keep cropping up. Our mental faculties are the pinnacle of a process of selective evolution, millions of years old. Our neurons compete for space and fuel, in the same way as our ancestors competed with one another for food and mating rights. Under this view, disparate mental phenomena such as consciousness, dreaming and even depression could be seen to have had an evolutionary purpose. The phenomena came into being, so the argument goes, in order to assist our survival. Their desirability in modern times depends upon their continuing ability to serve this purpose over changing circumstances – which is why traits like a depressive tendency and aggression are more likely to be labelled now as 'problems' than as strategies.

This is quite a compelling view, but it leaves a great swathe of human behaviour unexplained. Because every day, we humans do, say and think things that are not remotely helpful to our survival. We engage in practices that seem to be quite the opposite, in fact. And we could do no better than to look at the examples provided by our greatest writer for elucidation.

What is it about William Shakespeare that makes him such a genius? There is little doubt that the key to understanding his greatness is that he, more than any other writer in English, understood the workings of the human mind. From time to time, other writers from this and other countries, such as Chaucer, Montaigne and Dostoevsky, are cited as men who came closest to this aspect of Shakespeare's gift, but I suspect there is no one who so consistently portrays the full range of human consciousness. Whether we look at Lear, Prospero, Timon, Falstaff, Macbeth, Coriolanus, Richard II, Iago or Hamlet, we connect with those people at a very deep level, we know and understand them intimately, we empathize with them; we may not like them – indeed we may hate them – yet we laugh or cry with them, we fear for them, all because of the playwright's intuition. In Elizabethan and Jacobean times the English theatre flourished in a golden age, but none of Shakespeare's near contemporaries, I think, has this extraordinary understanding and the involvement with the breadth and range of the human mind. Christopher Marlowe was perhaps his closest competitor. On occasion, Marlowe may have written verse that was nearly as good but his 'leading men', Tamburlaine, Faustus and Barabas, the Maltese Jew, are exaggerated, mostly two-dimensional creatures in whom it is extremely hard to believe and who, consequently, only occasionally come to life. Ben Jonson was undoubtedly a gifted writer, but who has ever actually met a Sir Epicure Mammon (except very briefly at the Garrick Club) or thought like him? The characters of Subtle, the alchemist, and Volpone are convincing, but only at times; even at his best, Jonson's plays are largely peopled with caricatures rather than characters. This is good for farce or entertainment, but not so good for really changing the way we think about ourselves. The understanding of so many inner selves is what gives Shakespeare his towering reputation for intellect. And this understanding was a gift which augmented and stimulated the writings of others – as has been said of Shakespeare, he was 'not only witty in himself but the cause of wit in other men'.[79] And perhaps most remarkable is the fact that we do not know who Shakespeare really was. In spite of his brilliance in telling us about personality, we do not know him. We know virtually nothing about his own beliefs, beyond a few superficial likes and dislikes that creep out from the plays. Was he religious? Did he believe in God? What were his politics? Even his sexuality remains somewhat mysterious.

Many people consider that the single greatest play in the English language is *The Tragical History of Hamlet, Prince of Denmark*. This is because, within the context of this superb melodrama of revenge, Hamlet, the man, has a universal aspect to his character – he steps out of the play and exists almost independently. As an independent character, he succeeds where Pirandello's Father fails. *Hamlet* is the deepest look into the inner workings of the mind, a light shone into the psychic universe of the human brain. Though Shakespeare seems to have written sixteen plays after *Hamlet*, this is the drama that finds him at his apogee – all his writing is condensed in it, whether humorous, ironic, tragic or historical. And the poetry is among the finest he wrote, which is why, of course, so many phrases and quotations from it have crept into our language – so much so that we do not always recognize their origin when we read or hear them. What I find surprising and compelling about the play is that, every time I see it, and I've seen it more times than I can remember, it has me sitting on the edge of my seat. I know perfectly what happens in the duel at the end, but I still need to experience it because Hamlet himself transcends the play. Many essays, books, pamphlets and critiques have been written about this incontrovertibly great play – possibly more has been written about *Hamlet* than has been written about any other work written in the English language. Jan Kott, the Polish theatre director and critic, once remarked that the bibliography of *Hamlet* would fill two volumes of the Warsaw telephone directory. So it is with diffidence that I add my own inadequate understanding to so many previous analyses of *Hamlet*, and the way I believe it holds a unique, accurate mirror to human personality.

The man Hamlet

Hamlet has been described as a man who could not make up his mind, as a man eaten up with anger, who thought too much and was paralysed by his uncertainty, as a man who was self-absorbed and consumed by his troubles, as a man who was too good for his environment and times, as exhibiting a 'split apprehension of women and an attempt to salvage purity from an initial conviction of general debasement'. We have seen *Hamlet* interpreted as swashbuckling adventure, as a play about espionage, about vengeance, about the Oedipus complex, about subversion, or about totalitarian politics in a fascist state. Most of these are surely very short of the mark; others may be condemned as travesties.

The key to my thinking about Hamlet is this. It is inescapable that Hamlet repeatedly unnecessarily exposes himself to danger, and to unhappiness. And Hamlet is constantly wrestling with paradoxes within himself. The first puzzling thing about Hamlet is that he gets a clear commission when he meets his father's ghost on the battlements who spells out for him the nature of his death by murder. Whether the ghost is real, or whether this is Hamlet's own conscience or consciousness speaking, is open to interpretation (it is said, incidentally, that Shakespeare played the part of the ghost himself in the first production). But the position is clear: the ghost is telling the truth – indeed, Shakespeare's ghosts never lie. Hamlet is given instructions – he is to avenge his father's death but on no account to harm his mother, Gertrude:

> But howsoever thou pursueth this act,
> Taint not thy mind, nor let thy soul contrive,
> Against thy mother aught; leave her to heaven.

Yet this commission from his father (or his conscience) is met by his omission. He doesn't kill the King (until too late, at the end of the play when his own death is already encompassed), and, in spite of the ghost's instructions, he harms his mother hugely – for example, by what he says to her in the bedroom scene. Hamlet is certainly no coward, he is quite convinced of the truth of what the ghost says, he is a popular man of action, he has opportunity to kill the King, yet he fails to do what he has agreed. His indecision is, on a conscious level, motiveless.

A second paradox is that Hamlet is a 'good' man yet his

conscience reverses the normal rule – it expressly asks him to do something that is forbidden, to commit a murder. True, there is a conflict about killing the King while ''a is a-praying'. But if Hamlet has religious scruples, there seem none in executing the Ghost's command. And Hamlet could quite easily save himself from a conflict here. There would be no need for him to do the killing himself – he is clearly popular with his Danish subjects, he is the rightful heir to the throne and he can prove – in fact does prove – the case against Claudius. Yet his proof is private, only for his own 'satisfaction' and never intended as a cause for revolution or persuasion of others around him of the justice of his case. Indeed, he never argues the justice of his case, even to his closest friend, Horatio.

But the third set of paradoxes is the most revealing. The number of times that Hamlet consciously and deliberately puts himself at risk is striking. The play-within-a-play, 'Murder of Gonzago', memorably retitled 'The Mousetrap' by Hamlet when watched by Claudius, is a crystal example. Hamlet knows, and shows he knows, exactly what Claudius has done. He is certain that his father has been murdered, how the murder was done, where it was committed, and that the murderer has seduced his mother. Certainly, he has no need to observe the King's reaction to the play: 'the play's the thing, wherein I'll catch the conscience of the King'. His declared purpose is to study the King's facial reactions, yet earlier he admits the inscrutability of the King's smile: 'That one may smile, and smile, and be a villain'. The only possible outcome of this night at the theatre is to put Hamlet's own life in jeopardy. Hamlet increases his danger by visiting his mother's boudoir – no useful purpose can be served. He recognizes he is likely to be spied upon, and the killing of Polonius by stabbing him through the arras curtain compromises him further. He exclaims when seeing Polonius's dead body: 'Thou wretched, rash, intruding fool, farewell;

I took thee for thy better' – implying he thought it was the King. Yet only a few minutes earlier, in the immediately preceding scene, he rejects the chance to dispatch Claudius when he has the ideal opportunity. Yet anybody else alive behind the arras poses no threat to him. His treatment of Ophelia is inexplicable; it can only serve to make himself unhappy – and he knows that he is being overheard, so the tone of his conversation with her is most injurious to his own interests. Equally paradoxical is Hamlet's trip to England with his treacherous 'friends' Rosencrantz and Guildenstern, with letters that he knows contain instructions for his own death:

> There's letters sealed: and my two schoolfellows,
> Whom I will trust as I will adders fang'd;
> They bear the mandate; they must sweep my way
> And marshal me to knavery.

Indeed, in being 'marshalled to knavery' he has already agreed with himself to kill Rosencrantz and Guildenstern (even though he has not done the easier and more appropriate task – kill the King, the fount of the knavery). Above all, Hamlet agrees to fight a duel with furious Laertes, valiant and skilled a fighter though Ophelia's brother is. This is the man whose father Hamlet has assassinated, and whose beloved sister he has made mad. Hamlet knows clearly that foul play is likely – indeed, Horatio warns against the fight and offers to make excuses for Hamlet so that he need not turn up for the fatal duel. In switching foils during the fight, yet again Hamlet shows his suspicions. He knows intuitively the foil is poisoned, and when he sees the King warn the Queen not to drink, he makes no overt move to prevent her from taking poison. So though Hamlet avows he is against suicide and murder at the start of the play, by the end of it his actions have produced eight deaths (including his own) to no conscious advantage – the stage is littered with bodies.

So what is going through Hamlet's mind?

The portrayal of Hamlet demonstrates how profound was Shakespeare's insight into this particular aspect of personality, as into so many other facets of human psychology. Everything the Prince of Denmark does is geared to cause his own self-damage. The self-damage is at an unconscious level; the most likely interpretation therefore is that Hamlet is gaining a psychic reward. He achieves unconscious pleasure from his actions and his predicament. In this respect he exhibits an extremely common trait. This kind of behaviour, attempts at self-destruction or self-damage, frequently occurs among some individuals who though not in the slightest bit mad exhibit neurotic behaviour. Some psychoanalysts refer to this as psychic masochism.

Hamlet's behaviour, although neurotic, is certainly not, in my view, abnormal. Sigmund Freud (and we shall come to him in a moment, because he also attempted to analyse this play) would assert it is. To me, his behaviour is that of a perfectly normal individual who shows a common character trait and his reactions reflect only a rather more extreme version of how many of us might react in a similar situation.

Hamlet's reactions are merely towards one end of a common spectrum of self-damaging behaviour. His behaviour is near the end of the spectrum, because of the intense stress and disorientation of his position. But nearly all of us react in similar neurotic ways from time to time, only the reaction is less because our circumstances are not generally as extreme. Have you ever lost your temper when there was no conceivable advantage to you? Have you ever gained brief satisfaction by verbally attacking someone you love, only to feel deep remorse afterwards? Have you ever shown aggression to a colleague or friend when it might have been far better to have been temperate or forgiving?

Occasionally, our attempts at self-damage produce catastrophe, as happened in Hamlet's case. One most celebrated catastrophe of this kind in real life was that of Oscar Wilde. He should never have pursued his libel case; he knew his defence was far from secure. All Wilde's reliable friends advised him against this action in court, and the inevitable public exposure. The result was ruin, the absolute disgrace from which he never recovered. In Carson, the counsel for the Marquess of Queensberry, he was up against a contemporary from his days at Trinity College. Wilde knew that Carson and he were poles apart – Carson had always been a conservative, rigid, ambitious man with no time for the views of men like Wilde, who was soft, liberal, affectionate, more intelligent and more gifted. Wilde's replies in the witness box to a man quite prepared to destroy him eloquently point to his inability to avoid continuing with his downhill career of self-damage:

CARSON: May I take it that you think 'The Priest and the Acolyte' was not immoral?

WILDE: It was worse; it was badly written.

CARSON: I take it, that no matter how immoral a book may be, if it is well written, it is, in your opinion, a good book?

WILDE: Yes, if it were well written so as to produce a sense of beauty, which is the highest sense of which a human being can be capable; if it were badly written, it would produce a sense of disgust.

CARSON: A perverted novel might be a good book?

WILDE: I don't know what you mean by a 'perverted' novel.

CARSON: Then I will suggest *Dorian Gray* as open to the interpretation of being such a novel?

WILDE: That could only be to brutes and illiterates. The views of Philistines on art are incalculably stupid.

This highly articulate man, so capable of dissembling when he wants and manipulating the emotions of others, cannot resist a cheap victory. 'It was worse; it was badly written'; 'The views of Philistines on art are incalculably stupid'. A brief, cheap *frisson*, bought so expensively.

I wonder whether this 'reward activity' has some similarities with other risk-taking behaviour, or with behaviour that seeks pleasure from something which gives us a feeling of insecurity, horror or danger. Is this why *Alien* is one of the most popular films ever? Or why Mary Shelley's not particularly brilliant novel *Frankenstein* is one of the most universally known books and a bestseller for the last 150 years? We humans enjoy jeopardy. We get 'pleasure' from it.

Defence mechanisms

All of this is, of course, not so far from Sigmund Freud. Modern theorists working in the field of human behaviour frequently treat this nineteenth-century Austrian with a degree of scepticism – and with good reason. For instance, Freud placed too much emphasis on the sex drive, or libido, in the formation of the self. Others point out that his understanding was too heavily rooted in the very specific cultural background of nineteenth-century Viennese bourgeois society to be of much use to people in other places and eras. And a major problem with Freud, like my interpretation of Hamlet, is that his views were often assertions incapable of proof, and not subjected to statistical evaluation. Some critics would view his work as little different from phrenology.

Nonetheless, Freud's observations provided a bedrock for seeing various elements of human behaviour as survival strategies – helpful or otherwise. Studying cases of hysterical paralysis and fainting – which chiefly affected affluent young Viennese women with no outlet for their sexual or emotional energies – Freud was the first to coin the term 'defence'. In essence he argued that feelings were the same as other observable forces in the physical world, like magnetism or pressure. Energy, as we know from our early physics lessons, never goes away – it merely converts into different forms. Freud seems to have believed the same was true of emotional energy. That which was denied expression did not go away, it merely built up and found other, less obvious and often less helpful forms of expression. Perhaps the most helpful analogy is that of a fast-flowing stream. If for some reason the stream is dammed up, then it bursts its banks, flooding the surrounding terrain.

In cases of paralysis and fainting, Freud argued that, in order to avoid being overloaded by overwhelmingly powerful feelings, people unconsciously 'cut out'. They became ill in a manner that baffled the best physicians, because there was no physical cause for their disorder. What 'afflicted' them was, in fact, a survival strategy or defence mechanism. And in spite of the disruption this bout of fainting or paralysis caused to their lives, in spite of their consciously expressed hatred of the symptoms affecting them, this strategy nonetheless provided a form of unconscious gratification. By 'cutting out' – through paralysis or loss of consciousness – they obtained some temporary relief from their emotional distress. The focus of their concern, and the concern of everyone around them, became the 'illness' – not the underlying feelings which were too uncomfortable to look at. In addition, they were fussed over and treated as invalids – a way of ensuring they received the love and attention they might have been lacking in their daily lives. So in this way, we see how an apparently inexplicable, disruptive and unhelpful strategy can serve a clear purpose.

Other analysts, such as Melanie Klein, and Freud's

daughter Anna, have taken Freud's theory further, and argued that there are as many as ten 'defence mechanisms' in human behaviour – all of which have a common purpose of avoiding powerful and uncomfortable feelings. They also have in common the fact that they can cause as much trouble or more than they seek to avoid.

For instance, people employing the technique of *projection* take their uncomfortable feelings and project them onto others. A glance at the tabloid press will show how universal this particular trait is – asylum seekers, Islamic fundamentalists, warmongers and Michael Jackson all serve a particular purpose in being figures of hatred. By hating them, and by seeing them as forces of evil in the world, rather than as flawed human beings like ourselves, we can cope with our own powerful feelings of hatred – towards ourselves, and even towards people we also love, like our parents and children. We could argue that the lavish detail given in some newspapers to accounts of rape, torture, murder and sexual misdemeanour is a further example of projection. It recognizes impulses that exist within all humans, but by holding up certain individuals as scapegoats and roundly condemning them, it provides a means by which we can deal with these impulses.

Another strategy, one which might apply to the doomed Prince of Denmark, is that of *displacement*. Displacement occurs when we deflect feelings from their rightful object onto another. A classic example of this is the office hierarchy in which the boss, angry with his wife, rips into his undermanager, who proceeds to be nasty to his PA, who then goes home and kicks the cat. But sometimes we attack ourselves rather than the people we are angry with. And so we return to the Prince of Denmark. We could argue that Hamlet is actually furious with his father for dying, and with his mother for remarrying. But he knows that he should love and respect his parents, not be angry with them. The discrepancy between what Hamlet feels and what he knows he ought and ought not to feel gives rise to a great weight of uncomfortable feeling. So, unconsciously, Hamlet 'displaces' it by undertaking forms of action that attack himself. In his various ill-fated strategies, we have a clear demonstration of the way a defence mechanism can paradoxically do the very opposite of what it was meant to achieve. Perhaps the most appropriate metaphor might be of a rusty, ill-fitting suit of armour, which protects us from arrows, but cuts and chafes us when we wear it.

The psychoanalytic and the neurological interpretation of human behaviour are traditionally at odds with one another. For instance, a psychoanalyst might say that an agoraphobic – someone afraid of leaving the house – was actually afraid of what they might do if they left the house, or was trying to control the behaviour of their loved ones. A neurologist would instead pinpoint an overactive amygdala and an underactive frontal lobe. But it is faintly possible that the two disciplines are not that far apart. Psychoanalysts argue that the story of the human psyche is a story of conflict between the 'id', the impulsive, unthinking emotional element, and the 'ego' – which is the rational, thinking, society-conforming part of the self. It has been suggested that this sounds curiously like the struggle between the frontal lobes and the limbic system! Some people have proposed that this may be 'in the genes'; in the rare disorder Angelman syndrome, babies are born without any expression from certain genes on one maternal chromosome – chromosome number 15. The result is mental retardation, speech difficulties, changes in the EEG with epilepsy, and jerky, awkward movements. In its polar opposite, Prader-Willi syndrome, where babies are born lacking expression of the equivalent genes on paternal chromosome 15, the symptoms in later life include mental retardation, placidity, obesity and an underdeveloped sex drive. It has been argued, fancifully in my view, that the father's genes might be responsible for those regions of the brain that correspond to the 'id', while we inherit our ego-systems, including the rationalizing component of the frontal lobes, from our mothers.[80] Our personality, in this view, is nothing less than a competitive battle between male and female genes inside our developing brain.

Whatever you make of this theory, it seems clear that our brain can sometimes hinder us as much as it can help. This may become even more clear as I go on to discuss, in the following chapter, the ways in which our brain is wired to help us interact with one another.

chapter**eight**

The loving mind

In July of 1852, a 41-year-old wealthy and handsome Irish landlord named John Rutter Carden attended a dinner party given by the Bagwells of East Cork, another influential family in his social circle. At this sedate affair, he was introduced to Eleanor Arbuthnot, the attractive young daughter of some neighbours. Eighteen-year-old Eleanor was not exactly swept away with Carden's charms – but from the moment he laid eyes on her, he was in love.

Carden and Eleanor's family moved in the same milieu of landed folk who hunted, dined and weekended together on a regular basis. There were endless opportunities for the two to meet. Carden, whose love for Eleanor seemed to grow stronger at every meeting, approached her mother and formally requested her daughter's hand in marriage. Her mother, a redoubtable widow, did not have a low opinion of Carden, but she felt that her daughter was too young to consider matrimony, and she also knew that her suitor's feelings were not in the slightest bit reciprocated. Her daughter seemed mystified and embarrassed by all the attention. As gently as she could, she advised Carden to drop the matter and find a bride elsewhere. He shouldn't have had any trouble – his name had been linked in the past with numerous women.

But Carden was a tenacious individual. When he came of age he inherited the run-down Barmane Castle. The estate had been neglected by his military family; its Irish tenants had long paid no rent and were not about to pay now. The castle was energetically rebuilt by Carden to withstand the assaults from the tenantry that followed, and the castle was successfully defended in floor-by-floor combat. He thus acquired a formidable reputation as a ruthlessly efficient rent-collector on this Tipperary estate – fending off assorted rebellions by his tenants and earning himself the nickname 'Woodcock' for his famous ability to avoid bullets. Like this bird, he was hard to hit. On one occasion, he dodged the rifle-fire of two disgruntled tenants, captured the culprits and marched them straight to the local police station. They were subsequently hanged. Later, he aimed a swivel-mounted cannon at a mob of protesters from the battlements of his castle, dispersing them swiftly. The tenants agreed that 'Woodcock' Carden was a reasonable landlord; they just did not expect to pay rent. Clearly, he was not the kind of man to give up easily.

For two years, Carden pursued Eleanor at every opportunity. He wrote her notes, declaring his undying passion in the flowery terminology of the medieval troubadours. He developed the notion that she was equally in love with him but that she was prevented from showing it by her family because they opposed the match. He followed the girl from Inverness to Paris, remaining a spectral and irritating presence at balls and parties, staring mutely at the object of his obsession from across crowded rooms and halls. When she was at home, he kept a lonely vigil outside her house. When she dined in restaurants, he sat at a table nearby, occasionally approaching her, but just as often worshipping her from afar. Nothing would dissuade him from his romantic cause – not gentle pleading, or threats, or public disapproval.

Matters reached a head in July 1854. On Sunday 2 July, Eleanor, her sister and mother went to church in their carriage. Woodcock Carden was lying in wait for their return – with three hired henchmen. They overpowered the coachman and seized the horses. Carden, armed with a bottle of chloroform, attempted to drag Eleanor away. In his disturbed mind, this was not abduction – since Eleanor wanted to be with him. But for a girl who wanted to be carried off, she certainly made a great fuss about it. She screamed and fought him. Her governess, a Mrs Lyndon, attacked Carden with her fists until his face was unrecognizable. Nonetheless, his passion was so great that he succeeded temporarily in kidnapping the unfortunate girl. The authorities gave chase, and Carden – who had regularly dined and hunted with the influential lawgivers of the area – suffered the ignominy of arrest.

Carden was committed for trial at Clonmel – charged with attempted kidnapping, actual abduction and felonious assault. But popular support for him ran very high. Indeed songs were written about him. The Irish gentry – fans of his ruthless methods of dealing with tenants – backed him to the hilt. Women – those who had not experienced his stalking at first hand, at least – considered him a romantic figure. The press presented him as a passionate Celt, whose love had been cruelly spurned by a frigid English girl unworthy of it. It is a measure of his popular appeal that, although the charge then carried a possible penalty of transportation to Australia, Carden received only two years' hard labour.

He served the full sentence. Offered an early release if he promised to stay away from Eleanor and never make any contact with her again, Carden refused – saying that he would far rather die in the Crimea. He was still in love with her. What's more, he knew that she still loved him. When Carden heard news that a servant had been dismissed from Eleanor's household, he was convinced that this was because she had been caught smuggling out love notes from his beloved.

Two years in jail did nothing to change Carden's point of view. Upon gaining his freedom, he made formal applications to the Lord Lieutenant of Ireland, and to the ancestral head of Eleanor's family – Viscount Gough of Loughcooter. They refused to intercede on his behalf. Still convinced that she would one day escape the tyranny of her family, Carden turned Barmane Castle into a paradisiacal playground for Eleanor – installing a Turkish bath and furnishing the place in the height of mid-nineteenth-century fashion. Until his death in 1866, Carden remained in love with Eleanor – following her forlornly as she travelled around Ireland and Europe. His was a life shaped by a single, overpowering obsession. We might add that poor Eleanor's life was considerably shaped by it too. Despite her good looks and sizeable inheritance, she never married, and devoted her time to the education of her nephews and nieces.

Stalkers

If he were to seek help for his unfortunate condition nowadays, 'Woodcock' Carden might be told that he was suffering from 'erotomania'. Stalkers – people who obsessively pursue and harass others, sometimes for years – display this type of behaviour. Stalking often includes an element of delusional belief – such as thinking, without any grounds, that the object of the stalker's obsession is equally 'in love' with them. In Woodcock's case, he thought that young Eleanor wanted to be with him and was being held prisoner by her family. He skewed all incoming information to fit this faulty hypothesis – as in his notion that the maid had been sacked because Eleanor had asked her to smuggle out a *billet doux* to him.

Nowadays, particularly in big cities, it seems increasingly possible for people to live their existence alienated from the world. Sometimes only the media provide a minimal contact with society outside, and that contact falls short of reality. It is not surprising that the objects of erotomaniac obsession, then, are often public figures – such as actors, pop stars and politicians – whom the 'stalker' has never even met. John Hinckley Jr, who shot President Reagan, suffered from a mistaken belief that his actions would impress the actress Jodie Foster. According to some accounts, Hinckley watched the film *Taxi Driver*, in which Foster starred, several times a day. It is, incidentally, quite an unspoken problem for my own profession – a surprising number of doctors, mostly men, are persecuted in this way and usually their family suffers as well.

Erotomaniacs, or stalkers, are a risk to themselves and a risk to the people they pursue. They generally have quite a shaky sense of personal identity, they are given to sudden, violent outbursts, and they are very sensitive to real or imagined rejection. Some stalkers are prone to having brief, turbulent love affairs and often have quite frequent periods of intense depression. Erotomaniacs may have eating disorders, are prone to drug abuse and often show other self-destructive tendencies. Very little solid work has been done to understand the underlying causes but erotomania is possibly the result of a problem with brain chemistry. The evidence is poor, but in some cases erotomania is preceded by an injury to the brain, or by epilepsy – suggesting that it has a neurological origin, perhaps within the temporal lobes. Another view is that it is a variety of Obsessive Compulsive Disorder (OCD). Certainly a key feature of OCDs is what psychiatrists call 'intrusive thinking' – in other words, the sensation of being unable to stop having certain thoughts. In the case of erotomanic people, of course, the thoughts are centred around the person with whom they believe they are in love.

Brain scans of people with OCD have revealed a peculiarity of the neural circuits which pass from the basal ganglia, through to the anterior cingulate gyrus and orbitofrontal cortex. To quickly refresh the geography: the basal ganglia are believed to play a role in the planning and timing of actions, the anterior cingulate gyrus is the area that becomes activated when we are concentrating on something, and the orbitofrontal cortex, among other things, plays a part in guiding behaviour sensibly (remember Phineas Gage?). In OCD, we see a sort of 'vicious circle' – in which the basal ganglia tell people to act, the anterior cingulate ensures that they become fixated upon acting, the orbitofrontal cortex tells them that whatever they are doing is wrong, and this final part of the circuit sends them back to square one. This explains why people with OCDs may find themselves 'imprisoned' by the need to wash their hands hundreds of times a day, or unable to leave the house without checking, again and again, that they have switched the lights off, locked the doors, etc. As with so many disorders of the brain, OCD is really just a malfunction of a sensible system, devised to ensure that we carry out our plans with efficiency.

Is love embedded in our brain structure?

The tale of John Rutter Carden also highlights another point – that there is a fine line between the states of mind we class as 'sane' and 'insane'. At one level, his is no more than an extreme tale of unrequited love – something that many of us may have experienced to some extent at some point in our lives. Even when love is reciprocal, in the first flush at least, most people experience an element of obsessive thinking. Song lines like 'You are always on my mind' and 'I've got you under my skin' express this classic feature of attraction. When we are first in love, we think of our lovers many times a day, to the exclusion of other people, other concerns. Without this facet of human cognition, Vodafone and Orange would never find it worth running a text message service. So if love itself is more than an idea, could it be the result of a specific state of brain chemistry? If that is so, can we observe it and can we measure it? Would it have been worthwhile for Andrew Marvell to have put his coy mistress into an fMRI scanner?

To be human is to be involved in relationships with other humans, and love is an aspect of ourselves which seems deeply etched within the structure of our brain. We are created from a relationship, however transient, between two humans. Possibly more than any other mammal we are raised most typically in family units comprising two or more humans. We enter and become part of further networks of human relations – schools, workplaces, army battalions, cricket clubs and informal groups of like-minded friends. The human being is a social animal – and there is no human society known to anthropologists in which the norm is for people to live their lives in solitude. From vast cities to small nomadic bands of Kalahari bush people, the human existence is one conducted alongside others. The instinct to be a part of a group has evolved since we were vulnerable hominids, incapable of hunting successfully on our own or protecting ourselves from predators.

But most of us also need a bit of 'space' and solitude – although the length of time we are prepared to spend in our own company differs from one person to another. In general, though, it doesn't take very long before our more basic need for company takes over and we begin to feel uncomfortable alone. This trait was studied by Stanley Schachter in 1959. Dr Schachter from Columbia University was one of those larger-than-life individuals, interested in everything, provocative, entertaining and sometimes highly controversial. His interests were wide – how people communicate and work in a group, the effect of being the oldest or youngest child in a family, the causes of obesity, nicotine addiction, the power of suggestion, and the interpretation of pauses in human speech, for example 'ums' and 'ahs'. His arch announcement about the power of 'bubba psychology' ('bubba' is Yiddish for grandmother) was typical. Any grandmother, he stated, could out-predict an economist – and therefore the stock market – because a bubba understands that people are not coldly rational about investing their money. 'Bubba psychology' hit the headlines and became argot for a time in *The Wall Street Journal*.

Solitary confinement is one of the most extreme human punishments. Being a solitary castaway, or a prisoner of war or terrorism, can be agonizing and devastating – as Schachter pointed out, the hermit-like isolation can lead a person to withdraw so much that they become akin to a schizophrenic. So the role of the religious hermit seems almost totally 'unnatural'. Schachter wrote of isolation for religious purposes:

> The solitary [hermit] begins to see himself as he really is. The meanness, the crookedness of natural character begins to stand out in a strong light; those wounds to disobedience or conscience in the past have left festering, now give forth their poisoned nature. There may be terrible uprisings of lower nature, and more formidable yet, resistance of self-will to the straight-jacket into which he will thrust it. Here you get the reason for what look like eccentricities of asceticism, its bread-and-water fasts, scourges and the rest, even to midnight immersions practiced by Celtic solitaries. It is Master Soul, the rider, whipping Brother Ass, the body, into obedience . . .

Schachter conducted one experiment on isolation rather like an extreme version of the TV show *Big Brother*. Schachter found five male volunteers prepared to live alone. The men were given very basic, separate accommodation – a bed, chair, table and toilet. Their rooms had no windows; they had no access to books, newspapers, radio or television. One man lasted twenty minutes before declaring that he wished to leave. Three of the volunteers managed two days in the room. Even the fifth man, a resourceful individual who racked up a total of eight days alone, eventually began to feel extremely anxious. Regrettably, unlike the winning *Big Brother* contestants, this man did not receive £50,000 and a period of C-list celebrity status – although his efforts somewhat advanced our understanding of the human mind. Humans need some form of contact with others, and they begin to experience anxiety when they are socially isolated. Even activities like watching TV or reading a newspaper can serve to maintain an illusion of contact with the outside world – without it, we are lost.

Schachter's studies did not stop there. Like so many of these classical researchers, he would not have obtained ethical approval for this study now. He felt that if social isolation makes people anxious, then people who are anxious might experience an increased drive to mix with other people to reduce their anxiety. In one experiment he recruited a number of women undergraduates, mostly psychology students. When the subjects arrived for the experiment for which they had blindly volunteered, they were greeted by a researcher in a white lab coat in a laboratory surrounded by coils of wire, transformers and electrodes. The experimenter introduced himself as Dr Gregor Zilstein of the Department of Neurology and Psychiatry and told the volunteers that the proposed experiment required them to receive an electric shock. To make some subjects more afraid than others, the experimenter used two different descriptions of the shock they would shortly experience. To the group he did not want to make fearful, he offered reassurance. He stated that they would feel no more than a tickle or a tingle. To those in whom he wanted to induce severe anxiety, he said that the shocks would hurt, but that they would be helping humanity because of the value of the research. 'Of course,' he said, 'these shocks will be quite painful but they will cause no permanent damage.' Before the shocks were administered, these students were asked to complete a questionnaire about their feelings about participating. They were asked, 'How do you feel about being shocked?', to which they replied by choosing from a five-point scale, ranging from, 'I dislike the idea very much' to 'I enjoy the idea very much'.

The poor students were then told that there would be a ten-minute delay while he prepared the electrical apparatus. He pointed out that he could not set up the devices with everyone hanging around, watching. So the subjects were given a choice. They could wait in one of the small adjacent rooms, just big enough for one person but where they could relax looking at magazines in a comfortable chair. As an alternative, they were offered an uncomfortable classroom nearby where there was room for everyone and plenty of hard seats. The students also filled out a form indicating their preferences on a score of five points from, 'I very much prefer being alone' to 'I very much prefer being together with others'. Over 60 per cent of the women expecting a nasty shock chose to spend the waiting period with other people. The great majority of women expecting a mild shock preferred to wait in comfortable isolation. Schachter concluded that, just as solitude makes humans anxious, so anxiety motivates us to want to be with other humans.

Of course, no shock of any kind was administered. But the ethics of such an experiment would now be regarded as extremely dubious. Needless to say, one problem is the impossibility of getting properly informed consent. Also the researcher is causing quite severe distress by making some volunteers fearful. He is also effectively threatening the self-esteem of those who found themselves troubled by participating.

Fitting in with the group

Most of us – unless we are enviably accomplished – will have had some experience of what it meant at some time in our lives not to fit in and to lose self-esteem. I used to wait in vain as the team captains at school picked everyone but me. (I am convinced to this day that I was good enough to be a Second to my Boy Scout Patrol Leader.) And those excruciating Saturday-night hops, spent in the local youth club where everybody had a smarter pullover and a better haircut. All of us will remember the levels of discomfort and anxiety that this failure to fit in produces.

But my hugging the bookshelves at a party is hardly human evolution in action; it is probably some form of mildly phobic behaviour, maybe induced by being repeatedly rejected as a member of my classmates' tiddly-winks team. Social exclusion tends to happen for three other main reasons – because we break, or don't seem to know, the 'rules' of behaviour; because we are incompetent or clumsy; and because we are unattractive. Each of these factors is a threat to the survival of the group. Someone who doesn't know the rules, or who knows them and still breaks them, threatens the cohesion of the group. Someone who is incompetent is a drain on resources – they'll eat the food, but be useless on the hunt, for example. Someone who is unattractive – and more on this later – may have inferior genes or bring illness to the group. Faced with social exclusion, we experience anxiety, and anxiety, as we know, then makes us wish to belong. In evolutionary terms, this anxiety would motivate humans to try harder to win the group's approval.

Perhaps in one sense the human brain has evolved into a highly 'social' organ itself. From the moment of conception, the brain develops by dividing itself into ever increasing complexity. Neurons develop by forming connections with one another, and by passing communications along those connections. Neurons are fed and nourished by communication with their neighbours – the seldom visited neural network is the network that tends to die away, which is why we can lose skills and faculties that we do not practise. And the brain depends upon the input of the external environment in order to develop. As we have seen in earlier chapters, we need contact with other humans to learn language and to process emotions. A brain that grows without sight, sound or touch of other humans is severely compromised indeed. It has been argued with good reason that the size and complexity of the human brain is partly determined by our living in large and intricate social groups. Studies of humans and our near-relations in the animal kingdom have shown a relationship between the size of the cortex and the complex nature of the social group.

So the human brain is a brain geared to interaction with others. There may even be neurons within the amygdala that are only activated in response to displays of emotion by other humans. It seems that in monkeys there are certain nerve cells in their frontal lobes that fire when the monkey undertakes specific tasks with its hand – such as picking up food and putting it in its mouth. What is so intriguing is that these particular neurons also fire when the monkey being tested sees another monkey doing the same thing. These neurons have been called 'mirror neurons' and Professor Ramachandran argues that 'with knowledge of these neurons, you have the basis for understanding a host of very enigmatic aspects of the human mind: "mind reading" empathy, imitation learning, and even the evolution of language.' There is now good evidence that similar neurons exist in other primates, particularly humans. Professor Rizzolati from Parma, the scientist who first described mirror neurons, and his team used fMRI and reported that there may be mirror neurons in a region of the temporal lobes called the superior temporal sulcus.[81]

Mirror neurons, if they truly exist in humans, appear to depend upon the input of human-generated information. For instance, these areas of the brain appear to be activated only when we watch another human performing a task, not when we watch a robot performing it. They seem to be activated only when watching goal-oriented behaviour – such as shelling a nut – and not when watching mimicry or purposeless action. However, as few people yet have attempted to record electrical activity from single neurons in humans, it is difficult to prove that they do fire in the way I am speculating. Indeed, it would be intriguing to find out whether, if they truly exist, they fire in response to watching a monkey undertaking human-like actions. And if this were the case, it would argue the basis for our empathy with them.

The human brain is also wired to be able to identify and discriminate between other humans – even when the information provided is minimal. If you were shown a film clip of me walking round in complete darkness, with only little points of light indicating key areas of my body, you would be able to identify what you saw as the figure of a male human, walking. You might even, if you knew me, be able to identify me as Robert Winston. Studies have shown that, whatever the level of visibility, humans are able to identify others by the way they move.

An impressive array of brain systems are involved in the simplest of interactions between people. Take the example of going to a party. We use our ability to recognize faces, our memory to acknowledge the person who actually invited us, and we recall other people to whom we might wish to talk. We use our amygdala to judge by their emotions what sort of a mood they are in, and our frontal lobes to direct our behaviour towards them accordingly. We use our attention and skill at focusing to concentrate and conduct a conversation amidst the hubbub of cocktail party chatter. We use our motor skills to know how close to stand to other people, and to balance a plate of snacks on our knees as we sip a drink. And our nose is telling us meantime what an appallingly unpleasant Beaujolais it is we're drinking – you might even be casting furtive looks to see into which plant pot you can pour it.

If just one simple faculty within this system happens to be working awry, then the social consequences can be disastrous. Perhaps we have trouble recognizing faces, causing us to seem aloof and rude. Perhaps we find it hard to switch attention between one focus and another, causing us to seem slow or boring. Perhaps we have difficulty recognizing other people's states of mind, causing us to seem insensitive or selfish. Perhaps we rush up to other people and conduct conversations just inches from the end of their noses – causing us to seem overbearing or plain disturbed. Maybe we miss the flowerpot and soak the elegant Qum carpet. The outcome for all five is that we end up being excluded from the group.

A cocktail party might seem a trivial example. Admittedly, the human race would probably just about continue if these gatherings were outlawed tomorrow. But the skills we call upon in such a situation are essential to our survival. Without a social brain, how could we meet suitable partners, fall in love and raise children? How could we identify people potentially of benefit to us, or others who might wish us harm? Without an ability to interpret and 'enter into' the emotional states of other people, how could we develop morality, or exercise compassion and forgiveness? These skills are central to being human, and they depend upon brain activity that is largely outside our conscious control.

It must be love

In evolution, the choice of mate depends upon the ability of one animal to recognize qualities in another of its species that would lend potential offspring an advantage in the business of survival. An aggressive 'alpha male' chimpanzee is likely to be better at hunting food and at defending his partner and offspring from predators or other males. So, female chimpanzees prefer to breed with alpha males. Not only would they be useful to have around while the female is saddled with the responsibilities of caring for her brood, but the alpha male's 'successful' attributes would be passed, through the genes, to her offspring.

We can see Darwin throughout nature and the principle is at work with humans. Society might have evolved to the extent that attributes like big muscles and an aggressive personality no longer lend an evolutionary advantage. But nonetheless, human females look for other indicators of 'alpha male' qualities. As Henry Kissinger, perhaps no great whiz in the looks department, has famously pointed out, 'power is an aphrodisiac'. Signals like a smart suit, a costly watch, a flashy car or a brilliant mind communicate that the male in question can amply provide for the female and her offspring, and pass on his survival advantages to the next generation. So these things become included in what we see as 'attractive'. Similarly, men still value attributes in a woman that communicate her superior ability to produce and care for healthy offspring – hence the timeless appeal of the 'hourglass' female figure, with its childbearing hips and ample breasts.

Of course, it is clear that in many respects to be human is to transcend the limitations of our evolutionary past. We do not, for the greater part, evaluate potential love matches and partners on the basis of looks or wage-earning abilities alone. We are motivated by characteristics like kindness and sense of humour, and by common goals and experiences. The temptation is to argue that such aspects might convey little or no evolutionary purpose at all. But that gentle musician and actor-comedian Dudley Moore, a rather short gentleman with small eyes and a prominent nose, once said he 'laughed women into bed'. Humour is attractive. Scan the 'lonely hearts' column of any newspaper or magazine and you will see time and again the term GSOH – good sense of humour. A man who can make people laugh is probably also a man with higher intelligence, a clear survival advantage. He is also a man with a level of social ability – one that could be used for security and gain. A kind man might be of little use in the hunting ground, but what about playing an active part in raising children? These examples show that, however much the world we live in may change, we are still flesh-and-blood machines orientated towards survival and reproduction. So the science of attraction and arousal is one of the most fascinating areas of brain research.

Hey, good looking

The evidence suggests that, whether we live in mud-huts or Manhattan loftspaces, attraction works on the same principles as it did in our evolutionary past. Various studies have indicated that what both men and women rate as attractive, handsome, beautiful, sexually arousing, etc., is symmetry. A Canadian study in 1999 showed that the more symmetrical men's bodies were, the greater the likelihood they were to have a range of sexual encounters. Newborn babies also show an increased preference for symmetrical faces, and this continues throughout our development. The Darwinian thinking is that a symmetrical face is not merely more pleasant to look at, but is also an indicator of physical health and strength – and this is why we unconsciously regard it as more attractive.

A good deal of what I feel is 'soft' research in this area has been done on orgasm. Much of it probably should be taken with more than just a grain of salt. But for what it is worth, a study by the aptly named Randy Thornhill at New Mexico University suggests regular orgasms are up to 40 per cent more common in the female partners of men with symmetrical bodies. Given that orgasm is pleasurable and that failing to achieve orgasm may carry some perceived stigma, how do we know that Randy's subjects were telling the truth? Or did he stand by the bedside with a clip board? But if orgasm does have an evolutionary purpose, perhaps it increases fertility. A University of Cardiff study by Dr Jacky Boivin, announced at the British Association for the Advancement of Science in 1998, claimed that, when a woman experiences orgasm, or at least a heightened degree of pleasure in the sexual encounter, she retains more of her partner's sperm, thereby increasing her chances of conception. Women in their mid-thirties were asked to rate their last sexual encounter in terms of the degree of pleasure it had given them and were tested within three hours of having had sex. Of those women who had given the experience a low score, nearly half had a post-coital sperm count of zero. Of the high scores, only one in ten had a low post-coital sperm count. But I remain very dubious about these findings – post-coital tests of the number of sperms left in the cervix are notoriously inaccurate and unreliable.

But we are on more solid ground with the notion that humans seem to be hard-wired to be attracted to those partners whose bodies and faces most accentuate the characteristics of the opposite sex. So the ideal woman, in every society and culture, has smooth skin, large eyes, plump lips and a relatively unpronounced jaw, attributes which most identify her as a woman. In addition, factors like youth and body shape indicate the woman's ability to carry out the role of childbirth.

In early adolescence, as their ability to conceive and bear children is switched on, women begin to store fat around the lower regions of their bodies. Studies at the University of Texas, by Devendra Singh, show that this pattern of fat distribution has a powerful effect on the way males rate female attractiveness. By measuring the distribution of fat, and asking male volunteers to rate a range of female shapes, Singh's team came up with the finding that the ideal waist-to-hip ratio is one of 0.7, regardless of the weight of the woman. Researchers have compared Miss America winners over a number of years – and found the same ratio. It is thought that this ratio represents the optimal condition for childbearing – and it alters as women go past the menopause. Production of female hormones decreases at this stage and the way fat is distributed across a woman's body starts to become closer to that seen in men.

Women equally are attracted towards men with the greatest distribution of 'male' sexual characteristics. Men are on average taller than women, and height lent us a survival advantage in terms of fending off predators and in collecting food. So taller men are more likely to be rated as attractive by the opposite sex, and are therefore more likely to reproduce. A research study performed in Liverpool and Poland studied this factor in a group of 3,000 men. They found that men who married and had children were on average 3 centimetres taller than men with no children. The unmarried men were also markedly shorter than the married men.

The higher presence of testosterone in men gives their body a whole set of differences from the female form: for example, a 'harder', bony facial look, with prominent jaw and cheekbones, and minimal facial fat. Female faces, or feminine-looking male faces, remain more childlike, with features less pronounced, large lips and more facial fat in the cheek area. Theoretically, men with more 'masculine' faces should be men who offer the greatest evolutionary advantage, because testosterone offers survival advantages – increased courage, aggression and sex drive. A handsome face, so the theory goes, tops a healthy body. The craze for the ragged, emaciated look of 'heroin chic' runs contrary to good biological sense.

Bizarre as it may seem, a simple difference in facial characteristics can mean the difference between a lifetime of success and one of failure. Researchers have found, for instance, that the facial dominance – grooved, growly and meanly masculine-looking – of graduates from a military academy tends to predict their final rank at the end of their careers. It has also been shown that more feminine-faced men show an increased likelihood to be found 'not guilty' in certain criminal proceedings, or to be handed lighter sentences. But it is in the arena of sexual attraction that the face plays the most important part.

What women want

A study at St Andrew's University by Ian Penton-Voak showed pictures of male faces to women at the peak of their conception cycle.[82] When asked to select the type of man with whom they would be most likely to have a casual fling, the women chose the most macho-looking faces – think George Clooney – over more gentle, feminine or less conventionally handsome types of faces, such as Leonardo DiCaprio or Willem Dafoe. In other words, when they were most able to conceive, they chose men who seemed the most likely to offer their offspring a genetic advantage.

In a refinement of this first study, Penton-Voak's team asked women to state how highly attractive they rated themselves, and also asked them to select male faces with whom they would be most likely to have a long-term relationship. This is where I have to admit that I glossed over certain facts earlier on. Because in some respects, women's biology leads them to want two, sometimes contradictory, things out of the men with whom they mate. Their evolutionary programming should be telling them to have sex with men who most demonstrate good health and strong male characteristics. But they also seek men who will stick around and help to look after the children. Unfortunately, the closer a male is to the ideal masculine type, the more likely he is to have high levels of testosterone. Accordingly, he is likely to have a strong sex drive and to seek out sex with a range of partners. In other words, the 'good genes guy' won't be the sort of guy who wants to stay in at night with the Ball and Chain and the kids. And also, his similarity to the 'ideal type' means that other women will be more responsive to his advances.

In accordance with this, both women and men associate macho, high-testosterone looks with the 'wham, bam, thank you, ma'am' approach, and not with sticking around to help raise children. In another St Andrew's study, which involved a vast number of participants, images were shown of two male faces – one masculine and one feminine. They were presented alongside two dating adverts and the subjects had to say which face went with which advert. The only real difference in the two adverts was that one man was obviously seeking a short-term

374

fling, while the other was interested in a long-term relationship. The participants – both male and female – overwhelmingly linked the macho face to the advert seeking a quick fling.

So we see a trade-off in the way women select males for the purposes of sex and reproduction. Women who are in search of temporary sexual fulfilment will have no qualms about selecting the most rugged, macho men – because they are not concerned about whether he will stick around and help. Women who want babies have to make choices. What they lose in the 'good genes' department, they may gain in the 'nurturing partner' department. Their choices mainly occur at an unconscious level, but they are also affected by factors more open to conscious scrutiny, such as their sense of self-esteem.

The conventional wisdom would be that the most attractive, high-testosterone males would be the worst choice for a long-term relationship, because of their propensity to go to make babies with other females. But studies show that women who rated themselves as attractive still plumped for the macho males to have babes with. Does this suggest that such women are less concerned about a man's 'stick-around' qualities because they feel able to provide for their own offspring, and simply want males to provide good genetic material? Women who rated themselves as less attractive showed a heightened preference for non-macho-type males – suggesting that they are more attracted to signs of sticking around, rather than those of 'top-quality' genes. The 'more attractive' women's choice in picking the macho guys might also be because they rated themselves as being sufficiently able to get other men to help raising the children or persuade the father to stay put.

But, of course, we do not base our choices on looks alone. In an earlier series for BBC Television, *Human Instinct*, women were asked to sniff men's T-shirts and then match them to faces. The experiment tends to confirm that we are more attracted to people with a genetic make-up different to ourselves – thereby ensuring that our offspring get the best possible inheritance. It is now thought that what may influence women in the 'sniff test' of mate selection is immunity to different diseases and infections. In other words, the ideal breeding partner is one who is immune to different types of disease than those to which we are immune ourselves. The resultant offspring would then be immune to the maximum possible range of threats to its survival.

When it comes to the question of genes, most of us prefer a mixture of the familiar and the exotic. We all tend to seek sex with people to whom we are not related. All of us carry some defective genes; if our chosen partner carries similarly defective DNA, it could mean disaster for our children. Statistically, most of us carry at least twenty to thirty variations in our genes that could cause a hereditary disorder. The results of marriage within the same group are well illustrated by diseases like Tay-Sachs. Tay-Sachs Syndrome is a serious disease. The brain neurons become swollen with fat and these babies develop blindness, paralysis and dementia. These children die within the first few years of childhood. The disease mutation is very commonly carried in the DNA of Ashkenazi Jewish people – about 1 in 35 will carry a mutation that could cause the disease in a child if they marry another carrier. Ashkenazi Jews had ancestors originating from Eastern Europe or Russia and they have always tended to have children with people from the same origins, not least because they had little choice in the matter, and, of course, because of religious observance and social custom. But the consequences have been the increased risks associated with inbreeding.

So, however much we may be programmed to seek 'outside' genes, there is also a human preference for what is familiar and similar. Sigmund Freud is perhaps most famous for arguing that small boys and girls have an unconscious wish to sleep with their mothers and fathers, and an accompanying fear of punishment from the other parent. More recent research into human behaviour indicates that his theories may contain a grain of truth. Photo-imaging techniques have now made it possible to morph and merge photographs with great degrees of subtlety. After modifying photographs, one study showed that volunteers were more attracted to faces that resembled their parent of the opposite sex.

Hawaii is an ethnic melting pot. There, Polynesian, Oriental, Indian and European people live in roughly equal numbers. Studies there have shown that children born of mixed relationships tend to marry into the ethnic group to which the opposing sex parent belonged. Boys, in a sense, look for the genetic model presented by their mothers, and girls go for that presented by their fathers. Other studies have shown that factors like parental eye-colour and age affect our choices of mate. However, the likelihood is that these things influence us, not because we are seeking to mate with people who carry the same genes, but simply because our sexual identity is formed at a very young age. In its malleable plastic state, the human brain becomes stamped with the image of the ideal mate. These images would naturally bear more than a passing resemblance to our own parents, if we were raised by them. I have not come across studies that examine adopted children, but – if these observations are correct – one would expect that people adopted as babies would have similar tendencies.

What men want

Men are arguably less 'choosy' when it comes to mate-selection, and are more ready to have sex, regardless of the partner. I have previously written about the study in which an attractive female stranger approached men on a university campus and suggested they go to bed. Three-quarters of the men responded positively. In contrast, when a similar approach was made to women by a male stranger, none of the women agreed.[83]

This difference has nothing to do with morality – although we may consciously provide that as the explanation. It boils down to biology. A woman has to bear a child for nine months, and nurture her offspring. This means she is less able to find food for herself and her baby, and is more vulnerable to attack from predators. Accordingly, she has to make very careful choices about the men with whom she mates – evaluating, as I pointed out earlier, their good genes versus their stick-around qualities. She can also only reproduce relatively few times within her lifespan. By contrast, each male has the potential to create millions of offspring, with the only necessary investment being the length of time it takes him from the initial approach to ejaculation.

But men are not just more interested in sex and less in relationships – although there is a tendency within our culture to assume this. Lord Byron summed up this attitude most neatly when he wrote: 'Love of man's life is a thing apart, 'tis woman's whole existence.' Are men really wired to reproduce all over the place, while women need commitment?

An American study conducted by M.J. Montgomery and G.T. Sorrell in the 1990s showed that adolescent boys fall in love earlier and more often than girls.[84] It also asked the subjects to answer the following question: 'If you met someone who had all the characteristics you desired, but you were not in love with him or her, would you marry them?' The cynical male might assume that more of the girls would have said no than the boys. But the reverse was true. Almost two-thirds of the boys said no, they wouldn't marry someone if they were not in love. This compared to just one-third of the girls.

These results tie in neatly with a study conducted by the American sociologist William Kephart. Kephart questioned 1,000 college students about their experiences of being in love, and the age when this occurred. He noted that, while girls initially fell in love more often and began to do so at a younger age than boys, there was a distinct pattern relating to sexual maturity. By the age of twenty, girls reported fewer and fewer cases of infatuation, whereas in boys the number continued to increase. Kephart argued that these results indicated a clear pragmatism on the part of women. As they become physically able to bear children, they are less inclined to be persuaded by romantic notions of love, and more geared towards rational decisions about long-term relationships. As the Montgomery and Sorrell study indicated, women are more likely to marry without love being part of the equation.

So, however far we think we have come from our evolutionary past, our brain and body tell a very different story. Everything we know about the science of attraction, and the Darwinian forces at play behind it, begs one question – is there really any such thing as love? Fortunately, the answer is yes.

The thrill is gone

William Proxmire, who retired in 1989, was a Democrat senator for Wisconsin known for his high personal standards of integrity, dedication and, especially, frugality. Proxmire started his career in counterintelligence during the second world war and later entered politics, standing unsuccessfully for the Wisconsin State governorship three times. Finally he was elected to the Senate to fill the vacancy created by the death of the now infamous Joe McCarthy. William Proxmire spent his political life as the Senate icon campaigning mercilessly against wasteful government spending. Over the years he took credit for saving millions of dollars of public money. His campaigning in the US caused programmes targeted by central government to be curtailed, modified or cancelled. He is remembered in this context for his successful fight against American plans to finance a supersonic transport plane. But he is best known for his Golden Fleece awards. These awards put government officials on notice 'to prevent fleecing of taxpayers'. He announced his first award in 1975; this was an attempt to galvanize public opinion against the decision of the National Science Foundation to finance an $84,000 study about why people fall in love.

William Proxmire, the old conservative, said in defending his position that people 'want to leave some things in life a mystery . . . and right at the top of the list of things we don't want to know is why a man falls in love with a woman, and vice versa'. But I do not believe we scientists should be blamed for offending the senator, or anyone else. It is sometimes said that scientific knowledge overturns the mysteries of life and thus makes the world a poorer place. But these studies are genuinely important. Although some of man's greatest attributes, like heroism and altruism, may be reduced to what seems a genetic equation, the study of psychology has a huge value for all societies. So what if consciousness and language are portrayed as a set of electrical waves in the brain? Does it matter if some scientific evidence strongly supports the view that laughter, affection and even the transcendent nature of a religious experience may have their roots in the interplay of neuronal activity? This is no indication that scientists should halt their endeavours. Science will never reach a point where it can explain and interpret the whole mystery of human existence – and nor should it aim to. The journey counts more than any intended goal. Understanding the neurological and psychological basis of love adds an important dimension to what will always be an enduring mystery.

Senator Proxmire was not the only person to protest about love and sex. Doris Autkrystof in her *Amedeo Modigliani: The Poetry of Seeing*[85] describes how, on Monday 3 December 1917, a group of guests gathered in the gallery of Berthe Weill's, the Paris art dealer, for the debut of Amedeo Modigliani's only solo exhibition. But it was bad luck that the local gendarmerie was located exactly opposite the gallery. There were a number of large paintings of provocatively posed naked women hung in the gallery itself and, to attract the huge crowd that entered, one of Modigliani's big nude canvases was prominently displayed in the window. Berthe Weill was hauled in to see the Commissioner of Police, who ordered her to 'take down all that filth' because the naked women were, outrageously, displaying pubic hair. The public scandal that followed, and the police activity, closed the exhibition. But love can be what makes another human paint an extraordinary series of obsessive masterpieces.

Amedeo Modigliani's paintings of Jeanne Hébuterne started when they met in 1917. One is in the Courtauld Collection in London, another in the Guggenheim in New York, and several others are in private hands. They are testimony to a great and poignant story of love. Some of the many paintings of Jeanne are breathtaking, and in some, Jeanne's delicate, fragile beauty stands fresh after ninety years have passed. The thirty-three-year-old Modigliani met Jeanne Hébuterne when she was an art student of just nineteen. Modigliani had suffered with repeated chest infections since he was a teenager. In some ways their story is a little resonant with that of *La Bohème*, but in reverse. Besotted and passionate immediately, they lived together in Montmartre. Their public love – and the violence between them – became a Paris legend. The poet and critic, André Salmon, described what he saw on one occasion:

He was dragging her along by an arm, gripping her frail wrist, tugging at one or another of her long braids of hair, and only letting go of her for a moment to send her crashing against the railings of the Luxembourg. He was like a madman, crazy with savage hatred.

Some of Amedeo Modigliani's many pictures of Jeanne show exquisite tenderness, but others seem distant and indistinct, and a few make her look disinterested and passive, even blurred. I find it interesting that he painted her in outdoor clothes, in formal clothes, in a nightdress, pregnant, wearing a large hat – but as far as I know, he never painted *her* naked. Because of the Great War, by a year later, in early 1918, conditions in Paris had become so difficult that Modigliani travelled to join the many other artists in their haunts in the south of France. Jeanne went with him and soon became pregnant, but shortly afterwards she and Modigliani had a stormy separation. Jeanne's parents were Catholic and her mother strongly disapproved of the Jewish Modigliani. But their love proved overwhelming and they were gloriously reunited before their daughter was born. It seems that on the way to register the child Modigliani got totally drunk, and the baby remained officially fatherless until she was adopted by Amedeo's family in Livorno, in Italy. By 1919, Jeanne was pregnant again and they moved into their first proper home together in Paris, the only environment Modigliani really liked. And now he was financially more secure and becoming a real success; at an exhibition in London, the writer Arnold Bennett bought one of the paintings and said that it 'reminded him of his own heroines'. But Modigliani's chronic alcoholism had always been a problem. That, combined with his poor diet, was getting the worse of him and his health now rapidly deteriorated. Days after New Year in 1920 his neighbour and friend for fifteen years, the painter Ortiz de Zarate, hearing strange noises upstairs and not having seen the couple around, suspected there might be something wrong. He went up to call in and was greeted by a shocking sight. Modigliani was writhing around and delirious, their apartment was in a state of chaos. The bed was strewn with empty bottles, and sardine cans were dripping their oil onto the filthy sheets. Beside him on the disordered bed sat a motionless Jeanne, heavily pregnant. Extraordinarily, she had listlessly not bothered to call for help, or for a doctor. When medical help came it was already too late. It was obvious that Modigliani had acute cerebral tuberculosis and was dying. He remained in a coma and finally expired on 24 January 1920. Montmartre produced an enormous funeral, attended by what seemed its entire population. Hardly a distinguished artist, writer or poet of the time was not present. Just two days later, Jeanne, aged twenty-two, threw herself out of a fifth-floor window, killing both herself and her unborn child.

It seems unbelievable, but Jeanne's family refused to allow them to be buried together. It was only years later that, in recognition of the love they shared, her body was finally released from its solitary grave and she was reburied with Amedeo.

This tale, so often repeated in various ways, illustrates a point. It is possible to analyse much of what goes on in the brain when love is in the air, but we will surely never stop being fascinated by the many forms love takes, or by the heights of achievement and destruction to which it may drive us. For those of us temporarily more interested in the mechanics of the business, scientists have identified three separate and distinct categories of the love experience – lust, attraction and attachment. Anthropologist

Helen Fisher, of Rutgers University, has argued that each of these stages had a different evolutionary purpose. Lust evolved to get us looking for mates, attraction to help us focus our attentions on an ideal partner and attachment to stay with them and raise our offspring.

Some mechanisms for a deadly sin

An MRI scan of the loving brain reveals that each of these stages has its own separate set of brain activities. When we are in lust, the airborne chemicals, pheromones, are particularly active. As we have seen, in the animal kingdom pheromones may act to indicate when a potential mate is on heat and ready for fertile intercourse. In humans, the actions of these chemicals on the brain may be more subtle, but still seem to be important.

The chemical androstenol (which, ladies, is said to smell faintly of sandalwood) is found in male sweat and can increase women's social interactions with men. Incidentally, it can – for those seriously interested – be bought on the worldwide web for around $30. Dr Andrew Scholey and colleagues at the University of Northumbria exposed both men and women to this type of pheromone. When asked to rate the attractiveness of another person, based only on a character sketch, women gave a higher score when in the presence of the pheromone even though they could not always consciously perceive it. In one controlled study he and his colleagues investigated the effects of exposure to male axillary secretions.[86] Thirty-two female undergraduates, of whom half were on the pill, gave a rating of attractiveness to male character sketches and photographs of male faces. They were tested on two separate study days, corresponding to different phases of their menstrual cycle, while exposed to male axillary pheromones and, unknowing, under a control condition without a pheromone. Pheromone exposure increased the enthusiasm of the women's ratings of vignette characters and faces. Use of the contraceptive pill or menstrual cycle phase had equivocal effects on the response of these young women to the vignettes but not to ratings of photographs.

Similar findings were reported in a study at the University of Vienna, conducted by Karl Grammer. His team asked men to rate photographs of various women by how attractive they found them. They then asked the men to repeat the task, but this time after having smelt female pheromones known as copulins. The effect of the copulins was to make the men far less discriminating – such that they gave higher scores to women they had previously not fancied.[87]

Studies at the University of Chicago have shown that a pheromone with an undetectable odour, with the improbable name of Delta 4-16 androstadien 3-one, has a significant effect on glucose usage in the brain. Undetectable steroids like this can have differing effects on women and men. A woman's temperature may rise, while a male experiences a fall in temperature – but these measurements tend to vary depending on the person taking the readings and how sexually active this observer is.

We are a long way from understanding the full nature and extent of pheromonal action upon the brain. As humans, we are driven to seek shortcuts, and as a result, the newspapers will always be full of advertisements proclaiming the 'scientifically proven' benefits of various so-called pheromone sprays and aftershaves. Various studies have indicated that it is possible to synthesize pheromones, and that these ersatz hormones can have powerful effects on behaviour. In one American study, seventeen young men were given an aftershave that, unknown to them, contained a lab-manufactured pheromone. Of the group, 41 per cent reported a noticeable increase in their sexual activity.[88] I am tempted to say that the scientist who can finally patent a powerful but harmless pheromone scent will find themselves sitting on a goldmine. But, as with the use of subliminal advertising, any device that allows one human to alter another's behaviour without their consent, particularly in the arena of sexual relations, is highly dubious.

Like the pheromones that lead to it, love itself can be a dangerous thing. Novelists, poets and songwriters have long made a link between the properties of love and the experience of being addicted to drugs. And it seems that there is much scientific truth in this. A study by Larry Young at Emory University in Atlanta, Georgia, examined rodent sexual behaviour and the related activity in an area of the brain called the ventral pallidum.[89] The pallidum is an area particularly involved in reward and addiction. Young has suggested that the reward and pleasure systems of the brain exist primarily to ensure that we are interested in sex, and therefore reproduce.

The prairie vole, a shy, grey-furred little creature not unlike an English fieldmouse, has proved to be of particular interest to people involved in the research of sexual behaviour. This is because prairie voles are monogamous. They mate for life and, rather sweetly, they normally opt to remain celibate if a partner should die before them. Male voles also show an active concern in the rearing of the young. Studies have shown that these family-loving prairie voles have high levels of a particular receptor in the ventral pallidum. A study at Emory University bred the related prairie vole genes into mice – who, in contrast, are normally sexually promiscuous creatures. The mice began to imitate the prairie voles in behaviour – settling down and remaining with a single partner. Studies of another promiscuous creature, the montane vole, have shown that they have significantly lower levels of this receptor in their ventral pallidum.

This work does not prove that monogamy may be genetically determined, but certainly suggests an influence. Monogamous creatures may get enough pleasure from being with one partner, while their more promiscuous counterparts have to seek repeated 'affairs' in order to achieve the same levels of satisfaction from their less enriched ventral pallida. This area is also activated when addicted rodents obtain a fix of drugs, or even when they enter areas of their cages associated with receiving the drug. The picture is somewhat reminiscent of what we learnt in the previous chapter about extroversion and thrill-seeking behaviour – in that the brains of people with these traits seem to be chronically 'unrewarded' by their activities, and thus they go in continual search of ever greater stimuli. Sexual promiscuity in humans may be a sub-type of this behaviour.

Rather unsurprisingly, this is an area of research that has suffered from lack of funding. It has been difficult for scientists to convince funding bodies that they should part with money to allow them to perform brain scans on copulating couples and cocaine snorters. And it was ever thus. Wilhelm Reich, favoured student of Freud and at one time a leading light in the field of sexual behaviour, was effectively drummed out of the academic community when it was discovered that he was performing research on people as they engaged in sexual intercourse. It should also be pointed out that the reason so much sexual research involves rodents and monkeys is because it is hard to find human volunteers. There can, after all, be few situations less erotic than sitting in an MRI scanner with a team of white-coated scientists looking on – even if they are behind a glass screen in an adjoining office watching on an oscilloscope monitor.

High levels of pheromone production might account for what psychologists refer to as the rope-bridge effect. In a famous experiment at the University of Vancouver, two groups of men were asked to walk across two very different bridges. One was a very unremarkable pedestrian foot-bridge – no more perilous than the various stone and iron affairs that traverse the Thames and which many of us happily cross without batting an eyelid every day. The other was like a special effect from an Indiana Jones film – a swaying rope-bridge which dangled over a menacing 230-feet drop (but, I write proudly, not as half as terrible as the one across the chasm, 2,000 feet up, in South Africa, as seen in the plate section of this book). As they reached safety on the other side, the two groups of men were approached by an attractive female who – rather unfairly, even if it was in the interests of science – flirted with them a little and gave them her phone number while handing them the inevitable questionnaire to fill out. The men who had crossed the dangerous rope-bridge included more sexual content in their questionnaire answers. And a considerable number of them rang up the researcher later on. An explanation for this would be that fear (and possibly relief from it?) causes arousal in both brain and body, including feelings which lead us to over-rate the attractiveness of the opposite sex. And I thought that most people went bungee jumping just for the view.

The feeling of falling in love

Attraction can be defined as a form of focused attention and goal-orientated behaviour. Attraction is, of course, vital for the continuation of the species. For that reason, we experience actual pleasure when we see an attractive face. Analysis of the brain chemistry involved shows that some pleasure-and-reward regions of the brain – specifically the medial orbitofrontal cortex – are stimulated by the sight of attractive human faces. The level of stimulation is at its highest when the faces are looking directly at us, and when they are smiling. So, unromantic as it sounds, we humans are all wired to have a 'roving eye'.[90] Further to this, when we experience attraction to another person, our brains become fixated upon them. We ignore other people, we may even neglect our appearance, our work or our health. We have started to be in love. Tests taken at this stage indicate high levels of a chemical called phenylethylamine (PEA), structurally similar to amphetamines. This may account for the fact that some people describe this stage as a 'rush' of excitement, coinciding with other amphetamine-like symptoms such as increased heart rate, racing thoughts, loss of appetite, disturbed sleep and enlarged pupils. In the past, women sought to simulate some of the signs of attraction as part of their beauty routine. In medieval Italy, they put drops in their eyes extracted from the root of belladonna, which has the effect of dilating the pupils. It is no coincidence that 'belladonna' means 'beautiful woman'. In more modern times, studies have shown that increased pupil size has a powerful effect on both men and women. Show somebody two pictures of the same person, identical except that one of them has been artificially morphed to increase the size of the pupils, and the morphed picture will most often be identified as being more attractive.

The production of PEA is just one of the changes that take place inside our brain when we are in love. Thanks to modern technology, we can look inside the 'loving brain' and see exactly what is happening. Andreas Bartels and Semir Zeki of the Department of Cognitive Neurology at University College, London, undertook brain scans of seventeen subjects as they looked at pictures of their sweethearts, and then at pictures of friends who were the same sex as their beloved.[91] When the subjects were gazing upon their lovers, there was specific activity in the media insula and part of the anterior cingulate – areas we use to concentrate and block out other stimuli. There was activity in the caudate nucleus and the putamen, which are part of the brain's pleasure-and-reward circuitry.

As we've seen, feelings associated with love can also seem to have the quality of an obsession, or an episode of mental illness. People also show an increased propensity to think in irrational ways about the object of their affection – idealizing them and attributing qualities to them that they may not truly have. There is a grain of truth in the notion that falling in love is a little like a bout of madness. Donatella Marazziti of the University of Pisa argues that the evolutionary consequences of love are so important that there must be some long-established biological process regulating it. It is difficult to know how seriously to take statements like this – Darwinian pronouncements of this sort seem to be proliferating as more people attempt to sell their version of science, and they risk the work of others whose good science may not be taken seriously. But Dr Marazziti reports that she managed to find twenty subjects who had recently fallen in love in the six months before her study (possibly not too difficult in Tuscany) and she compared these people with twenty OCD patients not on drugs, and twenty normal controls. She measured the levels of the protein responsible for serotonin transport in circulating blood cells. Her findings suggest that subjects who were in the early romantic phase of a love relationship were similar to OCD patients in probably having reduced serotonin levels in the brain. Once the initial flush of being in love falls off, after twelve to eighteen months, Dr Marazziti notes that serotonin levels return to normal. Professor Sylvia Finzi at Pavia University, presumably one of her competitors, archly noted that the symptoms of being in love 'were usually easily recognizable' and that: 'Reducing love to a game of molecules belittles it . . . No neurotransmitter will ever be able to keep love in a cage.'

There are all sorts of risks in taking research of this sort too seriously. First, all the subjects in love were female – what happens to males? Secondly, how do you define 'being in love'? Thirdly, is a group of twenty people enough for statistical significance? And, among other objections, does measuring the level of a related chemical transporter really reflect what is happening in the brain? I put these arguments, not because I disapprove of Dr Marazziti's work or wish in any way to undermine it. But reports on these kinds of psychological issues generally get massive attention in the press and the reports sometimes misrepresent the importance of really solid insights into brain function.

However intensely we may feel the first rush of attraction, we probably do not need scientists to tell us that it does not last for ever. This thought has not stopped research by Cindy Hazan of Cornell University, who has written extensively about attachment.[92] One long-term study, which has involved 5,000 subjects in thirty-seven different cultures, has shown that attraction lasts for between eighteen months and three years. The suggestion is that by four years any child produced is sufficiently biologically secure for a couple to break up and move on. According to this theory we humans are serial monogamists. As the *Sunday Times* – always keen to report on intellectually tough scientific issues – pointed out in August 1999:

Hazan's findings offer a scientific explanation for many famous bust-ups, including that of the marriage of Prince Charles and Princess Diana, who fell out of love soon after the birth of their second son. Another celebrity cited as proving Hazan's theories is golfer Nick Faldo, 42, who dumped Brenna Cepelak, 24, bang on 30 months after their adulterous affair began.

Gwyneth Paltrow, who recently won an Oscar for her role in the romantic comedy *Shakespeare in Love*, said of her failed three-year

As I understand it, Dr Hazan has identified dopamine, phenylethylamine and oxytocin as the chemicals that produce what Elvis Presley famously described as 'that loving feeling'. Whether or not she actually measured them in her subjects was far from clear to me in what she has published. She is reported as asserting, 'Like a drunk grows immune to a single glass of alcohol, the effect of these chemicals wears off, returning people to a relatively relaxed state of mind within two years.'

According to Helen Fisher of Rutgers University, who at least has used fMRI, levels of dopamine and PEA also reach a peak during the earlier phases of love.[93] It is not clear why levels of these chemicals drop after this period, but it might be that the brain develops a kind of 'tolerance' to producing its own 'drugs', and eventually stops making them at higher levels than normal. This could explain why some people, sensing that – in the words of the old song – 'the thrill is gone', continually go off in search of new love affairs. Mistakenly, they believe that one day they will find the continuously stimulating partner – the ideal who will ensure them an everlasting supply of PEA.

Chocolate manufacturers have recognized the value of the publicity surrounding research into the neuro-transmitters associated with love. Chocolate, as it is sold for the consumer market, is an extraordinarily complex compound and contains various chemicals, among which are small amounts of PEA. The 'health benefits' of con-suming products from the plant *Theobroma cacao* (translated as the 'food of the gods') have relied on the information that chocolate, especially rich, dark choco-late, contains catechins, which may give protection against heart disease, cancer and numerous other diseases.

Chocolate manufacturers and those reliant on this industry are very ready to point out that eating chocolate boosts one's appetite, but does not cause weight gain – though how this works is a mystery to me – and that sugar in chocolate has a calming effect and reduces stress. It is also stated quite frequently that chocolate has analgesic properties – but I will continue to take paracetamol. There is some evidence that serotonin and endorphin levels are altered in the brain after eating chocolate. But for me, certainly, and a number of other people I know, the caffeine, phenylethylamine and theobromine elements in chocolate seem quite frequently to start a migraine attack. This probably happens as a result of the release of the neurotransmitter norepinephrine, which alters cerebral blood flow. So, far from being a case of 'chocolate is the food of love, play on', it might be more, 'Not tonight, dear, I've a headache.'

Accentuate the positive

The final phase of love is attachment. It is said that once PEA levels start to drop off, this is what humans are left with – attachment. It would seem to have a clear evolutionary purpose, because our children spend so much time being dependent upon their parents. Attachment is associated with raised levels of oxytocin and vasopressin in the brain – which have the effect of inducing bonding and reducing aggression. Bonding works for parents, but is also a powerful psychological force between them and their offspring. And oxytocin induces smooth muscle contractions (the muscles over which we do not have conscious control) and heightens nerve sensitivity. Its presence therefore enhances the pleasurable qualities of sex, and so it is not surprising that in both women and men, levels of the chemical rise dramatically from the first kiss to the post-coital cuddle. It continues to act throughout the reproductive cycle. It has a powerful effect on uterine contractions, and during labour and delivery when oxytocin levels are at their highest it is crucial for the birth of a baby. Because the release of oxytocin allows the milk ducts in the breast to contract, it enables breastfeeding. Solid research shows that blocking the production of oxytocin in rats prevents bonding between mothers and their young.

Reduced oxytocin levels may be just one of the factors involved in mothers who experience problems bonding with their new babies. This could even be why mothers who have had difficulty with their labour due to weak contractions, and those not having a labour at all (but are delivered by Caesarean section), occasionally feel emotionally removed from their new infant. I do not know of any good work to prove this, however, and admittedly there are likely to be many other factors which impair the relationship between mother and child after a difficult delivery. A University of Washington study showed that the act of childbirth itself normally triggers a certain neural pathway. Mice born surgically were found to have mothers who were significantly less maternal towards them. But if the mothers were given injections of the neurotransmitter norepinephrine during the surgery, their nurturing skills increased.

The hormone oxytocin also seems to play a role in the memories we form to aid our social interaction. Jennifer Ferguson at Emory University has rapidly made a name for herself in this field. In a series of interesting publications she and her colleagues have shown that mice bred so that they are missing the oxytocin gene do not develop social memory. These male mice were repeatedly introduced, along with other, normal mice, to a female who was unable to conceive. The normal mice, after repeated pairings, began to spend less and less time exploring the female. But the mice without oxytocin showed the same level of interest in her, no matter how many times they were introduced. But in spite of their lack of social airs and graces, these modified mice were perfectly capable of remembering other things – like finding their way through a maze, locating and remembering where a food was by its smell, and being able to seek out – wait for it – chocolate. The evidence suggests that oxytocin helps the brain to 'remember' socially important information and that the amygdala is a key in this process.[94]

Oxytocin seems also to have a role in maintaining feelings of comfort and well-being in close relationships. Dr Rebecca Turner at the University of California in San Francisco tested the idea that oxytocin is released in response to intense emotional states.[95] She asked twenty-six women between the ages of twenty-three and thirty-five (who were not lactating) to remember and re-examine a past event that caused them to feel a positive emotion about somebody, such as love or infatuation, or a negative emotion, such as a break-up. Because the well-being induced by massage may change oxytocin secretion, the respondents were also given a Swedish massage of their neck and shoulders for fifteen minutes. Blood was taken before, during and after the parts of the experiment and their levels of oxytocin measured. It is not surprising to me that the results did not show a great deal – measurements of oxytocin like this are likely to be a very crude reflection of what is happening in the brain. Nevertheless, the relaxation massage was associated with a rise of oxytocin and past negative emotion seemed to cause oxytocin levels to fall slightly; positive emotions had no demonstrable effect. There was marked variability between subjects. Some women who showed substantial fluctuations seemed to have more secure relationships.

The evidence suggests that male-brain activity is just as

important in the process of bonding and attachment. High levels of vasopressin are found in prairie voles and hamsters whose female partners have just given birth – and they exhibit correspondingly high levels of aggression in response to any threats towards their young. But while it increases aggression towards the outside world, vasopressin also stimulates father–child bonding. Hamsters – who do not traditionally play the role of 'doting dad' – when injected with additional vasopressin spent more time around their pups. A study at the University of Amherst, Massachusetts, has shown that an interaction between vasopressin and testosterone in male voles ensures they remain aggressive towards outsiders but are friendly and nurturing towards their own offspring. In the human male, testosterone and hormones like oxytocin or vasopressin have contradictory effects. The story goes – and the older I get, the less I believe it – that, in most men, testosterone production decreases with age. As a result, the oxytocin and vasopressin in the male brain may become the dominant chemicals. So some behavioural experts conclude that older men become less interested in seeking sexual activity with a range of partners, and more geared towards a single, stable relationship.

So the evidence suggests that, over time, the first mad fever of love becomes replaced by a more solid partnership, one more ideally suited to the raising of children. But some research has suggested that the most long-lived relationships are those that preserve a degree of the earlier, less rational phase. Dr Ellen Berscheid, Regents Professor of Psychology at the University of Minnesota, was the first to identify the 'pink-lens effect', whereby couples in love idealize their partners and make overoptimistic judgements about them. We think, for instance, that our lovers are 'brilliant' and 'wonderful'; we overestimate their intelligence, their honesty, their generosity and their looks.

As we saw above, this effect has been shown to decrease over time. In general we would expect that, as a relationship progresses, we become more realistic about our partner's strengths and weaknesses – we 'love them for what they are', rather than some idealized, rose-tinted view we may have picked up in the initial frenzy of attraction.

But a study at the University of New York, Buffalo, has also shown that the greater the intensity of the pink-lens effect, the greater the likelihood of the couples staying together. Another, more long-term study at the University of Texas, Austin, followed 168 couples who married in 1981. Once again, the couples who idealized one another the most at the outset were also those whose relationships were the most long-lived. So – here is the absolutely obvious solution to life's most complicated problem – it takes a degree of irrationality to make a happy marriage. But, certainly, an ability to 'accentuate the positive and eliminate the negative' is a great asset for any married couple – as I suspect it is in any human partnership or friendship.

Damned lies and statistics

Why, if the human brain is such a finely tuned device making continual choices based on our need to survive and reproduce, do we repeatedly make mistakes? How do people end up in abusive, damaging relationships? Why do certain people gain advantages by cheating and lying, and how is it that we are unable to spot them?

In evolutionary terms, there is much to be lost by making a wrong decision about another human being. A cheater, for instance, might employ the strategy of pretending to gather food for the group, but secretly keep the greater share for themselves. If, over time, cheaters develop strategies that successfully mask their dishonesty from other members of the group, these abilities would tend to enter the gene pool – ensuring that every group has a convincing liar within it who pays lip-service to the community but really only looks out for themselves. In times of scarcity, people like this would be dangerous to the survival of the group.

This is thought to be one reason why the human brain has developed a specific mechanism for assessing the mental states of other people. Work at the Institute of Neurology at University College, London, by Joel Winston in Ray Dolan's[96] group has pinpointed the importance of the superior temporal sulcus (STS) region of the temporal lobes in making such judgements. The STS becomes activated when we are observing the movements and gestures of other humans but it is also active when we view static images of the human face and judge their trustworthiness. Perhaps we use the STS to work out what people are thinking as much as what they are doing. Various studies using fMRI and PET scans performed while people are attempting to infer the mental states of others also show a 'loop' of activity, encompassing the medial prefrontal cortex, lateral inferior frontal cortex and the temporo-parietal junction. We have evolved a complex mechanism to ensure that we know what our fellow humans are about to do.

As already discussed, people with autism show a specific inability to 'read' other people in this way. And there is evidence to suggest that this aspect of autism is related to a particular abnormality within the brain. Uta Frith, of the Institute of Cognitive Neuroscience at University College, London, has conducted a number of important studies.[97] One involved taking PET scans of subjects' brains as they watched a sequence of moving shapes on a screen. The subjects – ten of whom had autism, ten of whom did not – were shown three sequences. In the first, the shapes moved in a random fashion; in the second, they moved in a goal-orientated fashion, chasing and 'fighting' one another. In the third sequence, the shapes' movements were more subtle, and were designed to elicit descriptions from the viewers that involved attributing mental states to the shapes – for instance, coaxing, teasing, tricking. In other words, in this third set of movements, the subjects were being asked to identify certain human traits of behaviour.

The study made a number of interesting findings. It became clear that the autistic people were much less accurate or consistent when it came to the third collection of shapes. Secondly, PET scans revealed that a common network of brain regions was active in both groups when they watched the third sequence. This network – composed of the superior temporal sulcus and medial prefrontal cortex – is one area where our brain assesses information about the intention of others. Crucially, in the autistic participants, a specific site – known as the extrastriate cortex – was highly active, but, unlike in the brains of the non-autistic participants, not in tandem with the brain regions around it. The researchers have argued that autistic people's inability to 'read' others may stem from this bottleneck in the brain's processing system – one bit of it is working fine, but it cannot communicate with the rest, like an axle which spins without turning the wheels.

In our evolutionary past, we needed to know who we could rely on. As the old saying 'blood is thicker than water' suggests, family members are more likely to be trustworthy. And psychological studies have shown that we are, in fact, more inclined to trust people whose facial characteristics resemble our own. At McMaster University in Canada, psychologist Dr Lisa DeBruine asked participants to play a simple trust game, in which they could not see their partners. The game involved asking them to choose between two options: dividing a small sum of money between themselves and their partners, or opting to trust their partner to divide fairly a far larger sum of money. In other words, the game tested how trusting people were.

Before each game, the participants were shown computer-morphed facial images which – they were told – were pictures of the people they would be playing against. The pictures were manipulated in order to resemble either the participants' own faces, or the faces of an unknown person. The results showed that when the participants believed they were playing against people whose faces resembled their own, they were more likely to trust their partners. This makes some evolutionary sense; people who resembled us were likely to be blood relatives and therefore more likely to behave unselfishly towards us. If your chief evolutionary objective is passing on your genes to the next generation, then assisting your relatives – however distant – gives at least some of your genes an elevated chance of being passed down the generations. Unfortunately, the business of living in increasingly large cities requires us to trust a circle of others much greater than the sum total of our blood relations. So it will be a long way to go before our brain will allow us to achieve the perfectly harmonious, cooperative society.

Lying to ourselves

It seems that, as far as assessing the behaviour of other, unrelated humans goes, our brain has predisposed us to be pretty cynical. A particularly interesting study at Cornell University, now known as the 'holier-than-thou' experiment, has shown how accurate we are at judging others' behaviour, and how poorly we judge ourselves.[98] In one test, people were asked to say how much of their fee for participating in an experiment they would give to charity.

One group of participants said they would give half of their five-dollar remuneration, but in practice gave an average of $1.80. Moreover, after meeting each other, they tended to estimate that they themselves would be more generous than the other participants in the study. A total of $19.89 was raised from this study; had participants given what they had promised, the charities would have received $31.72. These people were generally accurate in predicting how selfishly others would behave. But they were also consistently overoptimistic about their own levels of altruism. Perhaps this indicates how important cooperation is to the group. We depend upon the input of others in order to survive and this may explain why our ability for detecting selfish behaviour is more sensitive.

The need to behave in a manner most beneficial to group survival seems to have resulted in the evolution of mechanisms that signal to others when we are lying. We can see what's going on inside a lying brain – Daniel Langleben at the University of Pennsylvania has shown a pattern of increased blood flow to the anterior cingulate gyrus, which as we know becomes active when we need to concentrate on something.[99] But other brain regions are also important. A team led by Tatia M.C. Lee at the University of Texas at San Antonio recorded fMRI scans of participants who were told to do deliberately badly on a memory task.[100] The results indicated a complex loop of activity, involving areas of the prefrontal cortex and the parietal lobes. To lie effectively, it seems the brain has to first come up with the truthful response, then use further mechanisms to suppress it, produce a fresh, false response and monitor its own performance. Small wonder that we find it hard to lie continually.

We can spot liars from the way they smile. One of the first great neurologists to make detailed studies of the human face was the remarkable eccentric but phenomenal doctor Guillaume Benjamin Duchenne. He has been something of an icon to me ever since I became interested in the terrible disease to which he gave his name, Duchenne muscular dystrophy. This fatal disease, affecting only boys but carried by girls, was one of the first genetic disorders my scientific group and I attempted to detect in the human embryo in 1990.

Duchenne was something of a late starter – and as somebody who did not publish his first scientific paper until the age of thirty-two (very late indeed, these days), I empathize with that – as I do with the fact that he was, by all accounts, a remarkably undistinguished medical student. When he left provincial Boulogne for Paris, he was ridiculed for his coarse accent and had difficulty scraping a living. His father, like those before him, was a seafarer and Duchenne seemed to haunt the Paris teaching hospitals as an unpaid attachment; he was described as a lonely, mariner-like figure. But Duchenne had a most important quality for good scientific research – he kept immaculate records of what he saw. Moreover, he was persistent and would follow patients from hospital to hospital to gain insight into their condition. Over time, his neurological diagnostic skills came to be praised and he developed a particular interest in muscle diseases. This led to his attempts to stimulate muscles with electricity to try to preserve their function in wasting diseases like dystrophy, which included experiments to stimulate the facial muscles. His immaculate note-keeping, combined with his interest in photography, have left us an extra-ordinary resource, the record of his work dating back to 1860.

The Pinocchio factor

Duchenne eventually became internationally recognized for all kinds of research on brain function and for the identification of a number of important degenerative diseases. But relevant here is that he studied the mechanism of facial expression during emotion and recorded it in his atlas of photographs. This book is recognized to this day as being an enormous contribution to medical photography. Charles Darwin used a number of his photographs in his book, *The Expression of the Emotions in Man and Animals*. From Duchenne's pioneering work, we now know that we use four major facial muscles in the production of a real smile, but only two in a false, 'social' smile.

With the advantages lent to us by brain-scanning technology, we can confirm the mechanism of the false, social smile. The smile I fix to my face for a cocktail party (*Rictus ginantonicus*) – originates in the left brain. This communicates with the right brainstem, ordering the right lower facial muscles to respond. Simultaneously, the left hemisphere sends a command across the corpus callosum to the right hemisphere. The right hemisphere in turn causes the movement of the left lower facial muscles. An involuntary smile, which shows true feelings, on the other hand, will be betrayed by movement in the muscles around the eyes.

So an analysis of a simple human smile can tell us a great deal about a person's intentions. But facial expressions are not the only means by which a liar can be spotted. Liars also touch their noses more frequently when they are practising their devilish art. This 'Pinocchio effect' is thought to be because increased blood flow causes erectile tissue within the nose to swell, thereby making the hairs inside our nostrils stand up.

Dr Richard Wiseman at the University of Hertfordshire has conducted some of the key research into lying, and shown that true emotions tend to be registered on the face as fleeting 'micro-expressions' which last no longer than a quarter of a second. Wiseman, who began his working life as a stage magician and as a member of the famed Magic Circle, can certainly be said to know a fair bit about the business of deception. He has developed a list of some of the most glaring indicators of a fib. I spent a long session with him recently, which was filmed for television. I gave him two accounts of how I had spent successive evenings. One was strictly truthful, and one a fabrication. There was no way he could know beforehand which account was true. I pride myself at being quite a good liar – my children take pleasure in shaming their father by recounting how I used to cheat them from the age of six. I regret to say, I cannot bear to lose, even at Scrabble. One evening, I told Dr Wiseman, I had given a named lecture, received a gold medal, been publicly critical to a government official and had dinner with various members of the medical establishment afterwards. The following evening, I recounted how I had turned up slightly late for a dinner in Parliament, where I had heard a speech given by a notable football team manager about how his team had deservedly won the Premier League Championship. Moreover, I said, this appalling Scot had dared to suggest the club I support was 'rubbish'.

The TV producer and his team recording this farrago of nonsense were horrified because they thought they had wasted a valuable day's filming. There was no way, they felt, that Richard Wiseman would untangle my stories because I had sounded very plausible. But Dr Wiseman does not need a lie detector – indeed, he points out that these are useless. Moreover, he knows nothing about football, but this did not matter to him. He was looking for the involuntary signs we give out that reveal we are not telling the truth. These might include moving the head a lot when talking, repeated touching of the face, too much (or too little) eye contact, sweating and dilated pupils, remembering too few or too many details, delivering abrupt answers to questions and leaving long gaps, or no gaps at all, before answering questions. Repetitious 'ums' and 'ahs', too, are a bit of a giveaway. Building up a complete picture, together with voice quality, Wiseman has been invariably accurate with his assessments – as indeed he was in assessing which of my two stories was a fabrication. Dr Wiseman's research shows that even when a listener is blindfolded, the detection of a liar is mostly made with high accuracy by the blindfolded person. Voice quality changes. His research has led to the development of a software package called 'Verdicator', which checks for lying cues within the human voice. That, together with his records of how liars behave, means that you would not want to meet Dr Wiseman if you had just been arrested for robbing a bank.

Another big name in lying research is Paul Ekman of the University of California. Ekman has built a 3,000-item-strong database of universal human facial expressions – certain combinations or 'Action Units' which are used solely when we are telling lies. Ekman's Facial Action Coding System is now consulted by groups as disparate as the FBI and Hollywood animation studios. He maintains that everyone has a specific set of facial 'giveaway clues' – what poker players would call a 'tell'. When Mrs Thatcher publicly denied that she ordered the sinking of the *General Belgrano*, for instance, her eyelids engaged in a brief but rapid flutter. When Kim Philby denied being the 'third man' who tipped off Burgess and McLean that their cover as double agents had been blown, his right inner eyebrow gave a quick involuntary flicker.

TV footage of former US President Bill Clinton has, unsurprisingly, also undergone some quite thorough investigation by those interested in the science of lying. Alan R. Hirsch of the Smell and Taste Treatment Research Foundation in Chicago noted a range of changes in Clinton's body language when he publicly denied having sex with Monica Lewinsky. He compared this public interview with the one Clinton gave when he was sworn in as President. During the interview about Monica Lewinsky, Bill Clinton showed a 250 per cent increase in hand-to-face touching, and a 355 per cent increase in drinking and swallowing. Meantime his stuttering rate increased by 1,400 per cent, and other errors in speech by 1,700 per cent. While all this sounds impressive, there is a real problem of analysis. No matter how careful the investigators are, there is a huge difference between an interview given in highly favourable circumstances like a presidential inauguration, and one when a man in President Clinton's position is faced with an extremely hostile interviewer, in a situation which threatens both his public position and his private life.

But if our brain is built with this amazing equipment to assess the mental state of others, and if lying gives off obvious clues, how do liars get away with it? It may be more interesting to rephrase the question, and ask why do they get away with it? Could there be some evolutionary benefit in not detecting lies?

It is worth noting that, like my camera crew, we humans are not that good at spotting a *practised* liar, particularly one that knows how to obfuscate. One technology to improve the average person's ability to spot a fraud depends on being able to slow down and freeze video footage. In Dr Wiseman's studies, when he slowed video footage down to one frame per second, two-thirds of his 120 volunteers were able to spot the liars, which indicates that while we may all have the 'knack', we just can't use it much in daily life. Other studies have shown that ordinary individuals score only marginally worse than skilled professionals, like police officers and psychiatrists, in detecting liars. So why aren't we better at it?

One obvious answer is this. Just as we have evolved sophisticated brain technology to detect and signal lies, the liars themselves have been one step ahead of the game. In other words, cheats have evolved their own, increasingly sophisticated mechanisms for avoiding detection. And if you are a liar, the most deadly weapon of all is the ability to persuade even yourself that you are telling the truth.

Another possibility could be that the consequences for group survival in exposing cheats might be worse than the consequences of their cheating. For instance, given the example of the 'hunting cheat' above, perhaps it would be better for the overall survival of the community to have someone who contributed *something* to the group, rather than casting them out and thus being denied any contribution from them. We can see how accusations of lying, or socially harmful behaviour, have significant negative repercussions for any society. They do not merely result in one person or group of people being excluded, but create wider rifts and faction fighting in those who remain. Such damage could compromise the survival chances of a human group, particularly in harsh conditions.

We do not need to look at our ancestral past on the savannah to see that lies – or rather the facility to tell them and not be detected – are vital to smooth human relations. Take the example of a wife who asks her husband if he likes her new haircut. His actual feeling may be that he does not, or perhaps does not at that particular moment, but he will say, 'Yes, lovely, dear' because he knows this is better for his wife's self-esteem and for the continued functioning of their marriage. Likewise, it helps that his wife is able to believe what she most wants to hear, rather than root out an unflattering truth.

Similarly, when a father first glimpses his newborn child, his partner and everyone else around will almost certainly reel off a list of imagined similarities, such as, 'He's got your eyes', 'She's got your chin'. In fact, research has shown that very few facial characteristics of either parent are detectable in the faces of newborn infants. But it is in the interests of the family group for a father to believe that the child is his, thus ensuring that he sticks around to provide resources. Neither the mother nor the father believes a lie is being told – but if fathers did not have this capacity to 'gloss over' an obvious untruth, the social consequences could be disastrous. In a host of social situations, a 'white lie' goes undetected, because detecting it might simply do more harm than good. So does that mean that we are fundamentally without a sense of right and wrong? Not a bit of it – as I shall go on to explain.

The moral mind

We have seen that, in order to live alongside other people, our brains have had to evolve sophisticated technology for detecting and assessing the behaviour of others. We have also seen, time and time again, that the driving force behind so much of our mental make-up is the old evolutionary need to reproduce and survive. But we humans have the ability to move beyond the simple requirements of food, shelter and sex. Unlike most other animals – or what we know of them to date, at least – we are not bound by the immediate needs imposed upon us by our personal drives and the environment in which we find ourselves. Faculties such as the sense of self, and language, have given us the ability to move beyond the moment – to consider ideas in the abstract, to formulate plans, envisage improvements to the way we live our lives, and communicate them to one another, to create art, to have ideas.

A key area in which these for the most part uniquely human faculties play a part is in the development of attitudes such as morality, forgiveness and compassion. I have little doubt that, at one base level, morality is likely to be a hangover from our evolutionary past. The group requires rules in order to function safely, and sanctions in order to make people obey them. Commandments such as 'Thou shalt not steal' and 'Thou shalt not kill' are present in virtually every form of human society – probably because intra-group theft and murder would have been dangerous to group survival. I put stress on the term intra-group, because the same rules do not universally apply to other tribes or nations. In spite of the Ten Commandments, the ancient Israelites were not above killing Canaanites, or rustling their cattle, even though their treatment of the stranger, the foreigner, the captive and the slave was morally far advanced compared to the treatment the nations around them seemed to enforce on other peoples.

As I noted earlier, rule-breakers are rewarded with social exclusion. And this applies whether the rule concerns itself with the appropriate shirt to wear to a cocktail party, or not chatting up other men's wives. We use rules to ensure that our groups 'work' efficiently. And this drive to cooperate would seem to have a neurological basis inside our brains. In a study conducted by James K. Rilling at Emory University in Atlanta, Georgia, volunteers were asked to play a game called the 'Prisoner's Dilemma', in which players have to decide when to trust one another. MRI scans performed when the volunteers were cooperating together revealed a specific pattern of brain activity in the regions linked to pleasure and reward, like the caudate nucleus and nucleus accumbens. In other words, we may be hard-wired to receive pleasure from cooperation.

But human morality goes way beyond the simple survival needs of the group. What about, for instance, the principle that suggests we give up our seat on a crowded bus to an elderly woman? In evolutionary terms, there might be no conceivable value in helping the elderly. They are weak, they cannot hunt or bear children or defend us from predators. They are a drain on resources. Yet we accord them compassion and respect. Archaeological evidence suggests that 40,000 years ago our Palaeolithic ancestors – for whom survival must have been a far dicier business than it is for us – fed and looked after the elderly and the infirm.

What value could there be, for example, in forgiveness? The strictly deterministic, evolutionist point of view would suggest that we need to punish people who break our rules in order to prevent them, or others, from doing it again. So how could we have evolved codes of action that encourage us to forgive, to turn the other cheek?

From an anthropological point of view, one answer might be that these values only emerge as society becomes increasingly complex. In a simple society, there is relatively little specialization of roles. The men hunt, kill, make tools and build shelters. The women give birth, nurture children and prepare food. But as human groups become larger, we see an increased specialization. Larger groups allow for increasing amounts of food to be brought into the camp, so that there is a surplus to reward those who have not participated in the actual act of finding and killing it. These individuals can specialize in other tasks which benefit group survival, like making tools, or pottery. As time goes on, specialization increases, such that people become adept at tasks less directly related to the business of hunting and killing enemies. Classes of individuals like farmers, bakers, scribes and builders enter the ranks – all contributing something to the group, all receiving something in return for their contribution. More specialization means more surplus, more surplus means that survival becomes easier and easier – which in turn means that we can afford to show compassion to the weak and the elderly.

Not everyone would accept this view of the evolution of

morality. It has been noted, for instance, that the ability to acquire a surplus of produce, through techniques like farming, was the driving force in creating social divisions, rather than social harmony. A man who has surplus can pay others to perform tasks for him; he can also acquire possessions like larger houses, clothing, pottery and jewellery, which serve to accentuate his superiority to others. We start to see a class system, divided along the lines of power and wealth – quite the opposite of a compassionate society.

It might be of interest to take a look at the examples below.

> You are watching a view from a footbridge over a railway. You can see a runaway railway carriage, heading towards a forked junction. There is a lever in your hand, which you can pull to determine the direction of the railway carriage after it hits the fork. If you move the lever forwards, the carriage will kill one person. If you move the lever backwards, it will kill five people. Which way do you pull the lever?

> Now you are standing on another footbridge. This time you can only prevent the railway carriage from killing five people by pushing the person standing next to you off the bridge and into its path. Do you do it?

Did you answer 'forwards' to the first example, and 'no' to the second? If so, don't start worrying whether this demonstrates something bad about your moral character; this is how most people would answer both the hypothetical questions.

But in coming up with your two answers, you were probably using two different systems of your brain's 'moral network'. Joshua D. Greene at the Center for the Study of Mind, Brain and Behavior at Princeton achieved considerable publicity in the popular press with this study.

He gave sixty dilemmas to volunteers, and undertook fMRI scans of their brains as they weighed up the options.[101] The second of my examples involves a strong emotional content – because it requires us to imagine ourselves actively pushing another person to their death, regardless of the maths involved. Other dilemmas were like the first type, or had no moral component at all. Greene discovered that, in considering the 'emotional-type' dilemmas, there was significantly higher activity in three brain areas associated with processing emotions – the medial frontal gyrus, posterior cingulate gyrus and the superior temporal sulcus (STS). This latter region of the brain is thought to be the centre of our 'mentalizing' capacities – in other words, the skills we use to guess the mental states of other people. Brain scans of psychopaths often show lower levels of activity in the STS.

In contrast, the non-emotional type of judgements, as in the first example, showed more brain activity in areas like the dorsolateral prefrontal cortex and parietal lobes. These play a role in working memory. This pattern suggests that, when we have to deal with a dilemma with a strong emotional content, we put reality on hold for a moment and dip into our brain's little toolkit of social skills in order to decide how to act. There isn't some 'Organ of Morality' inside the brain as phrenologists might once have hoped, but instead there is a complex interplay between areas that control abstract thought and social behaviour. This implies, but does not prove, that morality evolved as a consequence of us living in increasingly complex social groups.

In another joint study, involving teams from the Universities of Sheffield and Manchester, brain scans were taken as volunteers read and made judgements on a number of social scenarios. The scenarios were designed to invite the reader to make a judgement invoking empathy or forgiveness. For example, an empathy-inducing scenario might be – *Your boss is not himself today. What do you think might be wrong?* A forgiveness-inducing scenario might be – *A young man in your street, whom you know to have recently lost his job, is visited by the police. What has happened?* The researchers found that very specific brain areas were active in the two types of scenario: one set of brain regions for empathy, and another for forgiveness.

We also know that humans mirror the action of others in order to understand their emotional states. Our amygdala registers the pattern associated with sadness when we see a sad face, and it meets anger with anger and fear with fear. In other words, the architecture of our brains predisposes us to interpret other people by feeling what they do, by putting ourselves in their shoes. It could be that concepts like compassion, empathy and forgiveness are just further refinements of this basic function – the add-on extras of the social brain.

In an earlier chapter I looked at the possibility that language was merely a further specialization of the brain areas that became enlarged as our ancestors made and used tools. It could be that the moral brain came as a later development along the same evolutionary river. The ability to conceive of oneself as a doing-I and a done-to-me, of a distinction between self and others, was central to the development of grammar and to the ability to build complex toolkits. It could be argued that the ability to verbalize these understandings also predisposed us to become moral. Your brain needs a level of abstraction to come up with a moral supposition such as 'I wouldn't like this if it was done to me' – one only possible with the fluidity lent to us by language.

chapter**nine**

The amazing mind: intelligence, creativity and intuition

Hermann and Pauline were worried about their sixteen-year-old son. He had always been a difficult boy – when he was born he seemed to have an overly large, misshapen head which, despite the doctor's assurances, never shrank down to an entirely normal shape. When his grandmother first saw him, she cried, 'Fat! Much too fat!' – and throughout his childhood, the boy was a heavy-set, rather lumpen figure, disinclined to running about or joining in the games of his peers. He did not learn to speak until he was nearly three, and was prone to bouts of terrifying anger. In one of these fits, he struck his violin teacher with a chair – and the teacher fled, never to return.

At school he was by no means unpopular, although – a little surprisingly for a Jewish boy in Central Europe – he struggled with foreign languages and teachers were exasperated by his ponderous way of thinking for a long time before answering questions, and even more by his habit of quietly talking to himself under his breath. The boy seemed happiest on his own, building card-houses of immense size and intricacy.

In 1894, when their son was only fifteen, Hermann and Pauline moved to Milan, to pursue business opportunities in the family's electrical engineering firm. Their son, to Pauline's regret, stayed behind in Munich to continue his education, lodging with family friends. The letters he wrote to his parents were sporadic, typically teenage – desultory and grudging in tone, reporting that he was doing extremely well at geometry, but telling little else of his life or his feelings. Then one day, he showed up in Italy. He and his form teacher hated one another, the boy reported. The teacher had said the school would be a better place if he weren't in it. His pupil quite agreed. So he had gone to see the family's old doctor and persuaded him to write a note – claiming that a decline in the boy's physical health necessitated him leaving school and going to live with his parents.

Not surprisingly, Hermann and Pauline were annoyed and distressed at this move on their son's part. He had lied for his own advantage and he had dared to quit school, a mere fifteen-year-old boy, without consulting them. But pleading and threats did nothing to change his mind. He had decided what he was going to do. He was going to prepare himself for the entrance exams for a highly prized place at the famous Zurich Polytechnical School. At the age of just sixteen, and without the benefit of specialized tuition, it seemed like a long shot. But impressed by their son's determination – and possibly also feeling a little guilty at having left him behind in Germany – his parents backed him.

Unfortunately, things turned out as they had feared. Their son did well in the mathematical and scientific papers of the entrance exam, but poorly in languages and history. The school suggested he needed at least another year of secondary education. Sighing at their son's stubbornness, Hermann and Pauline agreed that he should enrol for another year at a renowned Swiss school. But what was to say that he would cope any better there? How long would it be before he showed up in Milan again, citing personal antagonisms as the cause of his departure, and stubbornly insisting on some new and equally improbable plan of action?

In many respects, this little historical tale is no different from a drama played out between parents and children all over the world. The business of raising children is never easy, but some never seem to be happy from day one. They are difficult, at home and at school. They are wilful and impulsive, stubborn and argumentative. In the end, the only thing their loving parents can do is stand back and let them follow their own paths. Some of them do well – others, because of their indecision, or their stubbornness, their lack of social skills or their refusal to 'play the game', will always find life tough.

Need Hermann and Pauline Einstein have been so worried? Their son, Albert, did finally gain entrance to the Zurich Polytechnic School. And, of course, he went on to become part of history as the most renowned scientist of the twentieth century – the originator of the Theory of General Relativity, which radically altered the way we think about time and space. He is acknowledged as the possessor of an extraordinary intelligence, a genius, to rank alongside Mozart, Newton and Leonardo da Vinci, and countless others whose work has shaped our history.

But as we have seen, Einstein's childhood and adolescence were fairly unremarkable. His story, up until the point at which he gained admittance to the Zurich Polytechnic, was quite a familiar one of awkwardness and various tussles with authority. So what was different about Albert Einstein – what enabled him to turn from a difficult boy into a world-famous genius? Could it be that there was some quality, some aspect within his brain, which made him into a man of such extraordinary intelligence?

416

The contents of a Costa Cider jar

In 1955, Einstein died of an aneurysm. Not always a modest man, he had already made arrangements for his body to be cremated but his brain to be preserved for scientific research. Within seven hours of his death, the pathologist at Princeton, Dr Thomas S. Harvey, was dissecting his brain. Harvey preserved the unremarkable-looking object in formalin and then cut it into 240 blocks, each roughly the size of a sugar cube. Sections were cut from these blocks, and they were examined under a microscope. Neurologists all over the country eagerly awaited the results of Harvey's work. But his findings, when they came, were greeted with sighs of disappointment. Harvey said he could find nothing whatsoever that was remarkable or unusual about the structure of Einstein's brain. It was like any other damned brain, and it held no clue as to the incredible powers of its former owner.

So for a long time the brain lay pickled in a jar, but forgotten. In the mid-1970s, a journalist, Steven Levy, attempted to find the whereabouts of this treasured organ. Einstein's brain turned out to be still with Dr Harvey, now living in Wichita, Kansas. It was in two mason jars in a cardboard box that was marked with the words 'Costa Cider'.

Dr Marian C. Diamond was one of the first scientists to get hold of bits of Einstein's brain for detailed study. When the brain pieces arrived in her laboratory in Berkeley, California, they were now in a jar marked 'mayonnaise'. In 1985 she published a paper which gave news which was unimpressive if not frankly disappointing – Einstein's brain cells were no more and no less than those of any other average ageing man. Dr Diamond had compared the number of neurons with the number of glial cells in many areas of the brain and it was entirely average. Only in Brodmann's area 39 of the parietal cortex was there a difference. If you have ever hit the side of the head with your hand after saying something really silly, you now know you've got the right spot. And here there were actually fewer neurons in proportion – the ratio of neurons to glial cells was considerably less than average.[102] But like his ideas, Einstein's brain continued on its travel. In the 1990s, Dr Sandra F. Witelson and colleagues at McMaster University in Ontario, Canada, were given a unique opportunity. They painstakingly measured the size and shape of Einstein's 'organ of the mind' and in June 1999 the *Lancet* published their study.[103] The external surface of Einstein's brain was compared to those of the brains of thirty-five men of similar age. Each brain hemisphere was wider than average, even though the overall brain size was quite unremarkable. But Einstein's brain had an unusual pattern of both right and left parietal lobes, an area that is thought to be important for mathematical abilities and spatial reasoning. His brain also had a much shorter lateral groove, the parietal operculum, that was partially missing – and the parietal lobes themselves were 15 per cent wider than was seen in nearly all other brains.

Dr Witelson reports a similar structural variation in the brain of the famous mathematician Gauss, and the rather less well-known physicist Siljestrom. She tentatively concludes that she might have found the site of a specific kind of intelligence. But obviously the skills of the likes of Einstein and Gauss only provide evidence of one brand of intelligence. We do not only acknowledge some mathematicians and physicists to be geniuses. Clearly, the worlds of writers, musicians and artists, social theorists, actors and directors have contributed their fair share. And we also speak of high intelligence without any mention of genius. Among any ordinary collection of humans, there seem to be people who are 'brighter' than others. So what do we really mean by intelligence? Is a bright person someone who knows a lot, or just someone who has the capacity or the motivation to learn a lot? Or might it rest in entirely different abilities – such as the power to discriminate between useful and useless information in the swiftest time? But was Einstein's brain truly unusual? How do we account for intelligence – is it related to a specific feature of the brain, such as the findings from Einstein's autopsy might just suggest, or to our experiences and environment? Did Einstein's brain just develop in this way because of something in the environment? If the latter is the case, then perhaps we all have the capacity to be intelligent.

The measures of intelligence

In January 2003, a twelve-year-old boy from the Midlands won the finals of a television show called *Britain's Brainiest Kid*. When interviewed, he said that he'd been initially inspired by taking a test on the website of Mensa – a global organization for people whose IQ scores place them in the top 2 per cent of the population. He also said his interest had been sparked by watching an episode of the US cartoon show *The Simpsons*, in which child prodigy Lisa Simpson decided to join a high-IQ society not dissimilar to Mensa.

Little Lisa Simpson provides an interesting starting point for our journey down the lanes of human intelligence. For she is, undoubtedly, what we might call a child genius. She follows in the tradition of Mozart, who mastered the harpsichord at the age of three, was composing symphonies by the time he was six and spent his adolescence as the musical darling of the European aristocracy. Lisa Simpson does not enjoy quite the same benefits. Ignored by her beer-guzzling father Homer, teased mercilessly by her demonic brother Bart, patronized and belittled by teachers at school, Lisa's life is closer to what many gifted youngsters experience in reality – a hard time. It's not surprising that the parents of many of these child geniuses decide to educate their offspring at home, for better or for worse.

But, cartoon or not, it is notable that Lisa's background is that of blue-collar Middle America. Her surroundings are comfortable, not affluent. She attends a state school. Her parents are, for the most part, loving and kind, but far from being intellectuals – like Einstein's parents, in fact. Any skills Lisa has would seem to come from something innate inside her brain. And this returns us to one of the key themes of this enquiry.

We have seen much of the brain's adaptive, plastic ability to rewire itself continually – within limits – according to the environment in which it is put to work. So how, then, do we explain intelligence? Are child-geniuses merely the product of highly pressurized parenting, or do they possess some innate quality just itching to be teased into life? If this is the case, do we all have the capacity to free the genius inside ourselves – becoming more efficient at solving problems, more swift to remember, more creative? The short answer is yes, we do. While our chances of becoming Pulitzer-winning novelists or Nobel-winning scientists might be fairly slim, we all have the capacity to boost our intelligence.

But before we can explore the secrets of intelligence, it is necessary to explain what the term really means. This is not easy, as there are many different definitions to consider. Nonetheless, all theories come to an agreement on two things: first, that humans have a range of different abilities, and secondly, that intelligence can be equated with how high or low a person scores on a particular ability scale. The earliest definitions of intelligence came in the late nineteenth century from the French theorist

Alfred Binet, now thought of as the 'father of IQ'. Binet had been commissioned by the French government to identify children with special educational needs – and to do this, he developed a test that measured children's verbal, memory and mathematical skills.

Over time, Binet's scales of intelligence developed into two further concepts: mental age – which claims to indicate how far behind or ahead a child or adult is relative to others of similar age; and Intelligence Quotient (IQ) – calculated by dividing the mental age by the chronological age and multiplying it by 100. Needless to say, there is a problem with this means of 'measuring' intelligence. It rests on the assumption that tests of verbal and reasoning ability are the best markers of what makes someone 'clever'. This might be true in our particular time and culture – but what of other times and cultures where abstract puzzles were or are far from the norm?

IQ also ignores interpersonal skills – such as understanding other people's state of mind or knowing how to put people at ease – at which some people can be highly skilled, without having any ability at solving puzzles. A third problem with IQ tests is that they involve a skill, in the end, like any other. A student who has been doing dozens of dummy-runs before taking an IQ test will be familiar with the types of questions asked and perform higher than someone who has sat down to complete one for the first time. I remember as a small boy in the 1940s how important IQ was considered by my parents and by everybody else. Children like me practised IQ tests and, over time, certainly our scores improved, frequently enabling my middle-class suburban contemporaries and myself to get into particular schools. But were we, hot-housed small boys and girls, really more intelligent?

Intelligence, rather than being some specific ability – like being able to run fast over a defined distance – is a modular facility of the brain. In other words, there are different kinds of intelligence in all of us. Simon Baron-Cohen at Cambridge University has shown that many autistic people, for example, show high levels of intelligence in fields that involve 'closed systems' with predictable rules, such as mathematics or imitative drawing. But at the same time, they may have significant impairment in their social relationships.

It has been suggested, building on this idea of intelligence as a set of abilities, that we can think of two distinct varieties. Raymond Cattell has argued that *fluid intelligence* is our ability to see, interpret and manipulate relations between things – regardless of experience or practice.[104] A test of your fluid intelligence might be:

NOW is to NEVER
as
CLOSE is to a) FAR b) SELDOM c) NOWHERE d) WIDELY.

Crystallized intelligence, meanwhile, is the knowledge we acquire through experience. It has been defined as 'the extent to which a person has absorbed the content of culture' or 'the store of knowledge or information that a given society has accumulated over time'. Someone who uses crystallized intelligence is someone who puts his acquired knowledge into practice. A doctor who learns which of his patients likes to be greeted with a smile and a joke and which prefers to get straight down to business is using his crystallized intelligence. One who treats all patients the same is not.

These views of intelligence seem plausible – but are they really separate entities? Various studies suggest, for instance, that people with high fluid intelligence are more likely to employ crystallized intelligence. And conversely, how likely is it that someone who cannot see the connections between things is able to learn from experience? Another problem with this view of intelligence is that, once again, it relies for measurement on standard problem-solving tests. All manner of factors can influence a person's score when they are asked to sit down and complete a paper in exam conditions.

The intelligence of the footballer

In the 1980s, Howard Gardner, the distinguished American psychologist, challenged the traditional IQ model of intelligence.[105] He attempted to demonstrate that this was not an adequate measure of a person's intellectual abilities on its own. He proposed not two, but seven different arenas of intelligence – musical, verbal, mathematical/ logical, spatial, kinaesthetic (bodily control), intrapersonal (self-understanding) and interpersonal (understanding of others). The clear strength of this idea is in recognizing that people can display high intelligence in one or more of these arenas yet deficiency in others. However, it is difficult to support his argument fully. Paul Gascoigne, that remarkable English footballer who at his peak was one of the most mesmerizing players, clearly had outstanding kinaesthetic skills. The horrendous injuries to his knees, and the various operations that followed, were so serious that had you or I suffered them, we might never have walked normally again. All Arsenal supporters will remember with regret but admiration his astonishing curved free kick to score the winning goal against Arsenal in the Wembley FA Cup Semi-Final in 1991. And few can forget the irony of his reckless – dare one say, unintelligent – tackle in the subsequent Cup Final that damaged his leg so badly that he put his whole career in jeopardy. But in Gascoigne's case, such was his muscular development, his mental determination for fitness, the ability of his brain to help him understand where his knee was in space, and his ability to read instantaneously where other players would place the ball, that he was back playing international football. But it is doubtful whether his ability and body control could be seen as intelligence, gifted though he was. David Beckham, the England captain, has outstanding spatial judgement and physical agility. Yet he would be the first to admit that he is not the most eloquent of men, even though at the same time he has an admirable knack of putting other people at their ease and being highly and delightfully socially aware.

One advantage of Gardner's view of intelligence is that it does match much of what we know about the workings of the brain. Just as we saw that 'memory' could be fractionated into different forms that depend on different brain areas, we depend on different neural sub-systems for different arenas of our intelligence. It also implies that, with use and practice, we can make improvements. So, if intelligence is a set of skill arenas, would we be wasting our time if we tried to determine what an intelligent brain looked like, or what factors might cause it to be so? This has been a hot topic for research for many years. It builds upon a finding, way back at the start of the twentieth century, that most humans perform at consistent levels at a whole variety of cognitive tasks. In other words, that there seems to be some sort of general intelligence. Scientists have dubbed this *Spearman's g*.

The g-type

G can be measured by tests which fall into four rough categories: perceptual organization, working memory, processing speed and verbal comprehension. If we take processing speed, for example – it has been shown that people with faster 'inspection times' perform better on mental tests. The inspection time is how long it takes to register an image. If an image is flashed on a screen for 0.25 of a second, most people will have no difficulty in naming or describing it. But as the time gets shorter and shorter, only more intelligent people are still able to register what they see. Reaction times are also a measure of mental ability – if told to hit a button when they see a certain configuration of lights flashing, people will generally take a third of a second to register the correct cue (this is known as the decision time), and one-sixth of a second to move their hands (the movement time). It is thought that shorter decision times may be markers of higher intelligence.

More recently, intelligence researchers have suggested that an ability to overcome distraction might be a true marker of intelligence. Jeremy Gray at Washington University in St Louis, Missouri tested forty-eight subjects for g-type intelligence.[106] Using fMRI, the researchers measured brain activity as the participants performed a challenging mental problem. They had to keep a list of three words or faces actively in mind. Every few seconds, they had to add another word or face to their list, and drop the oldest item. But before they forgot the old item completely, they had to indicate whether the new item they were adding matched the oldest item they were dropping.

So for example, I might have the words FISH, CHAIR, BLUE in my mind. Then a new word would be given to me: ORANGE. I would say that this didn't match my oldest word: FISH, before dropping it. I'd then have CHAIR, BLUE, ORANGE in my mind. And so on to the next word: SEAT. And this does match the oldest word in my sequence. So I would confirm this before updating my new list: BLUE, ORANGE, SEAT.

The volunteers' brain activity was monitored as they got stuck in to this fiddly task. To complicate things more, the researchers threw in extra distractions – showing them items that did not match the oldest in their lists, but which did match one nearby in the ongoing sequence. Not surprisingly, the volunteers found these 'lure' items particularly confusing.

The study found that participants with higher g-scores were better able to respond correctly and swiftly, despite the interference from the 'lure' items. They appeared to do so by engaging several key brain regions more strongly, including the prefrontal and parietal cortex. The exercise seems complex, but the skill it was testing is similar to something we all call upon in daily situations. Imagine, for instance, trying to drive a car while conversing with someone in the back seat, or copying down a telephone number in the midst of a noisy party. Gray's study indicates we are able to concentrate better on a task in the midst of distractions the more intelligent we are.

So does this also mean that there is a specific site within the brain that houses our intelligence? PET scans have been performed on volunteers' brains as they act out simple and complex versions of three tasks – verbal, spatial and motor. Researchers discovered that the more complex the task, the greater the activity observable in the lateral frontal cortex – alongside the specific brain areas related to the problem. The results suggest that the frontal lobes might house a 'pool' of general intelligence, recruited in problem solving to assist more specific areas of the brain.

Intelligence born or made?

What creates individual differences in intelligence has been a question over which psychologists have argued for more than a century. And scientists are still attempting to tease out what part of our intelligence is due to 'clever genes' and what role our environment plays. It is known, as with Einstein for example, that there are structural differences in some intelligent brains. And people with a bigger brain, for instance, do seem to score higher in IQ tests most often – but overall, men do not score higher than women, even though the male brain is larger on average. Paul Thompson of the University of California, Los Angeles, has also performed scans on the brains of identical and non-identical twins, and suggested that intelligence might be related to the ratio of grey to white matter inside the brain.[107] His findings are powerful support for the notion that intelligence is, to an extent, genetically determined.

Thompson discovered that identical twins – who share the same genes – had exactly the same ratio of grey to white matter in their brains, and this corresponded to identical IQ scores. Non-identical twins, on the other hand, had varying amounts of grey and white matter in their brains, and they also scored differently in IQ tests. Furthermore, the more grey matter there was in the brain, the higher the IQ score tended to be. His MRI scans focused particularly on the two areas of grey matter controlling reading comprehension and speaking (Wernicke's area and Broca's area); these were highly similar in size in identical twins. Although these areas were similar in non-identical twins, the differences were greater than with identical twins, and fewer than in two unrelated individuals. This would suggest that our genes play a part in determining how much grey matter we have in our brains, and how correspondingly intelligent we are – at least if this is measured in terms of IQ. This is supported by the fact that identical twins who have been separated at birth tend to show as many similarities in intelligence as those who have been raised together.

But if we are born with intelligence, does that mean we are stuck with what we've got, or is it a faculty that can grow or shrink with age? Some considerable energy has gone into researching the relationship between intelligence and ageing. We have seen in earlier chapters that some degree of cognitive decline is inevitable in the ageing brain, although the evidence also suggests that we humans can do much to lessen its impact. Professor Ian Deary of Edinburgh University was interested in seeing whether intelligence was a faculty that changed over time. In this endeavour, he was aided greatly by the fact that every Scottish eleven-year-old sat a series of mental tests as part of a 1932 survey. Using his best detective skills in 1998, Deary tracked down a number of 77-year-olds who had taken part in these original tests. He discovered that most adults scored higher in their dotage than they had at the age of eleven – suggesting that intelligence improves with experience. He also discovered that those who scored in the higher ranges back in 1932 continued to show signs of higher intelligence in later life – indicating, happily, that we generally move forwards rather than backwards. He also found that change in thinking ability in these elderly people was related to a particular genetic background, further evidence of a genetic predisposition for intelligence.[108]

The 1932 records were also used in a study of intelligence and longevity performed by Professor Deary with Edinburgh University's Professor Lawrence Whalley.[109] He found that Scots who performed highest in the 1932 tests were also the most likely still to be alive in 1997. This might have something to do with the fact that an ability to use long-term planning is a key marker of intelligence. Arguably, people with lower intelligence might be more motivated by short-term gain, and would be thus more likely to engage in potentially harmful, but more immediately pleasurable activities. This could include a whole range of potentially life-shortening types of behaviour, such as risk-taking or the use of tobacco and alcohol. I personally find this research quite compelling. When I look around at the eighty- and ninety-year-old peers in the House of Lords who gained a peerage by appointment (perhaps through the law or another profession, presumably because in most cases they were particularly able), I am struck again and again how high intelligence seems to be a preservative. Many of these old men, though sometimes quite infirm, are extraordinarily articulate and many muster the best argument in debate. If ever there was evidence for not having an arbitrary retirement age, this must surely be it.

Professor Whalley's findings fit quite neatly with research recently conducted by Professor Merrill Elias of the University of Boston's Statistics and Consulting Unit. He has been evaluating the relationship between IQ and obesity. Men with a Body Mass Index of 30-plus (this would correspond to a man of 5 feet 8 inches in height weighing 14 stone), scored an average 23 per cent lower on IQ tests. The association in women is less convincing. It is not entirely clear why the connection only applied in men, nor whether a low IQ leads to obesity, or the other way round. Perhaps people with a lower IQ might be more likely to choose foodstuffs that are less healthy. People with a lower IQ might, for instance, end up in lower-income jobs, and therefore be inclined towards cheaper foods that are high in fat and sugar content.

Another long-term study of intelligence was undertaken by K. Warner Schaie in Seattle between 1956 and 1991, which involved testing adults every seven years.[110] Schaie's study used the model of fluid versus crystallized intelligence we discussed above. It found that there was a steady decline in inductive reasoning, spatial orientation, perceptual speed and verbal memory. But there was far less age-related decline in verbal or numerical fluency. The study also looked at outside factors that favoured the maintenance of mental abilities. The people who showed the least decline in their faculties had no cardiovascular or chronic disease, their personalities were flexible in mid-life and they lived with partners who also showed high levels of mental ability.

Boosting the intelligence of mice

So whatever your genetic inheritance, your way of life and environment can play a big part in determining your intelligence, and how well it lasts. Claire Rampon, of Princeton University, has shown how environmental factors can trigger the activity of intelligence-related brain systems.[111] Dr Rampon put groups of genetically similar mice into 'enriched' environments: basically a large box, with a floor covered with bedding material and items such as toys, birdhouses, small castles with multiple floors, a running wheel, paper tunnels – really all the sorts of things mice like to have available in their spare time. They were in this environment for three hours, six hours, two days or two weeks. She subsequently used gene probes to analyse 11,000 different genes in their brain, and to compare how well those genes were expressing (i.e. working) with littermate control animals kept in a standard environment. Even three hours made a difference. Dr Rampon discovered that the mice in the learning environment showed increased gene activity in over 150 genes associated with brain function, particularly those probably involved in neuron formation and structure, synapse function and brain plasticity and death of cells in the brain. What was particularly interesting was that some of these genes are known to be associated with learning and memory, and others are linked to age-related memory deficits such as Alzheimer's.

One of Dr Rampon's colleagues is Dr Joe Tsien from Princeton. His remarkable announcement from Princeton in 1999 caused a media sensation. His team had genetically modified mice to make them 'smarter'.[112] Dr

Tsien found that adding a single gene to mice significantly boosted their ability to negotiate a maze and to learn from objects and sounds around them. Moreover, they could memorize better. (The mice were called Doogie after Doogie Howser MD, a precocious character on American TV.)

First, the team tested the ability of the genetically modified animals to recognize an object. They put the mice into a box where they explored two objects for five minutes. Several days later, one object was replaced with a new one, and the mice went back into the box. Genetically modified mice remembered the old object and devoted their time to exploring the new one. Unenhanced mice spent an equal amount of time exploring both objects, indicating that the old object was no longer familiar. Gene-modified mice remembered objects four to five times longer than the normal controls.

Emotional memory was assessed by giving the mice a mild shock to their feet when they were put in another cage. At varying intervals later, the transgenic mice had more pronounced fear of the cage than the control mice. This was the case with a conditioned reflex – when the mice were taught to associate an audible tone with a shock, thus employing a different pathway in the brain. Spatial learning was measured by getting the mice to swim in a pool of water that had a hidden platform onto which they could climb to get out of the water. This is a very standard test, widely used in psychological testing. The transgenic mice found the platform after an average of three sessions, the control mice after six.

The gene Dr Tsien used, NR2B, helps control the brain's ability to associate one event with another by influencing the NMDA receptor, a receptor for the glutamate neurotransmitter. Dr Tsien not only gave mice extra copies of the NR2B gene, but he ensured that their activity increased as the mice aged, counteracting the decline of the natural gene. And as they got older, the brains of the mice retained physical features that usually characterize juvenile animals; in particular, they had a high level of plasticity and the ability to form long-term connections between neurons. These results could be of major interest to researchers trying to understand and treat human disorders that involve the loss of learning and memory. Clearly there is a commercial interest here: the NR2B gene – or what it is responsible for producing – could be a target for the pharmaceutical industry. There are, of course, extraordinary ethical and social issues raised by this discovery. Should genetic technology, if it was safe, be used to modify or enhance mental and cognitive attributes in people?

Boosting the intelligence of humans

Leaving aside the many ethical issues, safety is undoubtedly one key problem. Genetic modification is fraught with dangers to any offspring. The balance of genes and how they express in the brain would be altered at the peril of any human guinea-pig. So, while we are a long way from making superhumans by meddling with their genes, can we humans benefit by environmental changes? Obviously, the answer to this question must be yes, otherwise why would parents waste their money by sending their child to Eton or Cheltenham Ladies' College? But could a regime of brain-exercise for all of us or a simple change of lifestyle and eating habits give our grey cells a crucial kick-start? Encouragingly, some studies have suggested that activities as seemingly non-cerebral as chewing gum give you a mental workout. A University of Northumbria project tested volunteers for their ability to recall a series of words and pictures which they learnt after having chewed gum for three minutes. The gum chewers performed better – possibly because the act of chewing causes a surge of insulin in the body, and the hippocampus, seat of our long-term memory, is packed with insulin-responsive cells.

Three million people play bingo in the UK every year and you might think this is a mindless activity. Many people even bring their knitting to keep them occupied during the game, and some, such as a maiden aunt of mine, play the cards upside down to make it a bit harder. But anybody who has ever attended a bingo hall will have been amazed at the ability of certain old ladies to simultaneously listen to the numbers called, conduct a conversation, drink a port and lemon and play several game cards at once. Bingo is an activity that calls for high levels of attention, focus and efficient working memory. The players have to be able to check numbers off quickly and need rapid hand–eye coordination. And a recent PhD thesis submitted by Julie Winstone of the University of Southampton showed that experienced bingo players are helping to keep their brains in tip-top condition. She tested 112 women aged between eighteen and forty, and sixty and eighty-two, for faculties like speed processing, memory recall and scanning the environment for data. The study showed that the bingo players, old and young alike, scored significantly higher. Interestingly, in some areas of intelligence, the older bingo players were doing better than their younger counterparts.

Of course, some of the best examples for us to follow come from the redoubtable nuns of Mankato who keep their brains active well into old age by doing puzzles and participating in debates. We can all do the same for ourselves, but the thing to remember is that the brain adores a challenge. It is busy all the time, whether we are resting or working, but so much of that activity is routine. To build and strengthen new connections, the brain needs the challenge of fresh and unusual stimuli. So if you are doing the *Times* crossword in ten minutes flat (as my old history master at school used to do), you might not be giving your grey cells as much exercise as they deserve.

There's a lot of evidence to suggest that repetition is bad for brain health, and novelty is good. A study led by Robert Friedland, at the Case Western Reserve University School of Medicine, showed that people who had led mentally diverse and stimulating lives were less likely to develop Alzheimer's disease.[113] Arnold Scheibel of the University of California has conducted autopsies on elderly subjects and demonstrated that the neurons of those who had led more challenging and diverse lifestyles had a great density of receptor sites. So the message should be – make some changes while you can.

Laurence Katz, neurobiologist at Duke University Medical School, North Carolina, whose outstanding work on pheromones I've quoted before, suggests that unfamiliar challenges, however trivial or bizarre, are best for brain performance. So try brushing your teeth with the wrong hand. Make a cup of tea with your eyes shut. Read this paragraph backwards. It won't make you a genius, but it will give your brain a boost.

And when it comes to certain mental skills, training one part of the brain can be seen to have knock-on benefits elsewhere. But this effect has led to somewhat dubious research with equally questionable results. The Mozart Effect is a term coined by Alfred A. Tomatis some forty years ago. He alleged an increase in brain development in those children under the age of three who were made to listen to the music of Mozart. At Irvine, California, physicist Gordon Shaw and Frances Rauscher, a former concert cellist, studied the effects of listening to Mozart's Sonata for Two Pianos in D Major (K.448) on a number of college students. I blush in remembering that when I listened to it, aged nineteen, I fell asleep. However, Shaw and Rauscher found an enhancement of spatial-temporal reasoning, measured by the Stanford Binet IQ test. This effect was only brief and nobody else has been able to entirely duplicate their results as far as I am aware. Dr Rauscher has since moved on to study the effects of Mozart on rats. He claims that after he exposed rats in utero and then sixty days after birth to different types of auditory stimulation, animals that were exposed to the Mozart completed mazes faster and with fewer errors. Now he is removing their brains so he can slice them and see precisely what has changed as a result of this exposure to eighteenth-century Austrian culture. He speculates that this intense exposure to the music is a type of enrichment that has similar effects on the spatial areas of the hippocampus of the brain.

In 1997, Drs Rauscher and Shaw announced that they had proved that piano and singing are superior to computer instruction in enhancing children's abstract reasoning skills. They looked at three groups of pre-school children: one received private piano/keyboard lessons and singing lessons, a second received private computer lessons, and a third group received no training. Those children who received piano/keyboard training performed 34 per cent higher on tests measuring spatial-temporal ability than the others. They claimed that these findings indicate that music stimulated higher brain functions required for mathematics, chess, science and engineering. Certainly, Drs Shaw and Rauscher have stimulated an industry. They have also created their own institute: the Music Intelligence Neural Development institute (MIND).

To be scrupulously fair to Dr Shaw and Dr Rauscher, they argue that they have been terribly misrepresented. What there is, apparently, is 'a pattern of neurons that fire in sequences, and there appear to be pre-existing sites in the brain that respond to specific frequencies'. This is not quite the same as showing that listening to Mozart increases intelligence in children. However, according to various press reports, Shaw has not waited for the hard evidence before he helps parents to achieve their desire of enhancing their children's intelligence. Gordon L. Shaw's book and CD *Keeping Mozart in Mind* are available on Amazon.com priced $52.95 (or $41.50 secondhand). Apparently, the Governors of Tennessee and Georgia have been so impressed by this research that they have started programmes giving a Mozart CD to every newborn.

One important influence on a child's intelligence is the happiness of its environment. Some very recent research at King's College in London by Dr Terrie Moffitt and Dr Karestan Koenen of Boston University involved testing the IQ of British twins. No fewer than 1,116 pairs of twins at the age of five helped this enquiry. The mothers of these children had quite often been in an abusive relationship – 476 of them had suffered domestic violence and in 151 cases the violence had been classed as 'serious'. When the mother had been in an abusive relationship, the average IQ score of the twins was eight points lower than in children with happy, unabused mothers. It made no difference whether the twins were identical or not, demonstrating that this effect was not likely to be genetically determined.

A healthy mind in a healthy body

There can be no doubt, however: if you want to keep your brain healthy, you should also pay attention to your body. The Roman dictum *mens sana in corpore sano* would indicate that we have known for some centuries about the benefits of regular physical exercise for the brain. More recently, Dr Henriette van Praag, who studied opiate receptor function at Bar Ilan University in Israel, has worked on brain regeneration after injury and diseases at the Laboratory of Genetics at the Salk Institute at La Jolla, California. She examined the hippocampus – the site of the long-term memory – of mice who had access to a running wheel. Compared with mice deprived of exercise, the active mice had twice the number of cells in this region involved in memory.

And the same trends seem likely to be true for humans. Studies performed in the USA in the 1970s showed that children's reading scores were boosted considerably when they performed a short dance exercise every day for six months. A more recent educational study has shown that just five minutes of general jumping around at the start of the day results in improved concentration and more efficient learning of material. A simple explanation for this is that any physical activity raises our heart rate and increases blood flow to areas throughout the body, including the brain.

Different types of exercise can have varying effects on our mental health. A Harvard study in 1994 demonstrated that men who burn more than 2,500 calories a day in aerobic activity were 28 per cent less likely to suffer from depression. This might have something to do with endorphins – the natural opiates produced within the brain after physical activity. We've all of us experienced the warm, contented feeling that suffuses us after completing a physical chore, like digging the garden or climbing a steep hill. This may have less to do with our sense of achievement and more to do with the fact that our brains are flooded with endorphins. Sequenced movement activities like tai chi and hatha yoga have also been shown to have benefits for mental health – reducing stress levels and aggression.

Food for thought

Your brain is affected by what you put into your body as much as what you do with it. You may have been told as a child that eating oily fish, such as sardines, kippers, mackerel and herring, was good for your brain. And unfortunately for fish-haters, this is true. We saw in Chapter 6 how important fats are for the adequate development of a child's brain. And a particular type of long-chain fatty acid, found in breastmilk and oily fish, is vital for brain function. Norwegian and Danish research has shown that babies who were breastfed for less than three months were more likely to score below average for mental skills when they were five years old. It is interesting to note also that the lowest rates of dyslexia are found in Japan, where the diet includes a lot of fish.

When I was a small child it was wartime and, with rationing, many foods were in short supply. I remember I was forced to eat many repulsive foods. Reconstituted dried egg wasn't so appalling, but I found cheese completely repellent. I still remember meals at the age of two or three. I recall being seated on my nanny's knee in the kitchen being fed liquid cheese, while my parents were comfortably ensconced in the dining room. I still have a loathing – indeed almost a phobia – of all cheese to this day. No doubt the fat in it was good for brain development, but not nearly as good as the spoonfuls of cod-liver oil we were forced to eat. One swallow and a feeling of extreme nausea followed rapidly. But the Scandinavian research shows that my parents, along with many other wartime couples, were not so short of the mark and were being cruel to be kind. A child's brain contains a huge amount of fat, and fat intake is needed – especially during the first twelve months of life – for its adequate growth. And fish oils turn out to be remarkably good for brain development. Nowadays we can take them in more sophisticated ways – for example, in a pill capsule.

My family have always been convinced that fish helps make you brainy. My maternal great-grandfather was a rabbi and a rather bookish intellectual. He lived for much of his later life in Ramsgate, by the sea. In the belief that fish was really good for the brain, he eventually bought himself a large rod and spent many fruitless hours in often bitter weather on Ramsgate pier. My great-grandmother constantly teased him about his passion for fishing because he always came home wet but empty-handed. After one frustrating day spent in this useless activity this normally honest and highly moral man decided he would impress his family with his prowess. On the way home he guiltily slipped into the local fishmonger and furtively purchased two fish which the fishmonger quickly wrapped for him. On his return he proudly laid his prized catch on the kitchen table. The triumphant hunter-gatherer glanced round smugly at his expectant family. My grandmother, then a little girl, rushed to the table, picked up the fish and said, 'Oh, Dad, aren't you clever? Did you catch both these kippers?'

A recent experiment in Durham involved giving fish oil supplements to primary school children over a period of six months. Educational psychologist Madeleine Portwood selected a group of 120 children, aged six to eleven, who all showed evidence of learning difficulties, such as problems with coordination, reading, handwriting and spelling. Some of the results have been dramatic. After only three months of taking the supplements, one child's reading age improved by four years. For others, there has been a two-year leap in learning abilities.

Dr Portwood points out that, as a nation, we are not consuming nearly enough of the fatty acids so vital for healthy brain function. The government recommends eating one to two portions of fish like salmon or mackerel a week, but only around a third of us serve it up regularly.

In addition, she believes some children are deficient in fatty acids, not because of their diet, but because their bodies fail to make proper use of the fatty acids they are getting. It seems that these substances help neurons to grow and to form connections with one another. They may also be important in the brain's construction of myelin, giving the white matter its colour. This is the fatty sheath that covers neurons and conducts electrochemical messages through them.

Numerous studies have confirmed the importance of long-chain fatty acids in brain function. Laura Steve and John Burgess of Purdue University in the USA have discovered that children with higher amounts of the acids in their blood show better overall academic achievement and are specifically better at mathematics. Meanwhile, work being performed at Oxford University has shown that fish oil supplements can be useful in controlling the symptoms of Attention Deficit Hyperactive Disorder. In a six-month study, children attending a school for learning disorders were divided into two groups. One group took supplements for the whole period of the study, while another group switched after three months from taking a placebo to taking supplements. The results showed that taking the supplements led to a reduction in cognitive problems, behavioural problems and anxiety. Bernard Gesch, also of Oxford University, is taking this knowledge into the British prison system, where he has found evidence of a poor diet, low in some minerals and selenium, and high in the wrong kinds of fats.[114] He has found that young offenders in an Aylesbury institution showed a 30 per cent drop in instances of aggression when given fish oil supplements with their daily diet.

And if you come out in a rash when you eat fish, maybe you should try ginseng. So far there have been claims for its effectiveness in treating acidity, back pain, cancer, Crohn's disease, depression, impotence, chronic hair loss, urinary difficulties, chronic diarrhoea, headache, high blood pressure, insomnia, rashes of various kinds, hot flushes, diabetes, and as an aphrodisiac. Now it seems it improves intelligence. Dr Scholey, of the Human Cognitive Neuroscience Unit, University of Northumbria, Newcastle[115] undertook three studies involving mathematical tasks of varying difficulty, when a number of young men and women were tested for speed of problem solving. They were given different amounts of ginseng in a randomized fashion and were not informed whether what they had ingested was a placebo or the real thing. Different doses of ginseng improved accuracy and slowed responses. The most striking result was achieved with 320 milligrams of a ginkgo-ginseng combination. But the question remains: why did it take over 2,000 years for the Chinese to find out its cognitive value from a British scientist?

Evidence from Papua New Guinea indicates that there's one substance we should steer clear of if we want to keep our intelligence in tip-top health. Back in the 1950s, doctors noticed that among a certain tribe, the Fore people, there was a high incidence of a condition known locally as kuru, or the 'laughing death'. This degenerative brain disease causes its sufferers widespread mental decline, with depression, memory loss and impaired reasoning – these are, perhaps, the nicer of its symptoms. In addition, kuru sufferers were gradually robbed of the ability to walk, talk and even eat. It wiped out at least 10 per cent of the population. Initially, the colonial authorities thought that, since it was confined to one specific tribe, kuru was a genetic condition. They sought to prevent its spread by confining the movements of the Fore people to their homeland.

But anthropologists noted that there was something specific about Fore culture that might have been playing a part in the spread of the disease. To put it plainly, Fore villagers honoured their dead by eating certain parts of their bodies. It was considered a noble and respected action on the widow's part to prepare her deceased husband's body for consumption by steaming it in bamboo tubes. Furthermore, it became clear that kuru was most predominant in the women and children of the tribe. This custom stemmed from the belief that certain body parts, if eaten, imparted specific qualities to the eaters. And eating the brain should have helped them to gain wisdom and intelligence. We now know, of course, that kuru is almost certainly a prion disease, a disease of a disordered protein in the brain, causing damage similar to Creutzfeldt-Jakob disease (CJD), or BSE in cattle.

Drink not always a demon

You are unlikely to be in any situation where eating human brains is a possibility. But you might be interested in the benefits of a more common practice: visiting the pub. I have already dealt with excessive alcohol consumption and the risks of it causing irreparable damage to brain tissue. But it seems that the occasional tipple can be beneficial to your mental functioning. Research by Monique Breteler at Holland's Erasmus University has shown that light to moderate drinkers may be less likely to develop degenerative changes in the brain and possibly Alzheimer's disease. A further study at Milan University, by Alberto Bertelli, suggests that wine might be the ideal drink for brain-boosters. A chemical called resveratrol – found in vines, grapes and various drinks produced by fermentation – increases the activity and effectiveness of the MAP-kinase enzyme by up to seven times. In human tissue studies, this enzyme stimulates the growth of connections between neural cells, which may be good news for general brain health. Currently, research at the

University of Marburg is looking into ways to breed a resveratrol-heavy type of grapevine, having isolated the related genes. The stuff is most concentrated in red wine – one fluid ounce of a good claret contains around 160 micrograms of resveratrol – so don't be surprised if you don't get the same results from a Chardonnay. My old boss, Professor Murdo Elder, a Scot, used to creep into my office after hours to rummage around the drawers to, I regret to say, steal (or rather borrow, perhaps) the odd bottle – I've always kept a good malt there, purely for medicinal purposes as it's not uncommon to find the occasional member of staff in a dangerous state of collapse at the end of the working day. Professor Elder was adamant that my contribution in this respect was the most important I made to the academic health of the department. And he was convinced that he wrote far better scientific papers after a glass of scotch. He knew nothing, I am sure, about the research on resveratrol but it might just be that whisky has other properties.

What is very clear is that all the evidence suggests that the brain and the body are intimately involved with one another. Brain health affects bodily health, and vice versa. Increasing evidence suggests that factors such as stress, depression and mental attitude can have a visible effect on the behaviour of our bodies. As we shall see next, the idea of 'mind over matter' is beginning to seem less like a cranky belief confined to paranormal investigators and more like a basic fact of human biology.

Mesmerizing . . .

We come now to a fascinating and rather sad individual. Not many of us have made such an impressive contribution to human history that our surname contributes a new verb, noun or adjective. One thinks of 'galvanize' from Galvani, and 'Faradism' from Michael Faraday. Hans Mesmer was brought up on Lake Constance and qualified as a doctor at the late age of thirty-two. At the age of forty he became interested in the effects of magnets on the body and believed that he had discovered an entirely new principle of healing involving 'animal magnetism'. This 'animal magnetism' that he used was different from physical magnetism in that he believed that he could 'magnetize' paper, glass, dogs and all manner of other substances. Of course, had he but known it, much of modern neuroscience is founded exactly on this principle. The MRI scanner aligns molecules by their magnetic field, and dogs certainly fit the rule – though they are not often studied.

He set up his plate first in Vienna and then in Paris. Rather like the handful of millionaire plastic surgeons used now by the rich and famous in Hollywood, Mesmer's practice on the Place Vendôme quickly became a honeypot for the Parisian bourgeoisie. He had established his reputation in the years before through being apparently able to treat physical maladies using only the lightest touch of his hands. His basic notion was that people could be cured by the use of a magnetic field. So many of the deserving poor flocked to him for treatment that he resorted to mass methods of healing. He designed a magnetic *baquet*, a wooden tub nearly two metres across and around 30 centimetres or so deep, filled with water and iron filings. In this tub were fixed iron rods that his patients could grasp. In order to extend his practice, he 'magnetized' a tree, so that patients could be healed by holding ropes hanging from its branches. These devices 'worked' by inducing a 'crisis', normally convulsions or swooning.

Mesmer seriously believed that there was hard science behind the practice of what we might now call faith healing or the laying on of hands. He believed – building on

Isaac Newton's theory of a universal substance called the ether – that bodily disorders were the consequences of imbalances within the body's own magnetic fluids. Rather as we might even out a tube of toothpaste by squeezing it, these imbalances could be put right simply by touching, gentle palpations, or by passing the hands over the affected areas – a process which quite often resulted in the patients, as they were being 'cured', entering a trance-like state. There is an extraordinary cartoon of Mesmer from a French newspaper of around 1790. A fainting woman – clearly very well dressed – is collapsed in a chair and over her looms Mesmer, wearing a frock coat and making 'magnetic passes' which look sexually explicit. And Mesmer is wearing a huge head of an ass, with very long ears. He looks a little like Bottom in *A Midsummer Night's Dream* and presumably the implication is the same – that a cuckold is in the making.

For Mesmer, often regarded as the father of hypnosis, these trance-states were of minor interest, a mere side effect of his awe-inspiring treatment. So captivated was the public imagination by Mesmer's treatment that he began – not unlike some latter-day American evangelist preacher – to treat the 'sophisticated' people of Paris en masse, in stage shows carefully engineered with the use of elaborate costume, mirrors, lighting and music.

Some of the scientific community, instead of dismissing Mesmer as a side-show trickster, took his work very seriously. But possibly because he threatened its own lucrative treatments the medical establishment was very concerned about Mesmer. A Royal Commission into Mesmerism was set up which reported very critically. So Mesmer established a foundation – the Society of Honour – behind which he could hide. It consisted of a clinic, a teaching establishment and a register of qualified members who had received his training, and who paid for the privilege. From there he published his *Catechism on*

Animal Magnetism. At one point the catechism asks in what way should one touch a sick person in order for him to experience the effects of magnetism.

> First of all, one must place oneself opposite the patient, back to the north, bringing one's feet against the invalid's; then lay two thumbs lightly on the nerve plexes which are located in the pit of the stomach, and the fingers on the hypochondria [region below the ribs]. From time to time it is good to run one's fingers over the ribs, principally towards the spleen, and to change the position of the thumbs. After having continued this exercise for about a quarter of an hour, one performs in a different manner, corresponding to the condition of the patient. For example, if it is a disorder of the eyes, one lays the left hand on the right temple. One then opens the eyes of the patient and brings one's thumbs very close to them. Then the thumbs are run from the root of the nose [bridge] all around the socket.

Had his patients been put into a brain scanner after this therapy, it might well have been possible to prove an effect likely to be therapeutic. By 1791, Paris was in chaos. Being a foreigner, it was simplest for Mesmer to leave the revolutionary city. He retired at the age of fifty-four and settled eventually near Lake Constance. He had been fêted and famous in Paris, but now he was poor, obscure and forgotten. But he seems to have been contented enough with the quiet life, doing a little medicine, playing his glass harmonica, and remaining detached from the outside world until his death in 1815, at the age of eighty-five.

But Mesmer's rather fantastical *son et lumière* demonstrations created the bedrock for a new area of scientific consideration. The notion that physical symptoms could be alleviated by making suggestions to the brain spawned a century of scientific enquiry into the possible benefits of hypnosis. When Sigmund Freud began his forays into the unconscious mind towards the close of the nineteenth century, it was with a nod to the practices Mesmer had originated. Freud's earliest investigations centred upon the possibility that certain states of paralysis could be cured through inducing patients to relive traumatic experiences, which had become buried in the unconscious mind.

The brain *is* the body

However much scepticism we now feel towards the likes of Mesmer and Freud, medical professionals all over the world accept that there is a very powerful link between the physical body and states of mind. The placebo effect, which must have played a most important role in Mesmer's cures, is now firmly established as a healing entity. We also understand, for example, that some complaints can be purely psychosomatic – that is, arising from the mind. GPs now routinely look for symptoms of depression in those patients who repeatedly come to the surgery with aches and pains that remain resistant to any sort of treatment. In many cases, these patients are not time wasters or attention seekers. Nor are they are suffering from some well-hidden brain tumour or heart disease, but from a very elusive and disguised form of depression. It can, in many cases, be alleviated with antidepressant drugs.

Those inclined towards a more psychoanalytic view would say that people with this form of depression probably come from backgrounds where they were told to 'pull yourself together' and keep a 'stiff upper lip'. Becoming ill is therefore the only way that they can signal their emotional distress. But whatever perspective you take on this, when you have seen – as I have – people writhing in agony with a condition that has no bodily cause at all, you cannot possibly doubt the mind's power over the flesh.

Stress is perhaps the most classic, and common, example of the way a state of mind can affect the body. Consider the following scenario for a moment: You're in the office. The deadline is an hour away, and you've got a nagging headache. As each minute goes by, that pile in your In tray seems to grow ever more mountainous. Your phone keeps ringing. Each minute is punctuated by the little 'blip' sound telling you that more email has arrived in your computer. You spill your coffee, and when you go to fetch a cloth, you knock that carefully collated report onto the floor, sheets of paper flying everywhere. Your heartbeat soars, you come out in a sweat, your stomach churns and the headache intensifies. Your thoughts are muddled, panicky, incoherent. Stress, the demon of the modern urban lifestyle, is on the increase – and it can be a killer. High levels of daily stress are implicated in a raft of conditions from heart disease to cancer and even infertility.

Studies have shown that the ability to handle stress is vital to the functioning of our bodies. Robert Sapolsky of Stanford University has shown that rats raised in a 'gentle environment', which involved soft handling and regular touch and massage, produced increased amounts of serotonin, resulting in lower levels of aggression. As adults, they also had stronger immune systems and lived longer. But of course he is only demonstrating what all animal researchers and farmers know. Treat animals with courtesy, kindness and humanity and they will be healthier and much more gentle.

This phenomenon is equally true of humans. You may have noticed yourself that if you develop a cold or some other infection during a time of high stress in your life, it lingers longer and makes you feel worse. This is because substances such as cortisol and adrenocorticotrophin (ACTH), released when we undergo physical and mental stress, can compromise the body's ability to repair its own tissue and to fight infection. When a virus or bacteria enters the body, white blood cells divide and increase in number in order to fight the invasion. ACTH slows down the production of these white blood cells – thereby impairing our ability to stave off infection.

Dr Ronald Glaser of Ohio State University has demonstrated this effect by testing the blood of students before and after they sat examinations. The students emerging from the exams – in which we can assume they underwent a stressful experience – had significantly fewer white blood cells. Glaser also tested couples who were undergoing marital therapy. His results suggest that those couples who reported high levels of stress in their marriage had a less robust immune system.

We noted earlier that there seemed to be a link between high IQ and longevity. Research conducted at the University of California, Berkeley, has suggested that this could be related to the body's immune function. Dr Marian C. Diamond (who made the study of Einstein's nerve cells) noted that animals with damaged dorsolateral cortices – the area of the brain associated with complex cognitive functions – also had reduced levels of lymphocytes, which kill off viruses and neutralize toxins within the body. She proceeded to test this on humans by testing their blood before and after they had engaged in a one-and-a-half-hour session of contract bridge. Her results provided partial proof of the hypothesis – the bridge players had increased levels of CD4 T lymphocyte cells.

But why should this be? It is one thing to say that the brain can be tricked into suppressing or boosting the immune system, entirely another to suggest that group therapy or a tricky card game or falling in love can halt the progression of a disease. One explanation might be that the mind's ability to control the immune system is dependent upon attitude. For instance, our immune systems might not be impaired by stress per se, but affected by our personal attitude towards stress. Perhaps it matters just who is playing bridge, and with whom.

The field of psychoneuroimmunology is new, but one that is gaining some acceptance within the scientific community. It might be said to have gained its first foothold when a scientist called Robert Ader, of the University of Rochester School of Medicine in New York, experimented by conditioning mice to feel sick after they had drunk sweetened water. He enforced this conditioning by giving the mice a drug immediately after consuming the sweet water, which made them vomit. After a few repetitions, the mice were sick whenever they drank the sweet water, without needing to ingest the vomiting drug. But Ader's curiosity was alerted by the fact that these conditioned mice started to die, becoming susceptible to infections that they would normally have fought. When he studied the side effects of the vomiting drug, he discovered that it also suppressed the immune system of the mice. But its effects were short-lived; they ought to have reversed after the mice stopped taking it. The fact the mice were dying suggested one thing – it was their brains that were causing the suppression of the immune system.

Since this experiment was conducted in 1975, numerous studies have proved that mental stress can impair the body's ability to fight infections. More importantly, we are beginning to see that this is a process that can be affected by our own attitude. Given the right mental equipment, we can make our immune systems work more efficiently. A neat example of this is provided by the 'sherbet test'. It is known that adrenaline – though this too is released under stress – has the ability to boost the body's immune system. If we are given a dose of sherbet along with a dose of adrenaline, we can immediately see the adrenaline's positive effects on the body's white blood cells. But, rather like Ader's mice, our minds then become conditioned with repeated doses. If, later on, we are given a dose of sherbet without an adrenaline chaser, the immune system continues to behave in its adrenaline-boosted manner. Conceivably, then, the brain has the power to heal the body.

Medical professionals are beginning to make use of this knowledge in the treatment of disease. Dr David Spiegel and his colleagues at Stanford University, California, followed eighty-six patients with advanced breast cancer and secondary deposits.[116] Fifty of these women had psychotherapy each week for ninety minutes – the idea being that this would simply help them to cope with a terrible diagnosis – and thirty-six had routine oncological care. After ten years, only three of the patients were alive, and death records were obtained for the other eighty-three. Women in the group treated with psychotherapy had an average life span of thirty-seven months after diagnosis; the women in the control group only nineteen months. These findings are not conclusive: one of the problems is that to some extent the groups may have been self-selecting – there may have been something different between the people taking up the offer of psychotherapy and those refusing it, and this could introduce bias. Nevertheless, these results have continued to attract a great deal of interest.

More recently, Dr Spiegel's findings have been somewhat corroborated by Fawzy I. Fawzy, MD, a professor of psychiatry at the UCLA School of Medicine. Dr Fawzy recruited sixty-eight people with malignant melanoma skin cancer and invited half to participate in a ninety-minute education and support group once a week for six weeks. Compared with melanoma patients who received standard medical care, those in the group showed less fatigue and depression, and better coping abilities. Six years after Dr Fawzy completed this study, he learnt of Dr Spiegel's research and tracked down his participants to see if those in his group showed a better survival rate than the others. Despite the brevity of his group sessions (just six weeks), they had: among the patients who received only

standard care, ten had died. But among his group participants, there were only three deaths.[117] As this was a study
where the patients were randomly allocated to one or
other of the groups (and thus the researchers could not
influence what treatment was chosen), the results seemed
impressive, but recently Dr Fawzy went back to re-
evaluate the outcome more than ten years after the initial
treatments. Essentially the death rate was the same in both
groups – so the effect of psychotherapy, if it exists at all,
may be somewhat transient.[118]

Recently, Dr Christoffer Johansen's group from the
Department of Psychosocial Cancer Research, Institute of
Cancer Epidemiology in Copenhagen, made a thorough
review of the world scientific literature in this field.[119]
They identified forty-three randomized studies of the
effect of psychosocial intervention in cancer patients.
There were eight studies in which the survival of each
individual patient was assessed; of these, four appeared
to show some significant improvement. Of interest was
the general effect of psychotherapy on anxiety and depression; this was also inconsistent. It was disappointing
that psychotherapy and positive thinking only seemed to
be consistent in helping a few patients with their general
symptoms. As so often happens with subjects which are of
wide interest and newsworthy, it was obvious that there
had been much soft research and that enthusiastic stories
of the benefits of positive mental attitudes in the press
were based on quite flimsy evidence. Dr Johansen
and colleagues drew some important conclusions. They
commented that more large-scale studies with much
sounder methods were needed and that, in the meantime,
they felt that the question of whether psychosocial intervention among cancer patients had a beneficial effect was
unresolved. In too many of the studies, a number of essen-
tials were missing or incomplete; the failures of research
method included inadequate randomization of patients

allocated to different treatment groups, patients with different stages of disease being studied in each group, no assessment being made of psychological profile, and an absence of rigorous follow-up.

It goes without saying that stress is a broad term, used loosely. The stress of having a serious, possibly fatal, disease like cancer is different from the stress we experience travelling to work and different again from that experienced performing sport. Studies show that the way we respond to stress depends on the type of stress, and whether we feel able to control it. Short-term controlled stress like a bungee jump can actually boost the production of the body's own infection-fighting cells. Similarly, long-term controlled stress – such as a demanding job – can result in increased production of relaxant chemicals such as endorphins and enkephalins. This might explain why, when engaged in a period of high stress, we often fall ill after the pressure has eased off, rather than during it. Many actors report that they fall prey to colds and influenza just after finishing a job. 'I'm just resting' may not merely be a thespian's way of glossing over a period of unemployment after all.

And the reverse is also true: the most damaging kind of stress is the stress we feel we cannot control. In one experiment, rats were exposed to a series of shocks that could be stopped by pressing a metal bar. Another group also received shocks, but there was no mechanism by which they could escape them. Both groups were then injected with mitogens, substances which would normally stimulate the immune system. The rats experiencing stress they could not control had diminished production of lymphocytes. But the rats who were in control of their stress suffered no impairment to their immune function.

Similar experiments have been conducted with humans. In one experiment, healthy volunteers were exposed to noises over which they had control, and other volunteers to those they did not. In each case the noise was more than 100 decibels. Those who had no control over the noises were obviously increasingly helpless and began to feel rising anxiety. More importantly, when compared with those who could have turned the noise down if they felt they did not want to continue with the experiment, those with no control showed marked impairment of their immune system after the experiment. Studies of the families of Alzheimer's sufferers, and people who have lost partners, also confirm that various types of un-controllable stress cause most damage to the immune system.

So if this is the case, we should, by enabling people to reassert control, also help them to improve the response of their immune system. This is a possible reason why activities like group therapy and relaxation techniques may have a beneficial effect on the prognosis of disease. The more we feel we can control stress, the better placed our immune system may be to cope with its effects.

The genius inside

So we can, at least to some extent, boost both our brain and our body. But how does intelligence relate to that most elusive of human faculties – creativity? If we can make our brain stronger, can we also learn to unlock the secrets of creative genius, becoming better painters, com-posers, writers?

Before we can answer these questions, it is necessary to arrive at a definition of creativity and an understanding of the brain processes involved. Robert J. Sternberg, Yale University Professor of Psychology, suggests that creative thinking has a number of observable traits. First, there is an ability to switch between primary conscious and secondary unconscious thought processes. Secondly, and related to the first point, creative individuals show a low level of cortical activity. This could explain why children are more adept at thinking creatively than adults – because their frontal lobes are less developed. A child might, for instance, make an observation like 'those clouds look like mashed potato and that aeroplane looks like a sausage'. Adults generally refrain from making such statements if they don't want to seem odd. But those associations might be exactly what an artist or a poet uses in creating a painting or a poem. This brings us to Sternberg's third point. Creativity, he claims, is a right-brain activity – which entails making loose, or freeform associations between seemingly unrelated phenomena.[120] The right brain, it is maintained by some people, is better at grasping concepts in their entirety, at seeing the 'bigger picture' rather than the fine detail. This relies on a tendency to connect items into an overarching schema, rather than see them as individual and unconnected. Correspondingly, creative individuals are better at un-focused thought – at suggesting connections between ideas rather than in solving problems. One of the ways we can observe this is by using a device called the biofeedback machine – which shows users their EEG brain-wave patterns on a screen, and enables them to produce alterations in their internal brain states by concentrated effort. Creative individuals, asked to create a certain biofeedback pattern, are said to be markedly worse at this than their less creative counterparts.

Creative people seem sometimes to show a greater variability of response to external stimuli – factors like heart rate, galvanic skin response and cortical arousal increase more sharply in response to sensory input. We noted earlier on that this is also an observable trait in people of introvert-type personalities – so there might be some basis for assuming that creative people are likely to be more 'sensitive' to the world they experience. Arguably, this can explain why many artists, poets, novelists and the like tend to be introverts. They retreat from a world they find bewildering and potentially threatening to their sensitive sensory apparatus and seek solace in an 'inner world'. The fact that most creative people prefer solitude to do their work only reinforces their introversion.

Of course, if we took the arguments above to their natural conclusion, we might end up saying that all poets, composers, artists, etc., are right-brained introverts. And this is not the case at all – as anyone who has the slightest acquaintance with the life story of Ernest Hemingway can confirm. Certain activities, such as drawing and music for instance, may rely heavily on left-brain processes – which are arguably more developed in right-handed people. And novelists, playwrights, poets and the like also demonstrate a high degree of linguistic fluency – once again a left-brain activity. Guy Claxton, of the University of Bristol, suggests that the secret of creativity lies in an ability to switch modes. In other words, regardless of which side of your brain is dominant, if you are creative, you are able to suppress the elements you normally rely upon and dip into your opposite hemisphere. So a painter wishing to convey, for example, what the sky looks like to a child would switch off his analytical left brain and dip into his right to make the analogy of clouds with mashed potato. Certain neurotransmitters like acetylcholine may help this process and this – a simple difference of our chemical inheritance – might explain why some individuals find it easier than others.

Creativity appeared to be an entirely natural process for Mozart, who claimed that his intricate musical compositions came to him whole, stimulated by playing billiards, without needing any editing or even posing any disruption to his daily routine. Beethoven made endless sketches of what he intended, often hastily put together, before building up the final score of his great works. For Haydn, on the other hand, composition was a formal experience. Before sitting down to write he would dress in his best clothes, try things out on the keyboard and prepare himself for the meticulous craft of his trade. Brahms rose at 5 a.m., brewed monumentally strong black coffee and then sat down to write. Occasionally he could be seen, a solitary figure, lips pursed, sitting in a corner of his favourite coffee house during the later mornings. For others, creativity seemed to be linked to specific disorders of the brain. The Russian composer Dmitri Shostakovich claims in his biography that he was able to hear music inside his head simply by tilting it to one side. This might have been because Shostakovich had a fragment of shrapnel embedded inside his skull from his activities in the first world war. When he tilted his head, this shell splinter prodded and stimulated his brain's auditory cortex.

While Johannes Brahms seems to have shunned company and imbibed caffeine to achieve inspiration, other creative people have sought more extreme methods to engage their creative brain systems – and to disengage their normally dominant rational brain. Samuel Taylor Coleridge, for instance, wrote his epic poem *Kubla Khan* under the influence of laudanum. Not exactly a genius, fellow opiate-devotee William Burroughs is alleged to have created the manuscript for *The Naked Lunch* (1959) in such a drug-suffused state that he could not recall having written it. And frankly to me it reads like it. The painter Francis Bacon deliberately did without sleep and food in order to enter a heightened state of creativity.

Others eschew drugs and deprivation in favour of the fluid 'right-brain' thought processes of the dream state. The Nobel-winning physicist Niels Bohr famously arrived at his understandings of atomic structure at a point when his daytime ponderings were getting him nowhere. Exhausted and frustrated, he fell asleep and had a dream that featured a racecourse in which the horses ran in lanes corresponding to the orbit of electrons. The chemist Friedrich Kekulé describes how he dreamed two important observations on structural chemistry. The first is his description of the basic linking of the carbon atom:

> During my stay in London [1854–56] I resided for a considerable time in Clapham Road in the neighbourhood of the Common ... One fine summer evening I was returning by the last omnibus ... I fell into a reverie, and lo, the atoms were gamboling before my eyes! Whenever, hitherto, these diminutive beings had appeared to me, they had always been in motion; but up to that time I had never been able to discern the nature of their motion. Now, however, I saw how, frequently, two smaller atoms united to form a pair; how a larger one embraced two smaller ones; how still larger ones kept hold of three or even four of the smaller; while the whole kept whirling in giddy dance. I saw how the larger ones formed a chain, dragging the smaller ones after them, but only at the ends of the chain ... The cry of the conductor, 'Clapham Road', awakened me from my dreaming ...

Hardly the usual famous man sitting on the Clapham omnibus. A second important vision appeared to Kekulé in 1862, and led him to understand the benzene ring, the foundation of so much modern organic chemistry:

> ... I was sitting writing at my text book, but the work did not progress; my thoughts were elsewhere. I turned my chair to the fire and dozed. Again the atoms were gamboling before my eyes. This time the smaller groups kept modestly in the background. My mental eye, rendered more acute by repeated visions of this kind, could now distinguish larger structures, of manifold conformation: long rows, sometimes more closely fitted together; all turning and twisting in snakelike motion. But look! What was that? One of the snakes had seized hold of its own tail, and the form whirled mockingly before my eyes. As if by a flash of lightning I awoke; and this

time also I spent the rest of the night in working out the consequences of the hypothesis. Let us learn to dream, gentlemen, then perhaps we shall find the truth . . .

Bruce Miller at the University of California, San Francisco, is particularly interested in the ability of some autistic people to recall remarkable detail – like the British man Stephen Wiltshire, who was able from an early age to create fantastically detailed and accurate drawings, such as his remarkable rendition of the Victoria and Albert Museum. It seems that brain scans of these 'autist savants' revealed one common facet of brain architecture: they all had impaired function within the left anterior temporal lobe. Miller has argued that, because of this impaired left-brain function, savants have superior access to the creative resources of the right brain. Dr Allan Snyder, from Sydney, Australia, who has gained a great deal of attention in the press for his views, has taken these ideas a bit further in an attempt to see what happens if we can train a normal brain to shut down its analytical functions and draw more heavily on the right brain. He used a technique called Transcranial Magnetic Stimulation, which involves en-circling the skull with a series of powerful magnets. TMS is an experimental treatment currently being tested as an alternative to electroconvulsive shock therapy in patients with depression. Although not yet part of mainstream medical treatment, he claims some encouraging signs of its efficacy. When applied to a normal brain, the strong magnetic fields from the TMS device act to inhibit or shut down certain regions of the brain. Snyder asked volunteers, while fitted with a TMS device, to undertake a battery of tests, including drawing, mathematical reasoning, memory and looking for errors in texts. His experiments showed an average 30 per cent improvement on the tests. Subjects reported becoming more alert, more attuned to the fine detail of the world around them. Their drawings reflected this heightened awareness. Snyder's experiments have yet to unearth a mathematical or artistic genius, and as far as I am aware have not appeared in the peer review literature, so caution in interpreting these results is needed.

Not so rare at all

We all pick up cues from our environment without necessarily being fully conscious of what we are doing. Perhaps you, like me, have found yourself in a dubious part of a town and suddenly felt alert to a possible danger. This is a skill we all use in our daily lives. Driving along a busy stretch of road, we can tell – without exactly knowing *how* we can tell – which pedestrian is likely to suddenly sprint in front of us. Walking alone down a darkened street, we can tell by the body movement of the person approaching us whether they are likely to pose a problem to our personal safety. We have an innate ability to sense what is going to happen. And when we analyse our thought processes later on, it becomes clear to us how we reached our conclusion. Body language, a person's manner and speed of walking, the direction and intensity of their gaze – all of these are very good indicators of someone else's intentions, fair or foul. But what is interesting is that, in so many circumstances, we act on the information received *before* we have a conscious awareness of why we are acting. Our brain knows before we know.

We looked at this phenomenon briefly when dealing with the faculties of perception and attention. It has been shown that top sportsmen have only marginally faster reaction times than the general population. What lends them a unique advantage is their ability to anticipate, to modify their behaviour according to incoming information, without consciousness getting in the way and slowing it down.

The crucial feature of this ability for athletes is that it remains 'buried' in the lower motor portions of the brain. When we practise something new, as we have seen, the cortex and the basal ganglia work in unison. But over time, the higher, 'thinking' regions of the brain drop out of the equation, freeing themselves up for new activities. The process becomes embedded in the motor regions of the brain. If consciousness for any reason re-enters the equation, there is a slowing down, and an attendant lack of fluidity in the execution of the task.

Perhaps this has happened to you – you for some reason find yourself thinking about an activity you have been performing effortlessly for years. Suddenly, you lose confidence. You make mistakes, reinforcing your lack of confidence. Merely by thinking about how you do something, you forget how to do it.

Many sportsmen live in terror of this phenomenon, which some call 'the yips'. The darts-player Eric Bristow found his playing abilities paralysed for ten years due to an attack of the yips. It is thought that stress may cause performers to suffer a loss of confidence. Accordingly they become more self-conscious, thereby putting a once-fluid skill under deliberate scrutiny. Thinking, in short, seems to be one of the worst things you can do. The golfer Severiano Ballesteros was once asked what was going through his mind just before he struck the million-dollar shot. To which he calmly replied, 'Nothing.'

Hardly surprisingly, one of the ways sportsmen ensure that a skill remains firmly embedded in their motor cortices is by regular practice. As we know, the neurons of the brain fire in the same pattern whether we are performing a task or whether we are just going through the motions in our heads. For that reason, many sports psychologists now teach visualization techniques to their charges. The regular exercise of the related neural pathways cements the ability within the lower, unconscious regions of the brain.

Sportsmen are not the only people who rely on the ability to know without knowing. Devotees of detective fiction will know all about the intuitive 'hunch', that innate precognitive sense that keeps the hero pursuing his quarry, even when his gun and his badge have been taken away and he's been threatened with a lifetime of directing traffic. The television series *Columbo*, starring Peter Falk, was far more concerned with the hunch effect than it was with the solving of a crime. Each episode began with the murder taking place in front of our eyes – the satisfaction for the viewers lay in watching Columbo prove that his intuition was correct.

Fiction aside, cognitive scientists are convinced that intuition – the ability to direct our behaviour according to unconscious cues – plays an important role in the way professionals such as hospital trauma teams, policemen and firemen do their jobs. Cognitive psychologist Gary Klein has made a study of people who have to make split-second decisions in the course of their daily lives.[121] His research has led him to conclude that we often disregard the power of the intuitive hunch or gut instinct, preferring to teach people to rely on lengthy, slower, and sometimes far less efficient codes of action.

Klein was led to this conclusion, in part, through his dealings with a fire commander who insisted that he had extra-sensory perception (ESP). The commander related one key event that had convinced him to believe in his paranormal abilities. He and his crew had encountered a fire at the back of a house. Standing in the living room, he and his men doused the fire – which seemed to be in the kitchen – with water. But the flames did not recede. At this point, the commander reported that he had the sense that something was very wrong. Without knowing why, he ordered everyone to get out of the house. Just seconds after they did, the floor where they were standing collapsed into a burning basement. Had he not had this hunch, he and his men might all have died.

Seen in retrospect, this commander issued an order because the fire he was tackling did not match his expectations. If it was solely centred at the back of the house, why did it not recede when doused with water? There were other indications that all was not as it should be – the levels of heat where the firemen were standing were far more intense than they should expect from a small kitchen fire, again more consistent with a fire burning down below. Hot fires also make a lot of noise, but this fire was eerily quiet – because the real blaze was happening below them, muffled by the floor.

Klein continued to interview legions of veteran firefighters, and began to observe something crucial about the way human intuition works. Over time, the firefighters developed a personal 'archive' of different fires, how it felt and smelt to be near them, and how they could be expected to behave. Each time they encounter a new fire, they delve into this archive to find associations that match, and which then tell them how to behave. In other words, intuition is not some paranormal ability to see the future, but a technique of learning what to look for in a given environment, and of doing so without the conscious brain getting in the way. The supposedly psychic fire commander merely knew what signs to look for when encountering a fire, and when he detected an anomaly, he knew that it meant trouble.

The same ability has been observed in neonatal intensive-care nurses, who often have to make split-second decisions to save the lives of newborns. When premature babies develop an infection, the results can be devastating – sickness spreads like wild fire throughout the tiny body, resulting in rapid death. For that reason, skilled nurses have to become intuitively adept at spotting the early signs of infection so that they can act in time to save the baby's life. When interviewed by one of Klein's research colleagues, such nurses would say that they didn't entirely know how they knew, they 'just did'. But pressed further, it seems that these experienced nurses had built up a working archive of extremely subtle clues. For instance, the complexion of a premature baby in the early stages of infection would fade rapidly from pink to greeny-grey. They might cry frequently, yet suddenly become listless and lethargic. Their feeding patterns would become abnormal, their abdomens distended. But interestingly, half of the symptoms used by the neonatal nurses were not mentioned in any of the medical literature. They based their decisions on knowledge they had acquired themselves, and they applied it almost without knowing they were doing so.

Many doctors have had some experience of an intuitive decision taken under extreme pressure. One dire emergency I remember as a young surgeon in a rural hospital was when I was called to casualty to see a semi-conscious unmarried woman of twenty-two. She looked close to death and had a low, almost unrecordable blood pressure. She had been picked up by ambulancemen and there was nobody with her from whom to find out how she came to be in this state. She was deadly pale, her abdomen felt doughy and mildly distended. The likely explanation seemed that she was bleeding internally. And as I watched, she slipped into unconsciousness. Speed was clearly critical – no time to wait for tests here. There aren't many conditions that can put a young woman's life in such jeopardy in this way; an ectopic pregnancy is probably the most common. So simply on the odds, it seemed likely that she had a pregnancy in a fallopian tube – such things can bleed catastrophically. There was no time to lose if I was right, but it's a big decision to rush a person to an operating theatre and open the abdomen without even allowing time to scrub up properly. But the anaesthetist was doubtful about giving her an anaesthetic with her blood pressure, as it was by now unrecordable. So I opened the abdomen while he was still arguing – again not an easy decision to come to even on the spur of the moment. A huge amount of clotted and unclotted blood poured out from the open abdomen – so at least that part of my diagnosis was correct. But on looking at the fallopian tubes, both were entirely normal and there was no sign of a bleeding pregnancy anywhere in her pelvis. At the back of my mind was the thought that, very rarely indeed, an ectopic pregnancy can implant in the abdomen on the bowel. But this degree of bleeding was extraordinary for the very rare possibility of a bowel implant. I'd never seen it, and not consciously heard of it – but in a flash came the thought that with this degree of bleeding the liver might be the even rarer culprit. The incision was extended, a rapid search made – and there was the ruptured eight-week pregnancy on the liver. The pregnancy was quickly grasped and removed, a heavy pack was hastily inserted around the bleeding site, and miraculously we could immediately start to feel her pulse, at first ever so faintly.

The extra sense

So human intuition would seem to be less of a mysterious 'sixth sense' than a super-efficient means of evaluating information from the five senses we do possess. Nevertheless, the search for this elusive faculty goes ever onwards. Some researchers believe the sixth sense is a forgotten 'homing' instinct – still active in our animal ancestors, but now dormant in ourselves.

The argument for a human sense of direction comes

from the fact that there appears to be a mismatch within the allocation of our senses. Taste and smell could be said to act as a discrete pair – because they operate in the same way. Things to be tasted and smelt have first to be dissolved in water, usually the water in our mouths or nasal passages, and the actual processing of what they represent is done through chemical interaction. Meanwhile there is one difference between taste and smell – we have to put things in our mouths to taste them, but we can smell things at a distance.

We can draw the same parallels between hearing and touch. These operate through mechanical forces only – they both pick up meaning through motion. Touch, like taste, requires contact, whereas hearing, like smell, can be done at a distance.

Creating this set of parallels leaves sight on its own. We could argue that, like hearing and smell, it is a 'distant' sense – one not requiring contact for processing to take place. So could we be overlooking a natural contender for the sixth sense – one that works through contact? And if so what could it do?

Some scientists have pointed out that certain homing creatures, like butterflies and homing pigeons, find their way through space thanks to tiny magnetic-field-sensing materials inside their bodies. Pigeons find their way thanks to an in-built compass, a pod of magnetite crystals situated close to their skulls. Monarch butterflies are liberally doused with a fine, magnetically charged network of wing fibres. But we humans have no such faculty. If there is a sixth sense that equips some people to find their way more efficiently, it is very unevenly distributed throughout the human population, and seems to have no obvious markers in the body.

Others have suggested that the sixth sense may be more of a 'sense 5a', a refinement of the brain's olfactory powers. I have already mentioned the vomeronasal organ which contains receptors sensitive to chemicals that the nose itself seems unable to detect – like pheromones. A team led by Catherine Dulac at Harvard University has been able to isolate a gene in rats and mice that enables them to sense out pheromones.[122] Humans also have the gene, but with mutations that appear to render it useless. Yet, you will recall that humans can be influenced by pheromones, without being aware that they can detect them. This is thought to be one of the reasons why the menstrual cycles of women synchronize when they live in a communal situation. But if we haven't got the genes to detect pheromones, how is it that we can be influenced by them? Possibly, the human nose, though a crude device compared to the olfactory systems of many animals, is able to detect certain 'stronger' pheromones.

So there are interesting indications that we humans might, or indeed should, be endowed with a sixth sense of direction, or for detecting pheromones, but no concrete evidence that we are. That doesn't stop much extremely 'soft' research. From the days of Isaiah to the horoscope in your daily paper, we humans have been fascinated by the possibility that we can glimpse the future. Dean Radin, of the Boundary Institute in Los Altos, California, is currently engaged in a global study into ESP. Using online tests on his website *www.boundaryinstitute.org*, Radin has studied the results of over half a million trials, from 4,000 people in fifty-seven countries. His findings are certainly interesting, at least to the more gullible. On one particular test, where participants have to guess which of five cards conceals a picture image, top scorers get it right 48 per cent of the time over twenty-five trials. That may not seem terribly impressive – but the odds against that happening are, it is claimed, 2,669 to 1. A similarly small, apparently statistically significant result has erratically been shown when people attempt to influence the generation of random numbers with their thoughts.

I am extremely cynical about most of the stories of ESP, psychic ability and luck, as is the psychologist Dr Richard Wiseman. He describes how he conducted an experiment to see if there were individuals who could be genuinely regarded as lucky people.[123] Using the mechanism of a television show watched by around 13 million people, he asked the viewers to predict the winning numbers in the next UK National Lottery. During transmission, viewers who felt themselves to be unduly lucky or unlucky were requested to contact the TV show. Dr Wiseman expected just a few hundred to phone; an estimated million calls were received. With this huge number, only the first 1,000 callers could be sent a questionnaire. On the form they received, they were asked to state whether they classified themselves as naturally lucky or unlucky and whether they had psychic ability. The responses showed that, between them, these viewers intended to buy around a total of 2,000 tickets. When the researchers reviewed the data, they found that certain of the forty-nine possible numbers were frequently chosen by 'lucky' people, but avoided by 'unlucky' ones. Richard Wiseman states that his analysis showed the best shot seemed to be 1, 7, 17, 29, 37 and 44. After agonizing about the ethics of this, for the first and only time in his life, he says, he bought a ticket himself. Later that week, the draw was broadcast and the result was 2, 13, 19, 21, 32 and 45. Richard Wiseman had not managed a single number. Had the 'lucky' people done any better? Thirty-six of all the respondents had won money, four of them winning the princely sum of £58. The winnings were equally split between lucky and unlucky people. On average both unlucky and lucky people bought three tickets each, matched one number only and lost equal sums.

My cynicism about psychic ability might itself have something to do with the chemical make-up of my brain. Peter Brugger, of Zurich University Hospital in

Switzerland, took twenty believers in the paranormal and twenty confirmed sceptics, and asked the two groups to try to make out faces in a sequence that included both discernible and scrambled faces and other objects. He then repeated the exercise with words – some of which were real words, others scrambled versions of words and others just sounds. Brugger found that the believers were far more likely to say they had seen faces and heard real words when they had not. In an attempt to understand what might be the crucial differences in brain states between these groups, Brugger gave l-dopa – synthesized dopamine – to both. As we know high levels of dopamine are implicated in schizophrenia and mania – both of which can include a tendency to see the world as inherently meaningful and replete with coincidences. Brugger found that, when given dopamine, the sceptics replicated the results of the believers. Meanwhile, the extra dopamine had no discernible effects on the results of the believer. So perhaps we should talk less about the existence of the paranormal and more about people's differing capacities to believe in it.

Earlier on, in Chapter 3, I mentioned the findings of a Swiss neurosurgeon who, when operating on a patient with epilepsy, seemed to have found a possible neurological basis for the so-called out-of-body experience. This sensation of being outside one's body is particularly central to the accounts of some people who have, or claim to have, experience of what it feels like to die on the operating table. The near-death experience is a controversial area for brain science. If everything that we call 'mind' is a product of the working human brain, how can it be that some attributes of 'mind' can seemingly endure when the brain is no longer working?

Scientists and surgeons alike have been interested in this question ever since general anaesthesia has been in use. From full-blown accounts of meetings with the Maker to more mundane incidences of reporting conversations between the surgical team, there is a wealth of evidence to suggest that some aspects of consciousness cannot be put to sleep easily.

In one study, done in the 1980s, patients 'listened' to informative audiotapes while unconscious, and then they were tested for general knowledge after the procedure. Their scores rose from 37 per cent to 62 per cent subsequent to the unconscious learning experience. A number of scientists now believe that it's quite possible for the brain to absorb information while it is functionally unconscious. However, it has also been noted that any 'learning' that takes place seems to dissipate within a few hours after regaining consciousness. Also it seems that the deeper the level of anaesthesia, the patchier the memories will be. This supports the view that rather than there being some sharp conscious/unconscious switch within the brain, there is actually a subtler spectrum of differing states – ranging from fully alert at one end to dead at the other. The closer we are to waking consciousness, the more likely we are to absorb information.

But of course there is a great difference between these poorly controlled experiments and the near-death experience. By its very definition, this latter state is one in which the brain is not deemed to be at all close to waking. Professor Bruce Greyson, of the Division of Personality Studies at the University of Virginia, has subjected the near-death experience to extensive study, and by analysing the accounts of hundreds of individuals, has come up with a list of common attributes.[124] Near-death experiences, for instance, include a sensation of accelerated thought processes. People undergo a change in emotional state – most commonly reporting feelings of peace and bliss. Furthermore, they report abnormal spatial experiences, such as being outside of their bodies. In addition, they commonly report paranormal sensations, transcending the normal boundaries of time and space – such that they can see into the future or the past, or witness events outside the range of their sensory apparatus. But the mere fact that most near-death experience accounts are similar is convincing evidence for some that they merely arise from the chemistry of the brain.

Dr Susan Blackmore, psychologist at the University of the West of England, offers the view that the near-death experience is merely a product of the potentially dying brain. For instance, the commonly reported sensation of being in a tunnel and proceeding towards a pinprick of bright light simply represents the random firing of neurons within the visual cortex. The feelings of peace and bliss are a product of endorphins, released in response to the massive physical trauma of bodily death.

Scientists studying the near-death experience admit that their endeavours are thwarted by one obvious factor. Like consciousness, death is not some instant process, but a smudgier continuum of several stages, between which the boundaries are blurred. So how can we tell where the near-death experience fits in – does it occur as the brain loses consciousness, and if so, at which precise point during this loss-process? Does it occur while the brain is effectively 'dead', or when the brain starts to work again? Sleep studies, as we have seen, indicate that every night we shuttle backwards and forwards across the continuum between being close to waking and being 'dead to the world'. It might be the same for the dying brain – at certain points along its journey to absolute oblivion, there might be moments where it becomes more and less alert. There might also be differences due to individual peculiarities of neural wiring. As we have seen repeatedly throughout this book, no two brains will ever be completely alike. This in turn could mean that in some individuals certain brain functions, like hearing and memory, remain 'switched on' at lower levels of consciousness.

To date, research into near-death experiences has come up with no hard and fast answers. For every neuroscientist who steadfastly claims there can be no consciousness without a brain that is at some level still alive, there will be others who wish to storm the fortress, for reasons both noble and vainglorious. For everyone who draws comfort from the notion that every corner of existence can be rendered meaningful by the laws of science, there are others who wish to believe in mystery, in another dimension for which man's 'three score years and ten' is but a shallow rehearsal. And such divisions are evidence of the most enduringly beautiful aspect of the human mind. We've all got one, but how very differently each of us puts it to use.

But when we talk about creativity, we are essentially referring to a very rare subset of human abilities. If we were to sit down and draw up a list of people who had become famous for their extraordinary abilities, it would comprise only a tiny percentage of the total population. The focus of this book is the brain that exists inside all of us, not the rare exceptions. And there is much evidence to suggest that every brain is endowed with a myriad of extraordinary abilities. Each of us, in our daily lives, has the ability to anticipate what is about to happen, to unconsciously pick up cues from our environment and direct our behaviour accordingly. From the sportsman who knows which way an incoming ball will hit his bat, to the tourist who senses he may have strayed into a bad neighbourhood, to the fireman who knows when to get out of a collapsing building, to the artist communicating with us through his painting or his music, or the scientist solving the problems of the human mind, we are all truly remarkable.

References

1 Fried, I., et al., 'Electric current stimulates laughter.' *Nature* (1998), **391**:650

2 Gur, R.C., et al., 'An fMRI study of sex differences in regional activation to a verbal and a spatial task.' *Brain and Language* (2000), **74**:157–70

3 Lane, R.D., et al., 'Neuroanatomical correlates of pleasant and unpleasant emotion.' *Neuropsychologia* (1997), **35**:1437–44

4 Blanke, O., Ortigue, S., Landis, T., Seeck, M., 'Stimulating illusory own-body perceptions.' *Nature* (2002), **419**:269–70

5 Franzini, Louis R., and Grossberg, John, *Eccentric and Bizarre Behaviours* (John Wiley & Sons, 1995)

6 Blackmore, S., 'Crossing the chasm of consciousness.' *Trends in Cognitive Science* (2002), **6**:276–7

7 Sampaio, E., Maris, S., Bach-y-Rita, P., 'Brain plasticity: "visual" acuity of blind persons via the tongue.' *Brain Research* (2001), **908**:204-7

8 Leamey, C., 'Development and plasticity of cortical areas and networks.' *Nature Reviews Neuroscience* (2002), **2**:251-62

9 McCoy, N., and Pitino, L., 'Pheromonal influences on

sociosexual behavior in young women.' *Physiology and Behavior* (2002), **75**:367–75

10 Luo, M., Fee, M.S., Katz, L.C., 'Encoding pheromonal signals in the accessory olfactory bulb of behaving mice.' *Science* (2003), **299**:1196–201

11 Field, T., 'Preterm infant massage therapy studies: an American approach.' *Seminars in Neonatology* (2002), **7**:487–94

12 Livingstone, M.S., Rosen, G.D., Drislane, F.W., Galaburda, A.M., 'Physiological and anatomical evidence for a magnocellular defect in developmental dyslexia.' *Proceedings of the National Academy of Sciences USA* (1991), **88**:7943–7

13 Tallal, P., Merzenich, M.M., Miller, S., Jenkins, W., 'Language learning impairments: integrating basic science, technology, and remediation.' *Experimental Brain Research* (1998), **123**:210–19

14 Simons, D.J., and Chabris, C.F., 'Gorillas in our midst: sustained inattentional blindness for dynamic events.' *Perception* (1999), **28**:1059–74

15 Strange, B.A., Henson, R.N., Friston, K.J., Dolan, R.J., 'Brain mechanisms for detecting perceptual, semantic, and emotional deviance.' *NeuroImage* (2000), **12**:425–33

16 LaBerge, D., and Buchsbaum, M.S., 'Positron emission tomographic measurements of pulvinar activity during an attention task.' *Journal of Neuroscience* (1990), **10**:613–19

17 Ramnani, N., and Passingham, R.E., 'Changes in the human brain during rhythm learning.' *Journal of Cognitive Neuroscience* (2001), **13**:952–66

18 Cardinal, R.N., et al., 'Impulsive choice induced in rats by lesions of the nucleus accumbens core.' *Science* (2001), **292**:2499–501

19 Driver, J., and Frith, C., 'Shifting baselines in attention research.' *Nature Reviews Neuroscience* (2000), **1**:147–8

20 Schwartz, S., and Maquet, P., 'Sleep imaging and the

neuro-psychological assessment of dreams.' *Trends in Cognitive Sciences* (2002), **6**:23–30

21 Benedetti, F., et al., 'Influence of a functional poly-morphism within the promoter of the serotonin transporter gene on the effects of total sleep deprivation in bipolar depression.' *American Journal of Psychiatry* (1999), **156**:1450–2

22 Revonsuo, A., and Valli, K., 'The reinterpretation of dreams: an evolutionary hypothesis of the function of dreaming.' *Psykologia* (2002), **35**:472–84

23 Jung, C., *Memories, Dreams, Reflections* (Collins, 1962)

24 Sedikides, C., and Skowronski, J., 'The symbolic self in evolutionary context.' *Personality and Social Psychology Review* (1997), **1**:80–102

25 Baron-Cohen, S., 'The development of a theory of mind in autism: deviance and delay?' *Psychiatric Clinics of North America* (1991), **14**:33–51

26 Peskin, J., 'Ruse and representations: on children's ability to conceal information.' *Developmental Psychology* (1992), **28**:84–9

27 'Self-awareness and the frontal lobes: a neuropsychological perspective.' In Strauss, J., and Goethals, G.R., eds, *The Self: Inter-disciplinary Approaches* (Springer Verlag, 1991)

28 Pinker, S., *How the Mind Works* (W.W. Norton, New York, 1997)

29 Libet, B., et al., 'Responses of human somatosensory cortex to stimuli below threshold for conscious sensation.' *Science* (1967), **158**: 1597–600

30 Adolphs, R., Tranel, D., Damasio, H., Damasio, A., 'Impaired recognition of emotion in facial expressions following bilateral damage to the human amygdala.' *Nature* (1994), **372**:669–72

31 LeDoux, J.E., *The Emotional Brain* (Simon and Schuster, New York, 1996)

32 LeDoux, Joseph, *Synaptic Self: How Our Brains Become Who We Are* (Viking, New York, 2002)

33 Damasio, A., *Descartes Error: Emotion, Reason and the Human Brain* (Picador, 1995)

34 Dimberg, U., 'Unconscious facial reactions to emotional facial expressions.' *Psychological Science* (2000), 11:86–9

35 Rankin, A.M., and Philip, P.J., *Central African Journal of Medicine* (1963), 9:167–70

36 Calder, A.J., et al., 'Reading the mind from eye gaze.' *Neuropsychologia* (2002), 40:1129–38

37 Hooper, J., and Teresi, D., *The 3-Pound Universe* (Dell Publishing Co. Inc., New York, 1986)

38 Mayberg, H.S., et al., 'Cingulate function in depression: a potential predictor of treatment response.' *NeuroReport* (1997), 8:1057–61

39 Mongeau, R., Miller, G.A., Chiang, E., Anderson, D.J., 'Neural correlates of competing fear behaviors evoked by an innately aversive stimulus.' *Journal of Neuroscience* (2003), 23:3855–68

40 Rachman, S., and Seligman, M.E., 'Unprepared phobias: "be prepared".' *Behaviour Research and Therapy* (1976), 14:333–8

41 Öhman, A., and Soares, J.J., 'On the automatic nature of phobic fear: conditioned electrodermal responses to masked fear-relevant stimuli.' *Journal of Abnormal Psychology* (1993), 102:121–32

42 Select Committee on Science and Technology, House of Lords, *Cannabis: the scientific and medical evidence.* Session 1997–8, 9th report. The Stationery Office: HL paper 15

43 Hatzidimitriou, G., et al., 'Altered serotonin innervation patterns in the forebrain of monkeys treated with MDMA seven years previously: factors influencing abnormal recovery.' *Journal of Neuroscience* (1999), 19:5096–107

44 Hoffmann, A., *LSD – My Problem Child* (J.P. Tacher, 1983)

45 Damasio, H., et al., 'The return of Phineas Gage: clues

about the brain from the skull of a famous patient.' *Science* (1994), **264**:1102–5

46 McGivern, R.F., et al., 'Cognitive efficiency on a match to sample task decreases at the onset of puberty in children.' *Brain Cognition* (2002), **50**:73–89

47 Killgore, W.D., Oki, M., Yurgelun-Todd, D.A., 'Sex-specific developmental changes in amygdala responses to affective faces.' *NeuroReport* (2001), **12**:427–33

48 Newberg, A., et al., 'The measurement of regional cerebral blood flow during the complex cognitive task of meditation: a preliminary SPECT study.' *Psychiatry Research Neuroimaging* (2001), **106**:113–22

49 Lawrence, K., et al., 'Interpreting gaze in Turner syndrome: impaired sensitivity to intention and emotion, but preservation of social cueing.' *Neuropsychologia* (2003), **41**:894–905

50 LeVay, S., 'A difference in hypothalamic structure between heterosexual and homosexual men.' *Science* (1991), **253**:1034–7

51 Kuhl, P.K., 'A new view of language acquisition.' *Proceedings of the National Academy of Science USA* (2000), **97**:11850–7

52 Hollich, G.J., et al., 'Breaking the language barrier: an emergentist coalition model for the origins of word learning.' *Monographs of the Society for Research in Child Development* (2000), **65**:1–123

53 Huttenlocher, J., 'Language input and language growth.' *Preventive Medicine* (1998), **27**:195–9

54 Hasselmo, M.E., and Bower, J.M., 'Acetylcholine and memory.' *Trends in Neuroscience* (1993), **16**:218–22

55 Kida, S., et al., 'CREB required for the stability of new and reactivated fear memories.' *Nature Neuroscience* (1998), **5**:348–55

56 Ishai, A., Haxby, J.V., Ungerleider, L.G., 'Visual imagery of famous faces: effects of memory and attention revealed by f-MRI.' *NeuroImage* (2002), **17**:1729–41

57   Karni, A., Tanne, D., Rubenstein, B.S., Askenasy, J.J.M., Sagi, D., 'Dependence on REM sleep of overnight improvement of a perceptual skill.' *Science* (1994), **265**:679–82

58   Mednick, S., et al., 'The restorative effect of naps on perceptual deterioration.' *Nature Neuroscience* (2002), **5**:677–81

59   Zajonc, R.B., Wilson, W.R., Rajecki, D.W., 'Affiliation and social discrimination produced by brief exposure in day-old domestic chicks.' *Animal Behaviour* (1975), **23**:131–8

60   Brewin, C.R., and Beaton, A., 'Thought suppression, intelligence, and working memory capacity.' *Behaviour Research and Therapy* (2002), **40**:923–30

61   Genoux, D., et al., 'Protein phosphatase 1 is a molecular constraint on learning and memory.' *Nature* (2002), **418**:970–5

62   Logan, J.M., et al., 'Under-recruitment and nonselective recruitment: dissociable neural mechanisms associated with aging.' *Neuron* (2002), **33**:827–40

63   Inouye, S.K., et al., 'Cognitive performance in a high-functioning community-dwelling elderly population.' *Journal of Gerontology* (1993), **48**:146–51

64   Pervin, L.A, John, O.P., eds, 'A five-factor theory of personality.' *Handbook of Personality: Theory and Research* (Guilford Press, 1999)

65   Gazzaniga, M., and Heatherton, T., *Psychological Science* (W.W. Norton, 2003)

66   Johnson, D.L., et al., 'Cerebral blood flow and personality: a positron emission tomography study.' *American Journal of Psychiatry* (1999), **156**:252–7

67   Canli, T., et al., 'Amygdala response to happy faces as a function of extraversion.' *Science* (2002), **296**:2191

68   Boissy, A., and Boissou, M.F., 'Effects of androgen treatment on behavioural and physiological responses of heifers to fear-inducing situations.' *Hormones and Behaviour* (1984), **28**:66–83

69 Plusquellec, P., and Boissou, M.F., 'Behavioural character-
    istics of two dairy breeds of cows selected (Herens) or not
    (Brune des Alpes) for fighting and dominance ability.'
    Applied Animal Behaviour Science (2001), **72**:1–21

70 O'Connor, D.B., Archer, J., Hair, W.M., Fu, F.C.,
    'Activational effects of testosterone on cognitive function
    in men.' *Neuropsychologia* (2001), **39**:1385–94

71 Larkin, P., *Required Writing: Miscellaneous Pieces
    1955–1982* (Faber & Faber, London, 1983)

72 Taylor, S.E., et al., 'Portrait of the self-enhancer: well
    adjusted and well liked or maladjusted and friendless?'
    Journal of Personality and Social Psychology (2003),
    84:165–76

73 Taylor, S.E., and Brown, J.D., 'Illusion and well-being: a
    social psychological perspective on mental health.'
    Psychological Bulletin (1988), **103**:193–210

74 Hagen, M.C., et al., 'Somatosensory processing in the human
    inferior prefrontal cortex.' *Journal of Neurophysiology*
    (1988), **88**:1400–6

75 Larkin, P., *Required Writing: Miscellaneous Pieces
    1955–1982* (Faber & Faber, London, 1983)

76 Enoch, M.-A., et al., 'Genetic origins of anxiety in women:
    a role for a functional catechol-O-methyltransferase poly-
    morphism.' *Psychiatric Genetics* (2003), **13**:33–41

77 Bowley, M.P., Drevets, W.C., Ongur, D., Price, J.L., 'Low
    glial numbers in the amygdala in major depressive
    disorder.' *Biological Psychiatry* (2002), **52**:404–12

78 Nesse, R.M., 'Is depression an adaptation?' *Archives of
    General Psychiatry* (2000), **57**:14–20

79 Bloom, H., *Shakespeare: The Invention of the Human*
    (Fourth Estate, 1999)

80 Badcock, C., *PsychoDarwinism* (HarperCollins, 1994)

81 Iacoboni, M., et al., 'Reafferent copies of imitated actions
    in the right superior temporal cortex.' *Proceedings of the
    National Academy of Science USA* (2001), **98**:13995-9

82 Penton-Voak, I.S., et al. (1999), 'Menstrual cycle alters face

preference.' *Nature* (2000), **399**:741–2

83   Clark, R.D., and Hatfield, E., 'General differences in receptivity to sexual offers.' *Journal of Psychology and Human Sexuality* (1989), **2**:39–55

84   Montgomery, M.J., and Sorrell, G.T., 'Love and dating experience in early and middle adolescence: grade and gender comparisons.' *Journal of Adolescence* (1998), **21**:677–89

85   Autkrystof, Doris, *Amedeo Modigliani 1884–1920: The Poetry of Seeing (Basic Art)* (Taschen America Llc, 2000)

86   Thorne, F., et al., 'Effects of putative male pheromones on female ratings of male attractiveness: influence of oral contraceptives and the menstrual cycle.' *Neuroendocrinology Letters* (2002), **23**:291–7

87   Grammer, K., 'Androsterone: a male pheromone?' *Ethology and Sociobiology* (1995), **14**:201–7

88   Cutler, W.B., Friedmann, E., McCoy, N.L., 'Pheromonal influences on sociosexual behaviour in men.' *Archives of Sexual Behaviour* (1998), **27**:1–13

89   Young, L.J., Lim, M.M., Gingrich, B., Insel, T.R., 'Cellular mechanisms of social attachment.' *Hormones and Behavior* (2001), **40**:133–8

90   O'Doherty, J., et al., 'Beauty in a smile: the role of medial orbitofrontal cortex in facial attractiveness.' *Neuropsychologia* (2003), **41**:147–550

91   Bartels, A., and Zeki, S., 'The neural basis of romantic love.' *NeuroReport* (2000), **11**:3829–34

92   Kobak, R.R., and Hazan, C., 'Attachment in marriage: effects of security and accuracy of working models.' *Journal of Personality and Social Psychology* (1991), **60**:861–9

93   Fisher, H., et al., 'Review. The neural mechanisms of mate choice: a hypothesis.' *Neuroendocrinology Letters* (2002), **23** Suppl., **4**:92–7

94   Ferguson, J.N., et al., 'Social amnesia in mice lacking the oxytocin gene.' *Nature Neuroscience* (2000), **25**:284–8

95 Turner, R.A., et al., 'Preliminary research on plasma oxytocin in normal cycling women: investigating emotion and interpersonal distress.' *Psychiatry* (1999), **62**:97–113

96 Winston, J.S., Strange, B.A., O'Doherty, J., Dolan, R.J., 'Automatic and intentional brain responses during evaluation of trustworthiness of faces.' *Nature Neuroscience* (2002), **5**:277–83

97 Castelli, F., Frith, C., Happe, F., Frith, U., 'Autism, Asperger syndrome and brain mechanisms for the attribution of mental states to animated shapes.' *Brain* (2002), **125**:1839–49

98 Epley, N., and Dunning, D., 'Feeling "holier than thou": are self-serving assessments produced by errors in self- or social prediction?' *Journal of Personality and Social Psychology* (2002), **79**:861–75

99 Langleben, D.D., et al., 'Brain activity during simulated deception: an event-related functional magnetic resonance study.' *NeuroImage* (2002), **15**:727–32

100 Lee, T.M., et al., 'Neural correlates of response inhibition for behavioral regulation in humans assessed by functional magnetic resonance imaging.' *Neuroscience Letters* (2001), **309**:109–12

101 Greene, J.D., et al., 'An fMRI investigation of emotional engagement in moral judgment.' *Science* (2001), **293**:2105–8

102 Diamond, M.C., et al., 'On the brain of a scientist: Albert Einstein.' *Experimental Neurology* (1985), **88**:198–204

103 Witelson, S.F., Kigar, D.L., Harvey, T., 'The exceptional brain of Albert Einstein.' *Lancet* (1999), **353**:2149–53

104 Cattell, R., *Abilities: Their Structure, Growth and Action* (Houghton Mifflin, Boston, 1971)

105 Gardner, H., *Frames of Mind: The Theory of Multiple Intelligences* (Basic Books, 1983)

106 Gray, J.R., Chabris, C.F., Braver, T.S., 'Neural mechanisms of general fluid intelligence.' *Nature Neuroscience* (2003), **6**:316–22

107 Thompson, P.M., et al., 'Genetic influences on brain structure.' *Nature Neuroscience* (2001), **4**:1253–8

108 Deary, I.J., et al., 'Cognitive change and the APOE epsilon 4 allele.' *Nature* (2002), **418**:932

109 Whalley, L.J., and Deary, I.J., 'Longitudinal cohort study of childhood IQ and survival up to age 76.' *British Medical Journal* (2001), **322**:819

110 Bosworth, H.B., and Schaie, K.W., 'Survival effects in cognitive function, cognitive style, and sociodemographic variables in the Seattle Longitudinal Study.' *Experimental Aging Research* (1999), **25**:121–39

111 Rampon, C., et al., 'Effects of environmental enrichment on gene expression in the brain.' *Proceedings of the National Academy of Science USA* (2000), **97**:12880–4

112 Tang, Y.P., et al., 'Genetic enhancement of learning and memory in mice.' *Nature* (1999), **401**:63–9

113 Fritsch, T., et al., 'Effects of educational attainment on the clinical expression of Alzheimer's disease: results from a research registry.' *American Journal of Alzheimer's Disease and Other Dementias* (2001), **16**:369–76

114 Eves, A., and Gesch, B., 'Food provision and the nutritional implications of food choices made by young adult males, in a young offenders' institution.' *Journal of Human Nutrition and Diet* (2003), **16**:167–79

115 Scholey, A.B., and Kennedy, D.O., 'Acute, dose-dependent cognitive effects of Ginkgo biloba, Panax ginseng and their combination in healthy young volunteers: differential interactions with cognitive demand.' *Human Psychopharmacology* (2002), **17**:35–44

116 Spiegel, D., et al., 'Effect of psychosocial treatment on survival of patients with metastatic breast cancer.' *Lancet* (1989), **ii**:888–91

117 Fawzy, F.I., et al., 'Malignant melanoma: effects of an early structured psychiatric intervention, coping, and affective state on recurrence and survival 6 years later.' *Archives of General Psychiatry* (1993), **50**:681–9

118 Fawzy, F.I., Canada, A.L., Fawzy, N.W., 'Malignant melanoma: effects of a brief, structured psychiatric intervention on survival and recurrence at 10-year follow-up.' *Archives of General Psychiatry* (2003), **60**:100–3

119 Ross, L., et al., 'Mind and cancer: does psychosocial intervention improve survival and psychological well-being?' *European Journal of Cancer* (2002), **38**:1447–57

120 Sternberg, R.J., Kaufman, J.C., Pretz, J.E., *The Creativity Conundrum: A Propulsion Model of Creative Contributions* (Philadelphia, PA, 2002)

121 Klein, G., *Sources of Power: How People Make Decisions* (MIT Press, Boston, 1988)

122 Pantages, E., and Dulac, C., 'A novel family of candidate pheromone receptors in mammals.' *Neuron* (2000), **28**:835–45

123 Wiseman, R., *The Luck Factor: Change your luck – and change your life* (Century, London, 2003)

124 Greyson, B., 'Biological aspects of near-death experiences.' *Perspectives in Biology and Medicine* (1998), **42**:14–32

Glossary

ACTH – AdrenoCorticoTrophic Hormone is made in the pituitary gland; from there it enters the bloodstream, sending a message to the adrenal glands. The adrenals then produce various hormones in response, in particular cortisol.

Adrenal glands – These glands in the vicinity of the kidney are essentially a collection of cells specialized to produce the hormones cortisol, adrenaline and noradrenaline. These substances are important in the response to stress – fear, excitement, anxiety of various types, and in particular the classic fight-or-flight response to attack by a predator.

Alzheimer's disease – Dementia more common in advanced age, often with a genetic component, affecting about 20 per cent of people over eighty but occasionally much younger people. Usually involves wasting of brain substance and changes in the brain proteins. Loss of memory is a prominent symptom, as is impairment of language and comprehension. Psychiatric changes are frequent.

Amygdala – An almond-shaped structure in the medial temporal lobe involved in a range of emotional functions, including the detection of significant events in the environment, fear-conditioning and enhancing storage of emotional memories.

Via the hypothalamus, it can prompt release of stress hormones such as adrenaline into the bloodstream, giving a fright response and disrupting the control of rational thought. It is part of the so-called limbic system.

Anterior cingulate cortex – A region towards the front of the inside surface of the hemispheres. All its functions are not completely understood but it helps decision-making when the brain receives sensory stimuli that conflict, and it processes attention, concentration and pain perception. It is involved in controlling the autonomic nervous system. Morphine-like drugs decrease the activity in this region.

Autism – A condition characterized by lack of social awareness and failure to understand the emotions of others. Autistic children may be isolated, socially inept and aloof, and have narrowly focused interests.

Autonomic nervous system – This controls involuntary functions such as heart activity, digestion, blood-vessel constriction and relaxation, and sexual activity such as erection. It has two systems that work in opposition:

Sympathetic – speeds the pulse, constricts and dilates blood vessels, relaxes the bronchial tubing in the lung, reduces digestion. It is most active when stimulated by stress in response to the adrenals.

Parasympathetic – slows the pulse, increases gut motility, stimulates sexual arousal, increases salivation and constricts the pupils. It prepares the body to rest.

Axons – These are the long fibres of nerve cells (neurons). They can be thought of as the wiring of the nervous system.

Basal ganglia – A group of centres (or nuclei) in the centre of each hemisphere just above the midbrain. They play an important part in planning and coordinating movement and position of the body. They exert their influence over the networks that link the motor cortex to other cortical areas.

Brainstem – The part of the brain connecting its two cerebral hemispheres to the spinal cord. It is divided into three areas: midbrain, pons and medulla oblongata. It is the oldest part of the brain, where most basic, involuntary functions (such as breathing and control of heart rate) are regulated.

Broca's area – The area of the frontal lobe cortex – usually the left inferior frontal gyrus – described by Paul Broca as controlling speech.

Central nervous system – This is the 'controlling' part of the nervous system. It is covered by membranes called meninges. The brain, spinal cord and optic nerves are all part of the central nervous system. It is connected to the peripheral nervous system, which is made up of afferent nerves that carry sensory impulses from all parts of the body to the brain, and efferent nerves that take motor impulses to the muscles.

Cerebellum – This structure lies above the brainstem and behind the pons, and regulates balance and coordination of movements.

Cerebrospinal fluid – The watery fluid surrounding the brain and spinal cord. It contains glucose and proteins. It is derived from the bloodstream and is filtered by a membrane in the hollow structures inside the brain, the ventricles.

Cerebrum – The great bulk of the brain, divided into the two hemispheres, right and left, which are connected by the corpus callosum.

Cognition – The higher functions of the brain, which are often specifically human. They include: comprehension and speech, memory, problem solving, attention, calculation, and visual and auditory perception.

Coordination – Ordered, organized working of muscle groups or individual muscles to produce controlled movements.

Corpus callosum – The thick band connecting the two cerebral hemispheres.

Cortex – The outer layer of the cerebrum, which is rich in neurons and in which 'thinking' occurs. It normally has a layer of neurons about 4mm thick. But the word 'cortex' can be applied to the outer layer of many organs, including the ovaries and adrenal glands.

Corticotropin – See ACTH.

Cortisol – Hormone released by the adrenal glands during stress, but which is also necessary for the normal function of almost every part of the body. It regulates blood pressure, the use of sugars and proteins and causes breakdown of muscle protein into amino acids which can be used as an emergency energy source.

Dementia – Serious progressive loss of intellectual function, often associated with a disease such as Alzheimer's, but there are many different causes. There may frequently be a personality change, preventing a sufferer from engaging in any normal activity.

Dendrite – The thin part of a nerve cell that transports newly received impulses into the neuron.

Electroencephalography (EEG) – The commonly used technique that records the electrical activity in the brain, mostly from the cortex.

Epilepsy – An epileptic attack is a fit (or seizure), often with loss of consciousness, or occasionally changed consciousness, which may only be perceived by somebody who knows the person well. Severe attacks cause violent movements. It is due to an electrical storm arising from a focal point in the brain. Mild attacks in the temporal lobe may cause hallucinations or 'religious' experiences.

Evoked potentials – These are electrical signals recorded (usually on EEG) from the brain or scalp in response to a stimulus.

Frontal lobes – The largest lobes of the human cerebral cortex. The front part (anterior) is associated with learning, behaviour and personality. The back part (posterior) is the motor cortex controlling voluntary movements.

Glia – The cells of the brain making up most of its bulk. There are five times more glial cells than neurons. They are of various types, helping brain repair, brain nutrition, maintenance of cerebrospinal fluid and construction of the fatty myelin sheath of neurons.

Grey matter – The part of the brain, such as the cortex, where neurons are concentrated.

Hemisphere – Half of the brain's cortex (left or right), which is divided into four main lobes: frontal, temporal, parietal and occipital.

Hippocampus – Part of the temporal lobe immediately behind the amygdala, involved with memory and learning. It is part of the limbic system. The name means 'sea horse' – a reflection of its shape.

Hypothalamus – a collection of centres (nuclei) beneath the thalamus and part of the limbic system, which, through the pituitary gland, regulate hormones in the body and help maintain the body's internal state: for example, temperature control, food intake and heart activity.

Lateral geniculate nucleus – Part of the thalamus involved in processing visual stimuli and passing them to the cortex.

Limbic system – Structures forming around the border of the brainstem which are derived from a more primitive part of the cortex and are associated with learning, memory and emotional processing. The system includes the amygdala, the hypothalamus, part of the thalamus and the cingulate gyrus. Its utility as a concept is now hotly debated, because the range of functions it seems to undertake are done by independent sub-units of the system, and also by areas of the brain not originally thought part of the limbic system.

Magnetoencephalography (MEG) – Maps and localizes brain activity by recording magnetic fields produced by active neurons in response to stimulation. Magnetic fields are not distorted by passing through the scalp or skull, so it has advantages over EEG.

Medulla oblongata – Lower part of the brainstem which controls automatic functions and acts as a relay between the brain and the spinal cord.

Memory – Now considered as divided into a variety of types. Memory can be remote or episodic, such as long-time-past events from, say, childhood. Working memory allows you to recall events from a short time previously, such as whether the kettle is switched on. Prospective memory is the ability to remember a planned commitment. Procedural memory is associated with learnt motor skills and does not require intentional recollection – such as knowing how to ride a bike.

MRI/Magnetic Resonance Imaging – or nuclear magnetic resonance – is a method of imaging parts of the body using the magnetic properties of individual tissue types to build up a three-dimensional view. Functional magnetic resonance imaging (fMRI) allows the observer to see how the brain is metabolizing and can thus build up a picture showing what part of the brain is active during a given task.

Myelin – The fatty sheath that forms around axons and acts as a kind of insulator. Myelin greatly increases the speed at which nervous impulses can be sent.

Nerve fibres – These are axons that extend from the main body of the neuron and carry impulses to the nerve cell (afferent) or away from it (efferent).

Neuron – The individual nerve cell containing axons and dendrites and connecting with other neurons by synapses.

Neurotransmitters – Chemicals stored at the end of the axons in the synapses. When an electrical impulse reaches the synapse, neurotransmitters are released and cross the gap of the synapse. This initiates a message to the neighbouring neuron or cell, such as a muscle cell. There are around fifty different neurotransmitters, which may excite or inhibit transmission of an impulse. They include glutamate, GABA, acetylcholine, adrenaline and noradrenaline, dopamine, serotonin. Each of these transmitters has a different function.

Nucleus accumbens – Contains neurons that are part of the basal ganglia. It plays a role in regulating movement and complex motor activity. This structure is thought to be 'the reward centre of the brain' and an important neurotransmitter in its function is dopamine.

Optic nerve – The nerve that extends from the back of the eye – the retina – connecting it to the brain.

Paralysis – Inability to move a part of the body.

Peripheral nervous system – The nerves outside the central nervous system which run through the body to and from organs such as skin, muscles, joints.

PET – Positron Emission Tomography: a scanning technique using radioisotopes, such as radioactive oxygen, to identify blood flow or metabolically active brain areas.

Pituitary gland – The master gland at the base of the brain, connected to the hypothalamus, which controls other glands in the body, such as the adrenals, pancreas, thyroid, ovaries and testes.

Placebo – A sham or neutral treatment, often in practice a drug, which is inert and has no activity and therefore no intrinsic scientific value in improving or changing the medical condition. Because a person may expect to feel better after any treatment, placebo treatment may lead them to feel better and to have fewer symptoms from an ailment.

Pons – Part of the brainstem controlling respiration, and which is associated with arousal and sleep. It helps in control of the autonomic nervous system and is a relay centre between the cortex and cerebellum.

Proprioception – Awareness of position and movement in space.

Proteins – Compounds made up of amino acids that are responsible for all functions of cells. Proteins are produced as a result of the message produced by the DNA.

Pulvinar – Part of the thalamus involved in processing visual information.

Purkinje cell – A giant neuron with massively branching dendrites making many (perhaps up to 200,000) connections to other neurons. This cell is mostly found in the cerebellum.

Sensory cortex – Network of neurons (many of which are located in the parietal lobe) responsible for processing of sensations, including pain, smell, taste, temperature, vision, hearing, touch and proprioception.

Synapse – The gap connecting neurons to each other, controlled by the production of neurotransmitters. A typical neuron may have 10,000 such connections to other neurons.

Thalamus – The centre above the brainstem that acts as a relay station between cortex and sensory organs (e.g., retina, skin, inner ear) with the exception of some olfaction. It has many other functions, which include an influence on mood and some body movements.

Turner's syndrome – Loss of one of the paired X chromosomes in girls, occurring when the embryo is formed. It results in infertility, absent periods, short stature and cognitive changes.

Vagus nerve – The nerve which, among other functions, controls heart rate by carrying most of the parasympathetic signals from the brain to the body. When stimulated it slows the heart. It is a long nerve and is called a 'cranial nerve' because it is directly connected into the brain, in the region of the brainstem.

Ventricles – The four cavities in the brain containing cerebrospinal fluid.

White matter – The myelinated neuronal fibres and supporting glial cells making up much of the central nervous system.

Index

acetylcholine, 154, 161, 275, 304, 456
ACTH (adrenocorticotrophin), 448–9
Ader, Robert, 450
Adolphs, Ralph, 189
adrenaline, 62, 140–1, 450
adrenocorticotrophin, 62
agnosia, 119
air encephalography, 183
Alcmaeon, 20
alcohol, 18, 217–18, 222–3, 234, 236–7, 442
alexithymia, 193–4
Alien Hand Syndrome (AHS), 72–6
allomimesis, 201
alpha male, 369
alpha waves, 40, 152
Alzheimer, Alois, 33
Alzheimer's disease: alcohol effects, 442; development, 430, 434; islets of ability,

279; memory aids, 300; nicotine effects, 221; patterns of memory loss, 287, 297, 300, 303, 306; sense of smell, 104; stress, 454
amnesia, 279–82
amygdala: anger perception, 189–90, 208; appearance, 63; arousal, 139–40; behavioural approach system, 313; damage to, 189; in depression, 335; extroverts, 314–15; facial recognition, 124; fear perception, 139, 148, 189–90, 214, 249; laughter, 200; memory role, 144, 276–8, 297; nicotine effects, 220; REM sleep, 154, 160–1; response to emotions, 197, 198, 215, 366, 368, 413;

amygdala (*cont.*)
   role, 63, 124, 189–90,
   193; smell information,
   109
Anderson, David, 214
androstenol, 384
Angelman syndrome, 353
anger, 189–90, 191, 205–9
anhedonia, 212–13, 335
animal research, 50–4
anterior cingulate: cortex,
   91–2, 114, 175, 314,
   335–6; gyrus (ACG), 148,
   360, 401
anxiety, 155, 209, 226, 333,
   337, 365–6, 454
apoptosis, 80, 241, 242
apotemnophilia, 115–16
Arbuthnot, Eleanor, 355–8
Aristotle, 20, 21
arousal, 138–41, 312
Arvanitaki, Angelique, 49
Aserinsky, Eugene, 153
Asperger, Hans, 203, 204
Asperger's Syndrome, 203–4
attachment, 391, 393–7
attention, 134–8
Attention Deficit
   Hyperactive Disorder
   (ADHD), 61, 125, 147,
   221, 440
attractiveness, 365, 370–3
Aubertin, Ernest, 35
autism: abilities of autist
   savants, 459; gender
   issues, 253; inability to
   'read' others, 203, 254–5,
   398; orientation, 142–3;
   perception differences,
126; perception of eyes,
   204–5; PET scans, 398–9;
   Theory of Mind test, 166
Autkrystof, Doris, 381
axons, 58, 66

babies: brain development,
   241–7; emotional
   environment, 82–3;
   father's response to,
   407–8; language
   acquisition, 83, 246–7,
   261, 262–4; preference
   for symmetry, 244, 371;
   premature, 113, 463–4;
   self-awareness, 165–6
Bach-y-Rita, Paul, 104
Bacon, Francis, 457
Baddeley, Alan, 289
Balint's Syndrome, 142
Ballesteros, Severiano, 461
Baron-Cohen, Simon, 99,
   203–4, 253, 255, 422
Bartels, Andreas, 389
basal forebrain, 304
basal ganglia, 65, 93–4,
   141–2, 145, 225–6, 360,
   461
Baum, David, 244
Beckham, David, 424
Beethoven, Ludwig van,
   456
behavioural approach system
   (BAS), 313–14
behavioural inhibition
   system (BIS), 313–14
Bell, Andi, 300–2
belladonna, 389
Bem, Daryl, 257

Bennett, Arnold, 382
Beresford, John, 232
Berger, Hans, 39–40
Berger, Theodore, 276
Berscheid, Ellen, 396
Bertelli, Alberto, 442
beta waves, 152, 153
Binet, Alfred, 421
bingo, 433
biofeedback, 455
Blackmore, Susan, 103–4, 471
Blair, Tony, 195
Blanke, Olaf, 90, 93–4
blind people, 103–4, 112, 156
blindsight, 127–8
Blum, David, 61
body language, 194–5, 252, 460
Bohr, Niels, 457
Boivin, Jacky, 371
Bomford, Dr, 194
Boston Medical and Surgical Journal, 180
Bower, James, 28, 161, 275
Brahms, Johannes, 457
brain: in childhood, 208, 244–7; plasticity, 77–9, 80, 83–4, 103, 261, 264, 282–3, 291; regions, 62–6; size, 17, 37; surgery, 88–91; two sides, 66–9, 70; in the womb, 80, 241
brainstem, 65, 140, 153, 154, 160
Breasted, James, 18
breastfeeding, 245–6, 394
Brecht, Bertolt, 273–4

Bremner, Rory, 122, 195
Breteler, Monique, 442
Brewin, Chris, 298
bridge playing, 449
Bristow, Eric, 461
Broca, Paul Pierre, 19, 34–7
Broca's area, 19, 35, 238, 246, 264–5, 267, 268, 427
Brodmann, Korbinian, 32–4, 418
Bruce, Lenny, 226
Brugger, Peter, 468–9
BSE, 442
'bubba psychology', 362
Buckner, Randy, 304–5
Burgess, John, 440
Burroughs, William, 457
Bush–Gore election campaign, 136
Byron, Lord, 87, 378

Cajal, Santiago, 29, 30–2
Campbell, Ruth, 252
cancer, 451–3
Canli, Turhan, 314
cannabis, 218, 219, 223–7, 234
Capgras Delusion, 125
Caramazza, Alfonso, 269
Carden, John Rutter, 355–8, 360
Cardinal, Rudolph, 147
Carlson, Mary, 82
Carroll, Lewis, 154
Carson, Edward, 349
Carter, Rita, 120, 340
Caton, Richard, 39
cats, 65–6, 100, 106
Cattell, Raymond, 422

caudate nucleus, 390, 409
cerebellum, 28, 65, 145,
    225, 237, 284
cerebral cortex, 19n, 190, 226
Chabris, Christopher F., 136
childbirth, 394
children, 208, 242–7, 266–7
chimpanzees, 165, 201, 205,
    257, 267, 369
chocolate, 392–3
Chomsky, Noam, 267
Christie, Agatha, 280–1
Churchill, Winston, 322
cingulate: anterior, 91–2,
    114, 175, 389; anterior
    cingulate cortex, 314,
    335–6; anterior cingulate
    gyrus, 148, 360; cortex,
    141–2, 284; gyrus, 212;
    posterior cingulate gyrus,
    412
circadian rhythms, 158–9
Claparède, Edouard, 293
Claxton, Guy, 456
Clinton, Bill, 405–6
cocaine, 218, 224, 234–5
cocktail parties, 135–6, 368,
    402
Cognitive Behavioural
    Therapy (CBT), 81,
    198–9, 323, 334
Coleridge, Samuel Taylor,
    457
Columbo, 462
Comings, Kenneth, 61
COMT enzyme, 337
conditioning, 235, 450
consciousness, 90, 169–77,
    276, 461–2

Conway, Michael, 299
copulins, 385
corpus callosum, 66, 68–9,
    168–9, 252, 256, 295–6,
    403
Corteen, Dr, 135
corticotropin releasing factor
    (CRF), 140
cortisol, 448
Costa, P.T., 311, 321
Courchesne, Eric, 142
creativity, 454–60, 472
CREB, 287, 304
Creutzfeldt–Jakob disease
    (CJD), 51, 442
Crick, Francis, 162
cricketers, 173–4
Cro-Magnon man, 18
crying babies, 243
crystallized intelligence, 422
Curtis, Martha, 282–3

Damasio, Antonio: on
    'convergence zones', 278;
    on Gage's injuries, 237;
    memory study, 293–4;
    'storage sites' study, 120;
    on tumour patient, 191,
    192; on 'will' of stroke
    patient, 149
Damasio, Hannah, 120, 237
Darwin, Charles, 25, 259,
    369, 379, 403
Davidson, Richard, 211
Dawson, Geraldine, 82
deaf children, 112, 263
Deary, Ian, 428
DeBruine, Lisa, 399
deception, 166–7, 404–7

defence mechanisms, 135, 350–4
delta waves, 153
Democritus, 20
dendrites, 58, 274
DEP-1 gene, 338
depression: anhedonia, 211–13, 335; brain chemistry, 334–6; dopamine levels, 62; effects on personality, 334–5; as faulty sub-set of emotion, 339; genetic cause, 337–8; right-brain activity, 211; sleep difficulties, 155; types, 335; view of world, 74; visits to GP, 447
Descartes, René, 22–3
Dexedrine, 147
Diamond, Marian C., 417–18, 449
Dimberg, Ulf, 198
disgust, 146, 191, 197, 198
displacement, 352–3
dogs, 165, 201, 205, 339
Dolan, Ray, 143, 398
Doogie mice, 431
dopamine: alcohol effects, 223; COMT enzyme, 337; drug effects, 219, 229, 235; influence on character, 315–16; l-dopa, 469; in love, 392; nucleus accumbens, 146–7, 223, 235; role, 61, 193
dorsolateral prefrontal cortex (DPC), 91–2, 149, 154, 412

dreams, 153–4, 159–63
Drevets, Wayne, 335, 337
drug use, 217–20, 234–7
Duchenne, Guillaume Benjamin, 401–2
Duchenne muscular dystrophy, 402
Dulac, Catherine, 467
dyslexia, 126, 128–9, 270, 438

Ecstasy, 228–30, 234
Edelman, Gerald, 79–80
Edinburgh Review, 24–5
EEG (electroencephalography), 39–40, 143, 152, 172–3
Egypt, ancient, 18–19, 87
Einstein, Albert, 414–19, 427, 449
Ekman, Paul, 196, 405
Elder, Murdo, 443
electroencephalogram, 39–40
Elias, Merrill, 429
emotion, 68–9 190–1, 203–4, 252
empathy, 412
encephalitis, 202
endorphins, 62, 146, 223, 437, 453
Enoch, Mary-Anne, 337
epilepsy, 60, 69–73, 83, 87–8, 109, 202, 282–3, 469
epinephrine, 62
Erasistratus, 21
ergot, 230–1
Ernst, Richard, 46

erotomania, 359
ESP, 462, 467–8
Evans, Gaynor, 118
executive organization,
    147–50
exercise, 306–7, 437
extroverts, 310–14
eye movement desensitiza-
    tion and processing
    (EMDR), 333–4
eyes, 204–5, 254, 332–4,
    389
Eysenck, Hans, 310–12,
    321

fMRI scanning: attention,
    145; deception, 401;
    distraction, 425; epilepsy,
    91; love, 392; memoriza-
    tion exercise, 304; moral
    dilemmas, 412; P300
    response, 143; 'reading'
    others, 398; REM sleep,
    154
face recognition, 122–3, 138,
    262, 292–3
facial: characteristics, 244,
    368, 404–7; expressions,
    195–9, 248–9, 252, 403
family trust, 399–400
Farmer, Frances, 188
Fawzy, Fawzy I., 451–2
fear: amygdala response, 63,
    139, 148, 189, 197;
    communication of, 328;
    dreams, 161; facial
    expression, 197–8;
    impaired perception of,
    189; mesolimbic pathway,

146; phobias, 214–15,
    216; responses to, 213–14;
    survival strategy, 213–14
Feinberg, Dr, 72–3
Ferguson, Jennifer, 394–5
Field, Tiffany, 113
Fiery Angel, The, 86–7
fight-or-flight response, 109,
    139–40
Finzi, Sylvia, 390
fish oil, 438–40
Fisher, Helen, 384, 392
fluid intelligence, 422
Franzini, Louis R., 93
Frederick II, Emperor,
    239–40, 264, 267
Freeman, Walter, 185–8,
    213
Fregoli's Delusion, 122,
    124–5
Freud, Anna, 352
Freud, Sigmund, 160, 295–6,
    348, 350–1, 377, 446–7
Friedland, Robert, 434
Frith, Chris, 149
Frith, Uta, 398
frontal cortex, 141–2, 156,
    398
frontal lobes: agoraphobia,
    353; arousal system,
    138–41; development in
    childhood, 246, 247, 251,
    454; introverts, 312–14;
    lesions, 154; male ageing,
    253; 'mirror neurons',
    366–7; planning role,
    147–50; policing emo-
    tions, 202, 207–8, 214,
    328, 332; prefrontal edge,

335; self-awareness, 167–8, 247; in teenagers, 249–50
Fuller, Buckminster, 125
Fulton, John, 38

GABA (gamma aminobutyric acid), 60, 236
Gabrieli, John, 314
Gage, Phineas, 178–81, 189, 207, 237–8, 334, 340
Galaburda, Al, 128
Galen, 21, 50
Gall, Franz Joseph, 23, 24, 27, 35, 36–7
Galton, Francis, 98–9
Gardner, Howard, 423–4
Gascoigne, Paul, 423–4
Gazzaniga, Michael, 70–2, 168–9
gender differences, 68, 105, 204, 251–7, 318, 337, 379
genes, 99, 209, 256–7, 316, 337–9, 376, 432
geniculostriate pathway, 126–8
Genie, 84
Gesch, Bernard, 440
ginseng, 440–1
Glaser, Ronald, 449
glial cells, 57, 358
glutamate, 60, 431
Golgi, Camillo, 29–32
Goody, Jack, 260
Grammer, Karl, 201, 385
Gray, Jeffrey, 175, 313, 315
Gray, Jeremy, 425–6
Graybiel, Ann, 145
Greene, Joshua D., 411-12
Greig, Tony, 173

Greyson, Bruce, 470
Grossberg, John, 93
groups, 361, 365–8, 397–8, 401, 407–8, 410
guillotine, 22
gum chewing, 432–3
Gur, Ruben and Rachel, 68

Haier, Richard, 285
Halgren, Eric, 143
hallucination, 226, 231–2
Haloperidol, 212
Hamlet, 343–8, 352–3
Hammatt, Mrs, 185–6
Hammersmith cyclotron, 42, 43–4
hamsters, 396
happiness, 197, 198, 200, 211, 319, 334, 436
Harlow, John, 179, 180–1, 237–8
Harvey, Thomas S., 417
Hatzidimitriou, George, 230
Haydn, Joseph, 457
Hazan, Cindy, 391–2
hearing, 110–12, 466
Hebb, Donald O., 78
Hébuterne, Jeanne, 381–3
Hemingway, Ernest, 456
Herbert, George, 165
Herodotus, 240
heroin, 234–5
Herophilus, 21
Hindu mystics, 21
hippocampus: cannabis effects, 225–6; cingulum connection, 175; damage to, 288, 293; development, 242–3; exercise effects,

437; female tissue loss, 253; insulin-responsive cells, 433; limbic system, 63; memory role, 24, 30, 175, 221, 275, 276–8, 288, 293, 304, 433; posterior, 24

Hippocrates, 21, 241

Hirsch, Alan R., 405

Hobson, J. Allan, 160, 161, 290

Hodgkin, Alan, 48–9

Hoffmann, Albert, 231–2

Hollich, George, 262

homeostasis, 176–7

homosexuality, 256–7

Hornby, Nick, 255

Hubbard, Ed, 99

hunches, 462–5

Huttenlocher, Janellen, 264

Huxley, Aldous, 217–18, 233

Huxley, Andrew, 48–9

Hyde, Ida, 49

hypothalamus: arousal system, 139; behavioural approach system, 313–14; cannabis effects, 225; fear response, 214; hormonal centre, 109, 139, 319; limbic system, 63, 192; rat experiment, 212; role, 63–4, 155–6, 192; sleeping-waking cycle, 155–6

IKEA, 329–331

immune system, 376, 448–9, 450–1

inferior prefrontal cortex, 143, 398

insula cortex, 284

insulin, 432–3

intelligence: boosting, 430–42; born or made, 426–30; footballers, 423–4; G-type, 424–6; measurement, 419–23

intention, 149

intentionality, 170

introverts, 310–13, 314, 455–6

intuition, 461–5

IQ, 246, 269–70, 285, 419, 421, 427, 429, 436, 449

Isaacs Report, 55

Isabelle, 84

isolation, 362–4

Jackson, John Hughlings, 87–8

Jamner, Dr, 221

jogging, 306–7

Johansen, Christoffer, 452

Johnson, Debra L., 314

Jonson, Ben, 343

joy, 190–1, 328

Jung, Carl, 162, 295

Karni, Avi, 289

Kataria, Madan, 200

Katz, Lawrence, 107–8, 434

Kekulé, Friedrich, 458

Kelvin, Lord, 39

Kennedy, Rosemary, 188

Kephart, William, 379

Kesey, Ken, 189

Kissinger, Henry, 369

Klein, Gary, 462–4

Klein, Melanie, 351–2

Kleitman, Nathaniel, 153
Koenen, Karestan, 436
Kornhuber, Hans, 172
Korsakoff's syndrome, 293
Kott, Jan, 344
Kuhl, Patricia, 261
kuru, 441–2

LaBerge, David, 144
Laing, R.D., 189
Lancet, 418
Langleben, Daniel, 401
language acquisition and
    development: age of, 84,
    246–7, 261; brain areas,
    196, 246–7; evolutionary
    benefit, 257–9; hard-wired
    grammar, 262–70; left side
    of brain, 65, 263–4, 305;
    writing, 260–1, 269
Largactyl, 188
Larkin, Philip, 308–9, 310,
    313, 320, 322–3, 335–6
lateral pulvinar, 141
laughter, 199–203, 334, 370
learning, 159, 470
Leary, Timothy, 217, 232
Leborgne, Monsieur, 36
LeDoux, Joseph, 191, 277
LeDoux, Robert, 71
Lee, Tatia M.C., 401
left side of brain: corpus
    callosum, 66, 252, 296,
    creativity, 456; difference
    from right, 66–9, 74, 211,
    296; facial expression, 403;
    gender differences, 67,
    252, 318–20; happiness,
    211, 319–22, 322–3;

language, 65, 263–4, 305,
    456; memorizing role,
    305; role, 65–6
left-handedness, 253
left inferior gyrus, 36
LeVay, Simon, 256
Levy, Steven, 417
Libeskind, Daniel, 102
Libet, Benjamin, 172, 173–4
licking test, 310
Liddell, Guy, 341
Lima, Dr, 184
limbic system, 63, 100, 105,
    124, 146, 332, 353
Livingstone, Margaret, 128
Livingstone, Tessa, 296
lobotomy, 184–9
locus coeruleus, 156, 222,
    234
Loewi, Otto, 58–9
Lombroso, Cesare, 25
long-term potentiation (LTP),
    274–5, 298
love: appearance, 369–73;
    attachment, 393–7; brain
    activities, 383–7; feeling
    of falling in, 388–93;
    mystery, 380; survival
    qualities, 369–70; what
    men look for, 372, 377–9;
    what women look for,
    372–7
LSD, 217, 230–4
Luther, Martin, 20
lying, 166–7, 397–8, 400–8

Macbeth, 271, 291
McCoy, Norma, 106
McCrae, R.R., 311, 321

McCrone, John, 47
McDermott, Kathleen, 298
McGivern, Robert, 247–8
MacLeod, Peter, 174
magnetism, 444–5
Maguire, Eleanor, 302
mania, 150, 162, 229, 469
Mankato nuns, 81–2, 433
Maquet, Pierre, 154, 289
Marazziti, Donatella, 390–1
Marlowe, Christopher, 342–3
Marvell, Andrew, 361
mate selection, 369–79
Mayberg, Helen, 212
MDMA (methyl-dioxymethamphetamine), 228
media insula, 389
medial frontal gyrus, 412
medial prefrontal cortex, 398, 399
medial preoptic area, 251
meditation, 152, 162, 250
Mednick, Sara, 290
MEG, 47
melatonin, 22, 156
Meldrew, Victor, 327
memory: ageing and, 271, 287, 299–307; amnesia, 280–2; cannabis effects, 225–6; of early years, 242–3; emotion and, 272–6, 277; episodic, 279–80; exercising the brain, 303–7; false, 291–4, 296–9; flashbulb, 272, 275; long-term, 285–9; losing it, 230, 299–307; overload, 144, 271–2; procedural, 279–80, 282–6; psychoanalysis, 294–9; role of amygdala, 276–8, 297; role of hippocampus, 24, 65, 125, 225, 276–8, 279, 293, 298; short-term, 285–9; sleep effects, 159, 289–91; techniques, 300–3; trained, 24, 299–303; types of, 271–2; unconscious, 291–4
Merck, 228
Merzenich, Michael, 80, 129, 288
mescaline, 234
Mesmer, Hans, 444–7
mesolimbic pathway, 146
Method of Loci, 300–2
mice: exercise, 437; fear, 214; intelligence boosting, 430–2; long-term memory, 431–2, 437; oxytocin experiment, 394–5; pheromones, 106, 467; PP1 protein, 304; short-term memories, 286–7; vomiting, 450
Miller, Bruce, 459
Miller, George A., 285–6
Milne, A.A., 321–2
Milner, Peter, 212
mirror neurons, 198, 366–7
Mitchison, Graeme, 162
Mitford, Unity, 340–1
Modigliani, Amedeo, 381–3
Moffitt, Terrie, 436
Mohs, Richard, 306
Moir, J. Chassar, 231

Moniz, Egas, 181–4, 187–8, 213
monkeys: learning activities, 145; memory studies, 288; 'mirror neurons', 366; nucleus accumbens, 147; role of amygdala, 190; serotonin levels, 209–10, 230, 238
monogamy, 386–70
Montgomery, M.J., 378, 379
Montherlant, Henri de, 320
mood, 327–33
Moore, Dudley, 370
morality, 408–13
Mormons, 338–9
morphemes, 265–6
motivation, 288
Mozart, Wolfgang Amadeus, 416, 456
Mozart Effect, 434
MRI scanning: antisocial personality disorder, 207; development of technique, 34; loving brain, 384; meditation, 250; synaesthesia, 100; teenagers, 249; use of, 45–7, 91, 444
music, 81, 243, 282–3, 435, 456–7
Music Intelligence Neural Development Institute (MIND), 435
myelin, 57, 440
musticism, 217, 218, 229, 232, 233

Nabokov, Vladimir, 98
Nazism, 25, 27, 40, 59

near-death experience, 469–72
Nesse, Randolph, 339
Neural Darwinism, 79–80
neuron: calcium flood, 274; connections, 78–9; feedback, 78–9; first identified, 28–34; networks, 56–8, 78–9; number of, 56
neurotransmitters, 58–62, 147
Newberg, Andrew, 250
nicotine, 220–2
NMDA receptor, 431–2
noradrenaline, 62, 77, 161
norepinephrine, 62, 234, 394
novelty detection, 146–7
NR2B gene, 431–2
nucleus accumbens, 146–7, 202, 223, 315, 409

Observer, 228, 229, 336
Obsessive Compulsive Disorder (OCD), 359–60, 390
occipital lobes, 62, 94, 124, 154
O'Connor, Daryl, 318
oddball responses, 143
oestrogens, 287, 304
Öhman, Arne, 216
Olds, James, 212
optimists, 211–13, 322–3
opium, 224
orbitofrontal cortex, 200, 299, 360
orgasm, 371–2
orientation, 141–6
Ortiz de Zarate, Manuel, 383

out-of-body experience, 93, 469–70
oxytocin, 392, 393–5

P300 response, 143
pain, 113–14, 221, 225, 226–7
Panksepp, Jaak, 243
parietal: areas, 253; cortex, 94, 141–2; lobes, 63, 250, 412
Parkin, Alan J., 72, 75–6
Parkinson's disease, 61, 221
Parrott, Andy, 230
Pascalis, Olivier, 262
Passingham, Dick, 144–5
Penfield, Wilder, 88–90
Penton-Voak, Ian, 373–4
personality, 309–15
pessimists, 211–13
PET scanning: assessing mental states of others, 398; autism, 399; development of technique, 34, 42; false memory study, 298–9; frontal lobe study, 314; intelligence study, 426; procedural memory, 284; REM sleep, 154; use of, 42–5, 145
Phantom Limb Syndrome, 115
phenylethylamine (PEA), 389, 392
pheromones, 106–8 384–6, 434, 467
Philby, Kim, 405
phobias, 193

phonemes, 258, 261, 262–3, 265
phrenology, 23–7, 412
pineal gland, 21–2, 63, 156
Pinker, Steve, 170
'pink-lens effect', 396–7
Pirandello, Luigi, 324–7, 343
Pitino, Lisa, 106
pituitary gland, 64, 140, 192, 319
placebo effect, 447
planning, 147–50
Plato, 20
pleasure, 146, 386
Pod car, 207–8
Pollyanna, 322–3
pons, 154, 161
Porter, Eleanor H., 322–3
Porter, Lord, 227
Portwood, Madeleine, 439–40
Posner, Michael, 175, 284, 335–6
posterior hippocampus, 24
Prader-Willi syndrome, 353
prairie vole, 386, 396
prefrontal: cortex, 149, 288, 305, 337, 398, 401; leucotomy (lobotomy), 184–9, 238; lobes, 251, 288, 335
premotor cortex, 145
preparedness, 216
priming, 138
'Prisoner's Dilemma', 409
projection, 352
Prokofiev, Sergei, 86
prosopagnosia, 121, 124
Proxmire, William, 379–81
Prozac, 62, 158

psychoanalysis, 160, 294–6, 353
psychosis, 212, 213, 229
psychosurgery, 182–9
psychotherapy, 451–2
puberty, 208, 247–51
Purkinje cell, 28–9, 237
Purkinje, Johannes, 28
putamen, 390

Radin, Dean, 467
radioisotopes, 41–4, 250
Raichle, Marcus, 284
Rain Man, 126
Ramachandran, Vilayanur, 97, 98–9, 115, 366–7
Rampon, Claire, 430
raphe nuclei, 156, 222, 234
Ratey, John, 84–5, 125, 129, 142, 205
rats: cannabis effects, 225–6, 235; Ecstasy effects, 235; LSD effects, 234; Mozart study, 435; nicotine effects, 220; oxytocin levels, 394; play, 201; pheromones, 467; pleasure experiments, 212; REM sleep, 161, 289; sleep deprivation, 151; stress effects, 453–4; treatment of, 448
Rauscher, Frances, 434–6
recognition, 117–25
Reich, Wilhelm, 295, 387
REM (rapid eye movement) sleep, 153, 154–7, 159, 160, 161, 234, 289–90
resveratrol, 442

Revonsuo, Antti, 160–1
reward, 146–7, 223, 288, 386
right side of brain: corpus callosum, 66–7, 252, 296; creativity, 455; difference from left, 66–9, 74–5, 211, 296; facial expression, 403; gender differences, 67, 252; language development, 264, 305; memorizing role, 305; sadness, 211, 319–22
Rilling, James K., 409
ritual, 272–3
Rizzolati, Giacomo, 367
Robins, Lee, 235
Roediger, Henry, 298
Romanian orphans, 82, 113
rope-bridge effect, 387–8

sadness, 190–1, 197–8, 211, 319–22
Sagi, Dov, 289
Salimbene of Parma, 239
Sally-Anne experiment, 166
Salmon, André, 382
Sapolsky, Robert, 448
Schachter, Daniel, 298
Schachter, Gene, 279
Schachter, Stanley, 362–3
Schaie, K. Warner, 429
Scheibel, Arnold, 434
schizophrenia, 61, 221, 337, 469
Scholey, Andrew, 384, 441
Schwartz, Jeffrey, 81
Schwartz, Sophie, 154
self, 163–9

self-objectivity, 170
Seligman, Martin, 216
senses, 95–7
sensory overload, 101–2
sentience, 169–70
serotonin: anxiety levels, 209, 336; appetite regulation, 62, 228; cooperative behaviour, 209, 238; drug effects, 219, 228; in love, 390; in monkeys, 209, 238; mood regulation, 62; nucleus accumbens, 146; in rats, 448; wakefulness, 156, 158, 161
Seroxat, 158, 209
sex differences, *see* gender differences
sexual characteristics, 372–3
Shakespeare, William, 271, 342–8
Shapiro, Francine, 333
Sharma, Tonmoy, 212
Shaw, Gordon, 434–6
Shelley, Mary, 350
sherbet test, 450
Shereshevski, S.V., 144, 272
Shostakovich, Dmitri, 457
sight, 103–4, 466
Silva, Alcino, 287
Silvian fissure, 265
Simonides, 301
Simons, Daniel, 136
Singh, Devendra, 372
sixth sense, 465–7
Skuse, David, 252
Slater, Alan, 244
sleep: deprivation, 150–1, 157–8; disturbed, 155–7; memory and, 160–1, 289–90; need for, 151, 157–9; paralysis (SP), 156–7, 161; REM, 153, 154–7, 159, 160, 161, 234, 289–90; stages of, 151–3
sleepwalking, 155–6
SMA (supplementary motor area), 75
smell, 104–9, 375–6, 466, 467
smiling, 197–8, 200, 401, 403
Smith, Edwin, 18
Smith, Robert, 116
smoking, 220–2
Snowdon, David, 81–2
Snyder, Allan, 459–60
solitary confinement, 362–3
Sorrell, G.T., 378, 379
spatial location, 254
SPECT scanning, 250
Spector, John, 333
Sperry, Roger, 69–70
Spiegel, David, 451
spinal injury, 192–3
sportsmen, 173–4, 461–2
Spurzheim, Johann, 23–4
stage fright, 140–1
stalkers, 358–60
Sternberg, Robert J., 454–5
Steve, Laura, 440
Stickgold, Robert, 290
Strange, Brian, 143–4
stress, 221, 448–9, 453–4
Stuss, Donald, 168
subgenual medial prefrontal cortex, 337

subliminal messaging, 96, 137–8
substantia nigra pars compacta, 61, 226
Sunday Times, 391
superior temporal sulcus (STS), 367, 398–9, 412
Sur, Mriganka, 104
surprise, 191
symmetry, 244, 370–1
synaesthesia, 97–101, 242, 253
synapse, 59–60, 274
syntax, 266–7
systemizing, 254–5

Tallal, Paula, 129
tamoxifen, 287
taste, 109–10, 466
Tay-Sachs Syndrome, 376
Taylor, Shelley E., 323–4
tectopulvinar pathway, 126–8
teenagers, 208, 249–51
temporal: cortex, 143; gyrus, 270; lobes, 63, 124, 154, 253, 287, 314, 459
temporo-parietal junction, 398
tennis players, 174
Tennyson, Alfred, Lord, 322
testosterone, 248, 251, 253, 304, 316–19, 373, 374–5, 396
thalamus: anterior, 314; in depression, 335; dyslexia, 128–9; as input filter, 145; lateral pulvinar, 141, 142; limbic system, 63, 64–5;

posterior, 314; processing centre, 277; REM sleep, 154; sensory role, 64–5
Thatcher, Margaret, 405
THC (tetrahydracannabinol), 225–6
Theory of Mind, 166
theta waves, 152–3
Thompson, Paul, 427
Thornhill, Randy, 371
time lag, 172
Times crossword, 434
Tomatis, Alfred A., 434
touch, 112–17, 466
'trans-orbital technique', 186–7
Transcranial Magnetic Stimulation (TMS), 459–60
trust game, 399–400
Tsien, Joe, 430–2
Tulving, Endel, 279
Turk, David J., 168
Turner, Rebecca, 395
Turner, Victor, 272–4
Turner's syndrome, 252
twins, 331–2, 427, 436

Ungerlieder, Leslie, 287–8
unified experience, 170
Unwin, Stanley, 267–8
Urbach-Wiethe disease, 189

van Praag, Henriette, 436–7
variable access, 170
vasopressin, 393, 396
ventral pallidum, 386
ventromedial prefrontal cortex, 207, 333

Vicary, James, 137–8
visual cortex, 100–1, 103–4, 126–8, 148, 154
Vogt, Oscar, 33
Vomeronasal organ, 108, 466–7

Watts, Dr, 186
Weill, Berthe, 381
Weill, Kurt, 274
Weiskrantz, Larry, 127–8
Weixler, Petra, 201
Wernicke's area, 246–7, 265, 267, 427
Whalley, Lawrence, 428–9
whisky, 443
'white cut', 183–4, 185
Wilde, Oscar, 349–50
will, 149
Willatts, Peter, 245
Wiltshire, Stephen, 459
wine, 217, 442

Winnie-the-Pooh test, 319–22
Winston, Joel, 398
Winstone, Julie, 433
Wiseman, Richard, 403–4, 406, 468
Witelson, Sandra F., 418
Wood, Dr, 135
writing, 260–1, 269
Wundt, Wilhelm, 171

X-rays, 42, 182–3

Young, Brigham, 339
Young, Larry, 386
Yurgelun-Todd, Deborah, 249, 251

Zajonc, Robert, 292
Zald, David H., 332–3
Zeki, Semir, 389